An Introduction to
GROUND IMPROVEMENT ENGINEERING

198

An Introduction to GROUND IMPROVEMENT ENGINEERING

(A Useful Book for Researchers, Consultants, Engineering Students and Practicing Engineers)

Satyendra Mittal
B.E. (Civil), M.E. (Soil Dynamics), Ph.D., FIE,
MISSMFE, MIGS, MISET, MISTE, MIWRS
Department of Civil Engineering
Indian Institute of Technology Roorkee
Uttarakhand (India)

MEDTEC
An imprint of
SCIENTIFIC INTERNATIONAL PVT. LTD.
NEW DELHI (INDIA)

An Introduction to Ground Improvement Engineering

ISBN: 978-93-81714-36-2

First Published: 2013
Reprint: 2015

Published by: MedTec

An Imprint of:
SCIENTIFIC INTERNATIONAL PVT. LTD.
Publishers & Distributors
Head Office
4850/24, Ansari Road, Daryaganj, New Delhi- 110002
Ph.: 91-11-23287580, 91-11-23287584, 91-11-43512984
Fax: 91-11-23286096
E-mail: info@siplind.com
Website: www.siplind.com

Typeset at: SIPL

Branche Offices

House No. 31, 1st Floor, KKB Road, Chenikuthi, Opp. Sirdi Sai Bidya Mandir High School, PO-Silpukhuri, District-Kamrup, **Guwahati**-781003
Ph.: 09854311615, E-mail: debojit@siplind.com

7, Kohinoor Flats, Lukes Lane, Ambujavilasam Road, **Thiruvananthapuram**-695001
Ph.: 09895092553, 0471-4000057/58, E-mail: kerala@siplind.com

127/G, Manicktala Main Road, Kankurgachi, Near Yogodyan, **Kolkata**-700054
Ph.: 09433335049, 033-23204041, E-mail: debashisdas@siplind.com

Resident Representatives

70, 8th Main, 3rd Cross, Thigalarapllya, Hoodi, Mahadevpura PO. **Bengaluru**-560048 Ph.: 09886285836, E-mail: satyajitmondal@siplind.com

Plot No. 18, Door No. 3/378A, 6th Street, Kubera Nagar Extn., Madipakkam, **Chennai**-600091, Ph.: 09444017978, E-mail: krb@siplind.com

Printed at:
Chaman Enterprises, Delhi-110002

Dedicated

to

My most beloved parents

Late Shri Ramesh Chandra Mittal and Mrs Laxmi Mittal,

who showed me the path I have walked on .....

and

to those

who could have survived, if their houses were built on safe and sound foundations

Preface

This book is basically a compilation of my lectures which I had been delivering in IIT Roorkee and many parts of the world. I do not claim that by reading this book one shall be expert of ground improvement engineering but I can confidently say that this shall help in understanding the basic concepts of ground improvement engineering and also know its different techniques. No single technique is sufficient for improvement of ground as the measure adopted for improvement of the ground may differ from site to site. The literature available in this area is very scanty and the different methods are not compiled at one place. Therefore, it was thought that the material be collected and compiled in a comprehensive form so that the professionals may get fairly a good idea of ground improvement techniques.

Some chapters in this book are based on research work of my students namely Meenal Gulati, Arup Bhattacharjee, Sefali Biswas, V. Kannan, N. Sweta, Kanwar Singh, etc. It also contains outcome of my own research and consultancy works. Material for the book has also been collected from good books by Purushottam Raj, R.M. Koerner, CJFP Jones, NWM John, Swami Saran and S.K. Shukla besides several journals, manuals, etc. for the benefit of readers. Ground improvement engineering is now a compulsory subject in most of the technical institutes, hence, students and teachers will find this book very useful as all the relevant information shall be available to them at one place.

I am thankful to Strata Systems, Maccaferri Environmental Solutions, DuPont Typar® SF and M/s Garware Wall Ropes Limited for permitting me to use some of the pictures from their literature. My very sincere thanks are due to Neetu Kotnala for complete DTP work, Shri Sunil K. Panda (Editor) for preparing manuscript of this book and to my well wisher Mr. J.P. Shukla, for precious help at various stages of preparation of this book. Thanks are due to Kartikey, Nandita and Neeta for all time support. I have no suitable words to express my gratitude to Mr. Rajan Jain for publishing the book in a short span of time.

The practicing engineers and consultants shall also find this book very useful to get the design tips readily. I feel the first edition of this book shall certainly give an overview of different ground improvement techniques. I hope the readers will find this book very useful in their day to day professional life. I shall, however, anxiously wait for useful suggestions to improve this book.

Satyendra Mittal

Contents

Introduction and Earth Reinforcement

CHAPTER 1

1.1 INTRODUCTION

Ground improvement is the most imaginative field of geotechnical engineering. It is the field in which the engineer forces the ground to adapt to the project requirements by altering its natural state, instead of alter in design in response to the ground's natural limitations. The results usually include saving in construction cost and reduction of implementation time.

Solving transportation problems has been one of the most important social duties of Civil engineer, dating back to the days of early settlements. The opening of new frontiers, development of civilisations, interconnection of societies and defence of land all require the establishment and maintenance of functional network of transportation system and in turn require an input from the geotechnical engineer.

The planned development efforts over the last five decades have considerably enlarged and improved the road network in India, increasing it to manifold. The network, which now forms an important component of the infrastructure for transportation, provide comfort mobility to a large section of the population, stimulates their economic and social activities, and draws their great participation in the national life. The increasing level of economic level of activity has resulted in a large increase in the volume of traffic on the entire road network. Further, the axle loads have also intended to increase. To meet the demand for the increased traffic volumes and intensities, a quantum jump in highway construction is in offing. This expansion of the road network involves construction of expressways, highways feeder roads and the road bridges in widely depressed area with totally different soil conditions, each of which possessing unique problems to road scientist and engineer. The problem faced by the engineer in the construction of road bridges in the coastal and delta areas of the country arise primarily out of the low shear strength and high compressibility of the soft clay sub soils. Many embankments of these roads on embankment have to pass through a marshy area with extremely poor soil condition. Hence, it is necessary to improve load response behaviour of the soft soil to build the embankment economically and with high serviceability. If the depth of soft soil is more, then it is difficult to replace the virgin soil with suitable soil to provide support. As a result, road embankment may fail or experience long-term settlement. To overcome this problem at site, ground improvement is needed.

1.2 GROUND IMPROVEMENT

With the rapid industrialisation and increase in population, a lot of construction activities are taking place. Hence, an engineer is compelled to accept all types of land whether soft or good. Gone are the days when people used to reject any site on account of being waterlogged or full of soft soil or otherwise. Now the technology is available to improve all types of soils.

The concept of earth reinforcement is not new. The basic principles are demonstrated in nature by animals and birds and the action of tree roots. The fundamentals of techniques are also described in Bible covering the reinforcement of clay or bricks with marsh plant or straw for the construction of dwellings. Constructions using these techniques were existed in 5th and 4th millennia BC. Renowned sage Gautam Budha with his favorite disciple Anand was once passing through a town. He was told that the people were homeless in a nearby village. Budha suggested Anand to collect torn off and unused clothes from the people of town. After collecting the clothes, Anand cut the collected clothes into small pieces, as instructed by his preacher and mixed them into clay, built houses for homeless villagers. The clothes evidently increased the tensile strength of soil and thus the homes built by Budha survived for a long time. The earliest remaining examples of soil reinforcement are Great Wall of China and Agar-Quf Ziggmat which is situated 5 km north of Baghdad. This was constructed with clay bricks varying in thickness between 130–400 mm, reinforced with woven mats of reed laid horizontally on a layer of sand and gravel at vertical spacing varying between 0.2 and 0.5 m. Reeds were also used to form a rope (100 mm F) which pass the structure and act as reinforcement. This structure was 45 m tall (originally it is believed to have been over 80 m high) and is 3000 years old. The Great Wall of China is another example of reinforced soil where mixture of clay and gravel was used with tamarisk (bushes grown along sea side) branches.

The Romans are also known to have used this technique. Pasley (1822), a colonel in British army constructed retaining walls by reinforcing the backfill with brushwood, wooden plank.

Reinforced earth finds application in construction of dams, railway lines, river training works, dykes etc. In 1930s, French scientist constructed a ladder wall wherein retaining wall consisted of mass of granular filling unified by a row of tie members each having a small end anchor, together with a thin cladding membrane, as shown below (Fig. 1.1).

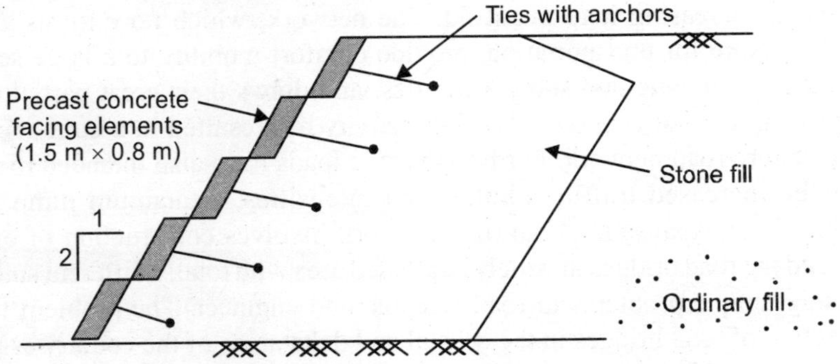

Fig. 1.1: Ladder wall

The ground improvement is the technique to improve soil characteristics to improve its engineering performance as per the project requirement by altering its natural state, instead of having to alter the design in response to the ground natural limitations.

The ground improvement techniques are used to:

- Reduce the settlement of structures.
- Improve the shear strength of soil and thus increasing the bearing capacity of soil.
- Increase the factor of safety against possible failure of embankment.

- Reduce the shrinkage and swelling of soils.
- Increase the liquefaction resistance.

1.3 HISTORICAL EVIDENCE

With the invention of wheel in the land of Sumer about 35000 years ago, the mobility of man was greatly enhanced leading to the establishment of settlement and the interconnection of societies. In those days settlements and their transportation system were not controlled by engineering considerations, but rather by religious, trade, agriculture and military object. In those days builders often faced great difficulties in the road construction over the marshy and unsuitable land without the facility of modern equipments.

Fig. 1.2: Technique adopted by Romans

Soil reinforcement technique was applied for the construction of roads over soft soils, using wooden logs by the builders of 2400 BC. The same technique was also later adopted by Romans and it has been shown in Fig. 1.2.

The Sumerians also used the concept of consolidation in Mesopotamia. At the Agarquf Ziggmat (1400 BC) in Iraq, reed were bound together to form cables that acted as horizontal drains for dissipation of pore water pressure build in the soil mass as earth structure increased in height.

Fig. 1.3: Compaction technique

Romans used timber piles as reinforcement to provide the load—support, to allow the construction of roadways over very soft soils, as all along the Vin Ostiense in Italy. The concept used by the Romans when crossing marshes along their 800 km network of roads is shown in Fig. 1.3.

The early accounts of densification go back to ancient India, where elephants were used for compaction of soils. The use of animals such as cows, oxen and calves for soil densification was specified in the old Hindu set of building rules expressed in Manasara Shilpashastra, written in Sanskrit.

The first use of dynamic compaction goes back to the days of Roman Empire. In chapter III of Book III of his De Architectura, on Roman architecture, Vitruvius writes, if, the soil is not firm, it should be made firm by dropping a weight. Ground densification methods were described by the Chinese in the Dung Dynasty code of 1103 AD. Some of these methods are illustrated in Fig. 1.4.

Fig. 1.4: Ground densification (1103 AD)

Chemical stabilisation was effectively used by the Romans for the construction of their famous roads. They improved their road building capabilities by introducing natural pozzolanic soils, lime and even furnace slag as construction material. Many examples were found in ancient road by the Romans in Italy and England.

1.4 METHODS OF GROUND IMPROVEMENT

Henry Vidal (1965) explained the concept of ground improvement in a very simplified manner. The same is discussed below.

Drop a handful of sand from certain height.

You will get a leap of sand [Fig. 1.5 (*a*)]. Keep your foot over it, sand will disperse in all directions. Measure this dispersion as 'X' [Fig. 1.5 (*b*)].

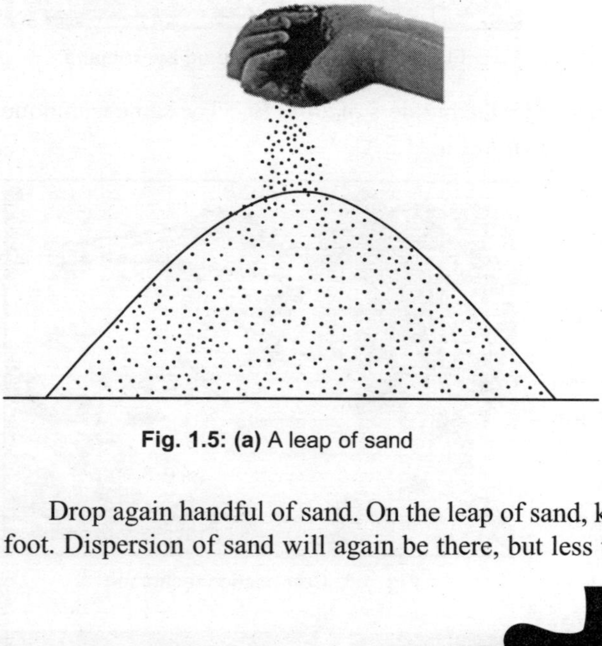

Fig. 1.5: (a) A leap of sand

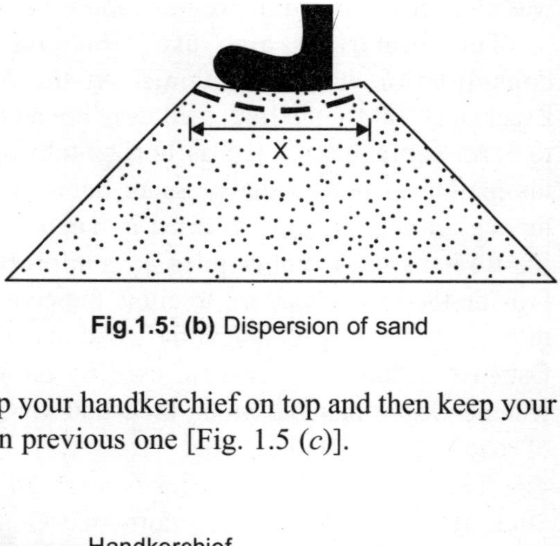

Fig.1.5: (b) Dispersion of sand

Drop again handful of sand. On the leap of sand, keep your handkerchief on top and then keep your foot. Dispersion of sand will again be there, but less than previous one [Fig. 1.5 (*c*)].

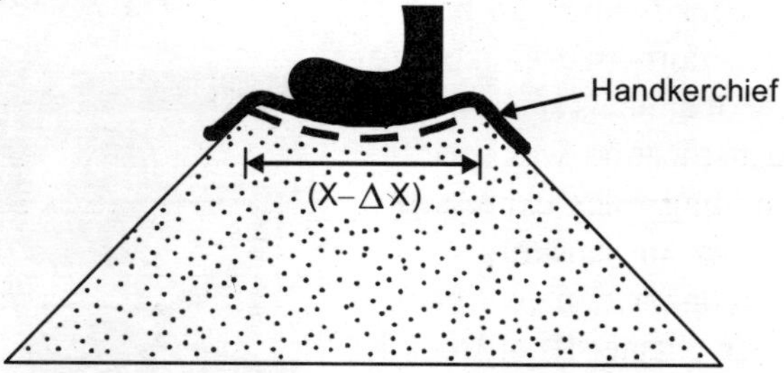

Fig. 1.5: (c) Less dispersion of sand after putting handkerchief

Do you know the reason? Magic?

No, it is mechanics!

Let us do one more experiment to understand further.

(*i*) Take a wooden box with its all four walls openable through hinge [Figs. 1.6 (*a*), (*b*)].

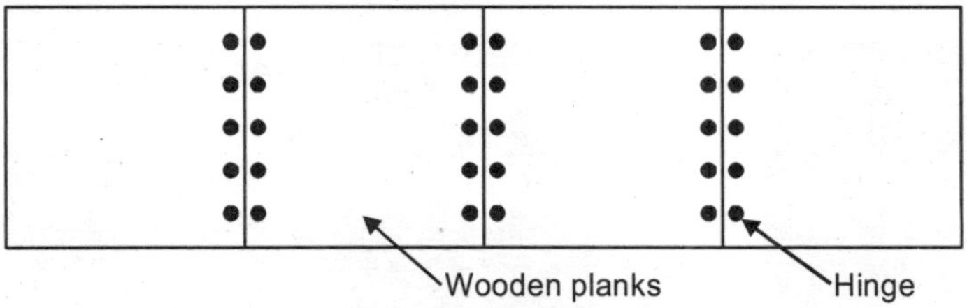

Fig 1.6: (a) Wooden box with all four walls opened

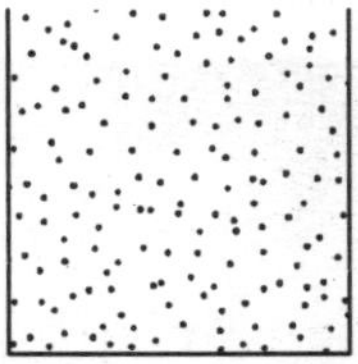

Fig. 1.6: (b) Wooden box with all four walls vertical

(*ii*) Fill the box with aggregate up to its top and ask a man to stand on the aggregate. Remove the four walls of box. There will be collapse of aggregates [Fig. 1.6 (*c*) to (*e*)].

Fig. 1.6: (c) A box is filled with aggregate

Fig. 1.6: (d) Compaction of aggregates

Fig. 1.6: (e) Collapse of aggregate after removal of box walls

Fig. 1.6: (f) Insert geogrid at various intervals with aggregate filling in layers

(*iii*) Empty the box and re-fill the box with aggregates, but in 3 to 4 stages. Firstly, put a geogrid [Fig. 1.6 (*f*)] and pour aggregates. After filling aggregate in one layer, put another strip of geogrid (explained in subsequent chapters) and fill another layer of aggregate, followed by another strip of geogrid and so on till the box fills up full. Ask a man to stand again on top of aggregate. Remove the walls of box. The aggregate layer will not collapse now even after removal of box walls. [Figs. 1.6 (*g*) and 1.6 (*h*)].

Fig. 1.6: (g) The aggregate layer doesn't collapse even after removal of box walls

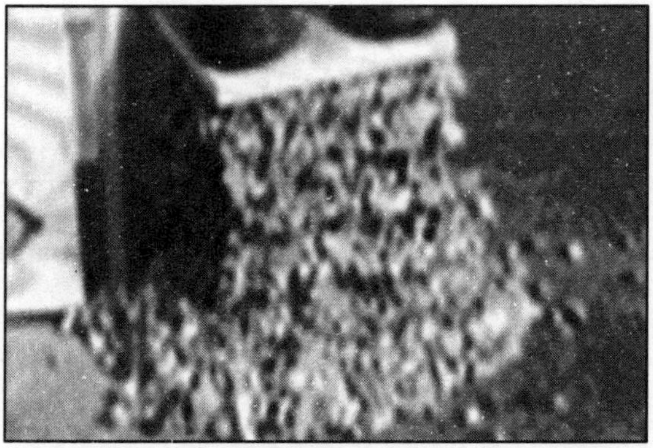

Fig. 1.6: (h) The reinforced soil mass weight

Magic again? No, it is mechanics again.

(*i*) Check if aggregate, which has been reinforced with geogrid (without using any cement or binder material) can take heavier loads. Put wheel of a car. The aggregates mass does not collapse now also Figs. 1.6 (*i*) and 1.6 (*j*)].

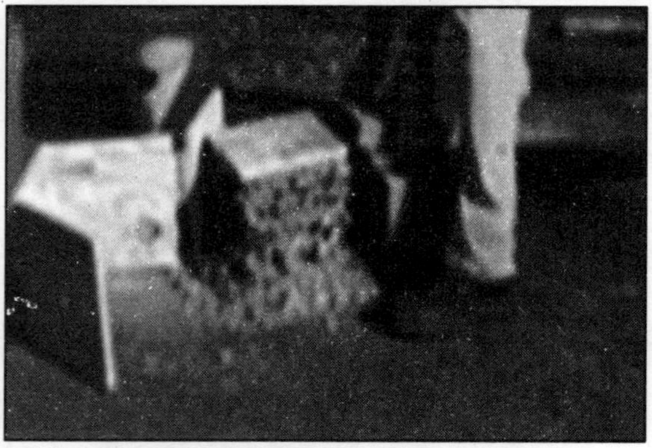

Fig. 1.6: (i) Check if the reinforced mass can take heavier load

Fig. 1.6: (j) The aggregates reinforced with geogrid can take the load of the wheel of a car, showing the power of geogrid. This experiment shows that strata grid can work as to sustain heavy loads

Fig. 1.6: Concept of reinforcement using strata grid (Courtesy: Strata Geogrid)

The reason is obvious. There is a friction developed between handkerchief (in previous experiment) and soil due to which adhesion took place between cloth and soil particles and thus soil particles did not flow away. Same thing happened in second experiment, where bond was developed between geogrid and aggregates.

If the virgin soil is soft and replacement of existing soil is not feasible, in that case it is recommended to treat that soil at site itself. If, the surface soil is granular, adopt the technique of densification. If the soil is cohesive, the ground improvement techniques applied may include consolidation, weight reduction or reinforcement using vertical reinforcing element. If the weak soil is very close to the surface, densification with heavy vibratory roller, chemical stabilisation with admixture or reinforcement with geosynthetics (explained in subsequent chapters) can be used. Ground improvement methods can be divided into eight main categories (Table 1.1).

Table 1.1: Methods of ground improvement

Sl. No.	Methodology	Nature of work
1	Removal of soft soils	Removal of unsuitable soil
2	Pre-compression or Preloading of soils	Preload the existing soil prior to construction
3	Stage construction	Construction of embankment in parts/stages
4	Densification	Static compaction, vibrofloatation, dynamic compaction, stone columns, blasting and compaction grouting
5	Consolidation	Preloading, mwick drain, vacuum consolidation
6	Weight reduction	Use of Fly ash as light material, wood chips, slag, geofoam, tyre chips
7	Soil Reinforcement	Use of geosynthetics, Soil nailing, micropiles
8	Chemical treatment	Jet or fracture grouting, admixtures mixing, deep mixing.
9	Biotechnical stabilisation	Brush layering, brush matting, composite system

1.4.1 Removal of soft soils

Soil replacement is conventional method for improving the soil when the thickness of clay layer is small. The clay is removed by excavation. It is filled with murrum and compacted at desired density.

1.4.2 Pre-compression

The technique of improving soil properties by pre-compression consists of placing of surcharge load on the ground. It is very effective in compressible soils and also those sites where water table is at shallow depth. The uniformly distributed surface load is placed prior to the construction of the intended structure. A part or all of the surcharge may be removed before the construction commences, depending on the requirements. A surcharge (overload) is placed temporarily and after a predetermined time lapse the intended structure can be built with occurrence of little or no additional settlement. The greatest advantage of preloading, as compared with other method of improving ground is its much lower cost. If it is done with vertical drains, it can be only 10–20% as expensive as other methods.

The vertical drains facilitate transmission of pore water pressure by capillary action under the action of preload. Another advantage of preloading is that it allows an immediate and direct assessment of its effects. By measuring ground movements, and in some cases pore water pressures it is possible to assess quantitatively the extent of ground improvement in terms of shear strength and predict its future behaviour. Preloading is essentially attractive when fill material is subsequently used at the same project for site preparation.

Preloading needs space. The area taken for preloading usually extends to about 10 m or more outside the perimeter of the proposed structures. This technique also requires large quantities of fill material. Granular fill is always desirable as it does not turn into mud when it rains. Also it can be used for embankment construction after the end of preloading. [Figs. 1.7 (*a*) to (*d*)].

Fig. 1.7: (a) Plain site

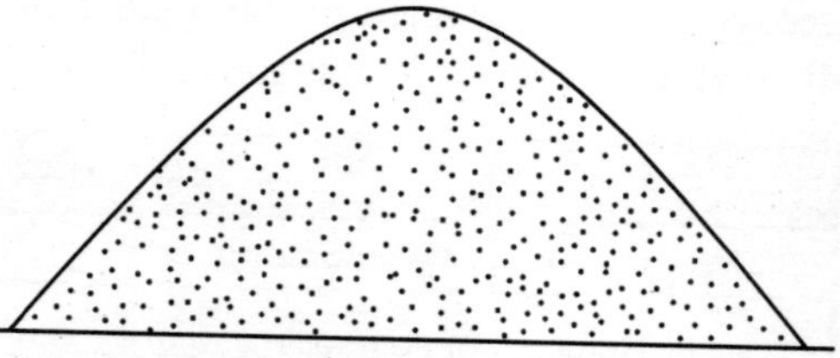

Fig.1.7: (b) Preloading by soil mass without drains

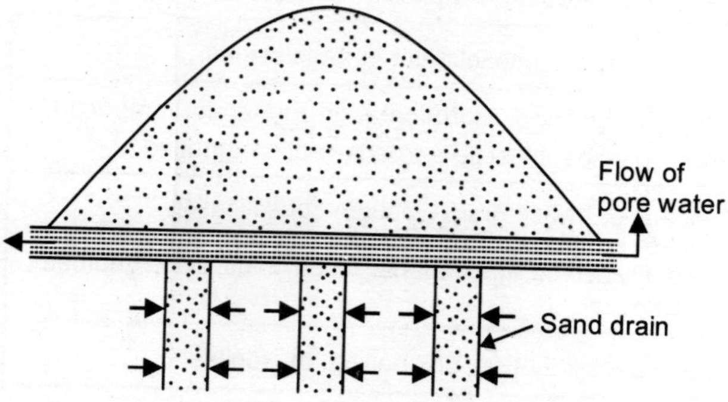

Fig. 1.7: (c) Preloading with vertical sand drains

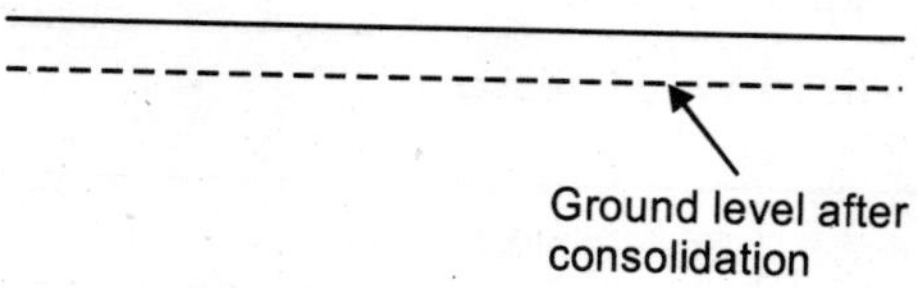

Fig. 1.7: (d) Improved ground at site

1.4.3 Stage construction

Stage construction is employed mainly as a means of gradually increasing the shear strength of soft clay, which would otherwise be adequate to carry the intended embankment load without failure. Stage construction is particularly economical technique in the construction of highways fills on soft clays. Since, the material can stay in place and need not be relocated.

In the stage construction process construction of an embankment is generally done in stages. In this process a part of structure is constructed to consolidate the existing soil for that load.

1.4.4 Densification

Soil densification is the process by which soil particles are constrained to pack more closely together through a reduction in air voids by the application of compactive efforts.

Compaction is measured in terms of dry density of soil. The principal factors of compaction, which influence the dry density of soil are:

(*i*) Moisture content
(*ii*) Energy input
(*iii*) Volume compacted
(*iv*) Soil characteristics

The standard laboratory tests used for evaluation of MDD and OMC for various soils are:

(*a*) Standard Proctor Test
(*b*) Modified Proctor Test

- The maximum dry density (MDD) and optimum moisture content (OMC) depends on the degree of compactness.
- The higher the energy of compaction, the higher is the maximum dry density.
- The higher the energy of compaction, the lower is the optimum moisture content.

- No portion of compaction curve can touch the zero air void line.
- The maximum dry density of compaction and the corresponding optimum moisture content will vary from soil to soil.

Using the results of laboratory compaction dry density versus water content, specification can be made for the compaction of a given soil in the field. In most of the cases, the field engineer has to achieve a relative compaction of 90% or more of specified laboratory test

$$\text{Relative compaction} = \frac{Y_{d\,\text{field}}}{Y_{d\,(\text{max})}}$$

where, ($Y_{d\,\max}$) is maximum dry density achieved in laboratory at its optimum moisture content.

1.4.5 Field compaction

The selection of proper compaction equipment is vital for mechanical stabilization. The compaction equipment are generally 8-10 ton static rollers, vibratory equipment with variable frequency and amplitude, sheep foot rollers for clayey soils and plate compactor etc. Ordinary compaction in the field is generally done by roller. Several types of rollers have been used for compacting the soil in the field. The most common roller are given below.

(*a*) Smooth wheel rollers
(*b*) Pneumatic rubber-tired rollers
(*c*) Sheep foot roller
(*d*) Vibratory rollers

1.5 CONSOLIDATION

"Consolidation is the process by which expulsion of water takes place from the void spaces by the application of the external load".

As per the fundamental concept of soil mechanics, the stresses induced by external loading on fine grained soil, the instantaneous pore water pressure increases. In due course of time this excess pore water pressure will be squeezed of the soil and a process will continue until the excess pore water dissipates and the soil skelton becomes strong enough to carry the new load on its own. Clayey soils have low permeability with long drainage path, hence full consolidation of soil skelton takes a very long time. The philosophy behind the use of vertical drainage is to shorten the drainage paths in order to accelerate the consolidation process.

Large consolidation settlement are expected whenever the high embankment is constructed over soft or normally consolidated clayey soils. In such cases pre-compression may be used to minimize the post construction settlement (section 1.3.2). To accelerate the consolidation settlement, sand drains, band drains, stone columns are being used these days.

The idea of using the vertical sand drains as means of accelerating the rate of consolidation was developed in USA. Since then vertical sand drains have been frequently utilized in order to reduce the time of consolidation of fine grained soils. However, some disadvantage connected with the use of sand drains is that sand must be chosen with extreme care in order to have good drainage property. Generally, for installation of sand drains pre boring of about 160 to 500 mm dia may be required.

In order to avoid the boring/drilling process for sand drains, sand wicks or 'Band Drains' with a diameter of only about 50 mm may also be used so as to cause least disturbance to ground.

1.5.1 Ground improvement by band drains

In the last 25 years, prefabricated band drains have almost entirely replaced conventional sand drains as the preferred method to speed up the consolidation process of soft cohesive soils under preloading embankment fills.

The prefabricated band drains, has an obvious advantage over the sand drains in its quality and of its small cross-sectional area. The mandrel for their installation can be made very slender and, therefore, the disturbance effects during installation can be reduced.

The first attempt to develop band drains was by Geotechnical Department of the National Swedish Road Board, later initiator and head of the Swedish Geotechnical Institute, established in January 1944. Fig. 1.8 shows the different types of band drains (Table 1.2).

Table 1.2: Dimensions and material of various band drains (Hansbo, 1977)

Drain type	*Width (mm)*	*Thickness (mm)*	*Filter sleeve*	*Country make*
Cardboard wick	100	3	None	No longer manufactured
Geodrain	95	4	Loose	Japan
Alidrain	100	6	Loose	Canada
Castleboard	93	3.2	Fixed	Japan
Colbond	300	4	Loose	Holland
Flodrain	100	5	Loose	Indonesia
Mebradrain	100	3.5	Loose	Holland
PVC	100	1.6	None	Japan

Fig. 1.8: Installation of band drains at site (Vishakapatnam, A.P.)

1.5.2 Construction

Sites, where the band drains are to be installed, quite often have the low bearing capacity. Consequently there may be need to reinforce the ground surface for the installation machines. Usually it is enough to use drainage blanket as a working platform, the thickness of which can be chosen with due regard to the bearing capacity of soil.

Mostly the drains are installed in an equilateral triangle pattern or in rows with a drain spacing equal to the spacing between drain rows. Finally the drained area is loaded in accordance with the loading schedule. It should be kept in mind that preloading time is governed by the soil layer which shows the lowest rate of consolidation. The stipulated time of consolidation mainly depends upon the surcharge to be applied and spacing of the band drains. Its design can also be done on the basis of consolidation theory.

The cost of extra loading whenever needed may add considerably to the total cost of the project. By using so called vacuum method, which is an alternative to preloading by the use of fill embankment.

According to the vacuum method, an under pressure is created in the drainage blanket, and consequently also in the vertical drains by means of a vacuum pump. The main drawback with this method is creating air tight seal of the drainage blanket.

1.6 WEIGHT REDUCTION METHOD

Utilization of fly ash as light weight material: The fly ash has low density and thus is useful for structural fills and highway embankments. Since the fly ash has lower density for the same fill height, the effective vertical stress will be lower than those seen in conventional soil materials. As a result it is very useful for fills over compressible soils, reclamation sites, as pre-consolidation pressure developed shall be low, thus settlement will be controlled.

1.7 SOIL REINFORCEMENT

The use of soil as a construction material has been practiced by man since ancient times due to easy and cheap availability in enormous quantities. But the use of soil in geotechnical applications has been limited due to its inherent weak (generally) tensile strength and susceptibility to change in characteristics with change in moisture content. To overcome this problem various methods have been used including soil reinforcement with reeds etc. However, use of natural material were restricted due to degradation problems of such materials.

1.7.1 Soil reinforcement by geosynthetics

With the advancement in the science, high strength geosynthetics have been introduced to provide an excellent solution of the ground improvement problems. Geogrid can be used with locally available soil as a construction material. The geosynthetics being made of high density polyethylene and polypropylene are virtually inert to the environmental effects.

The most common use of geosynthetics is in highway engineering especially for road constructed on weak sub grade. Geosynthetics like geotextiles, geogrid, geomembrane and geocomposites can lead to less deformation and high strength to the road and can reduce the thickness of the pavement (discussed in subsequent chapters). Geosynthetics based solution results in overall saving in the cost and natural material leading to environmental and natural resources protection (Table 1.3).

Table 1.3: Function of geosynthetics

Functions	*Geotextiles*	*Geogrids*	*Geomembrane*	*Geocomposites*
Separation	✓	—	—	—
Filtration	✓	—	—	✓
Reinforcement	✓	✓	—	—
Drainage	✓	—	—	✓
Erosion control	✓	—	—	—
Slope protection	✓	—	✓	—

Since the early 1970s, various type of geosynthetics have been used to reinforce soil in the construction of embankments built on soft soils and also construction of retaining walls. In case of reinforced retaining wall granular material are used as back fill. The geosynthetics provide the reinforcement to the soil. In the construction of embankment over soft soils, layer of geosynthetics can serve as reinforcing materials or can accelerate the process of consolidation of the soft sub-grade. The former function usually aim temporary increase of the safety factor of embankment which is associated with a faster rate of construction or the use of steeper slope that would not be possible in the absence of reinforcement.

1.8 CHEMICAL TREATMENT—DEEP MIXING WITH LIME, CEMENT AND LIME & CEMENT

Use of lime and cement columns of stabilized soil/lime mixture was developed in Sweden and Japan in 1970s. The predecessor to cement soil mixing column developed was use of mixed soil cement columns by intrusion in USA in the 1940s. The improvement of very soft clays and silts by mixing lime and /or cement with a rotary tool to form treated columns is being used in many parts of the world. The method involves advancing a mixing tool by drilling to a depth equal to the designed length of the column. Dry unslaked lime (CaO) and cement is delivered by compressed air to just above the mixing tool. Soil is mixed with lime or cement. Lime-cement columns have been used primarily to reduce settlement in deep deposits of very soft soils due to new construction loading. Another purpose of deep mixing is in improving global stability associated with new roadway embankment.

With deep mixing, there is little build up of soil stresses in adjacent matrix soils. The end product may be relatively rigid with cement or cement and lime columns in granular matrix soils or somewhat compressible soils when installed in fine grained soils. Deep mixing with lime only, produces a compressible column especially in reinforcing fine grained soils.

1.9 BIOTECHNICAL STABILISATION

A non conventional ground improvement technique used for embankment construction involves biotechnical stabilisation using live branches and stems as reinforcing element. In this system the soil and vegetation interact in three scenarios as follows.

Friction is developed along the interface between the soil and the live stems and roots, in a manner similar to that of mechanically stabilized fill.

The growth of roots along the buried stems has the effect of the artificial cohesion developed in the fibre reinforced soils. The woody vegetation acts as a hydraulic drain, resembling the action of wick drains (Figs. 1.9 & 1.10).

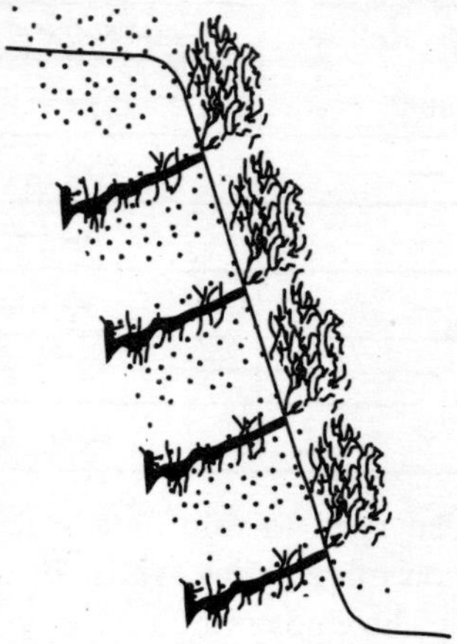

Fig. 1.9: Biotechnical stabilisation on slopes

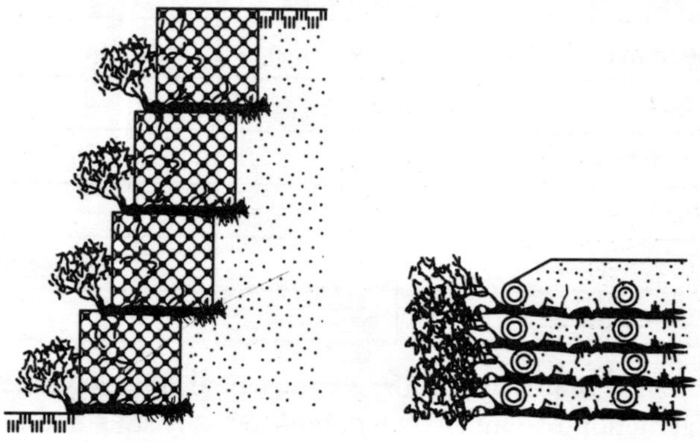

(a) (b)

Fig. 1.10: Biotechnical stabilization **(a)** slopes **(b)** embankment

1.10 MODERN STRUCTURE

The modern concept of earth reinforcement was proposed by Casagrande who idealized the problems in the form of weak soil reinforced by high strength membranes laid horizontally in layers. The modern form of earth reinforcement was introduced by French scientist Vidal in 1960s, wherein he gave concept of composite material formed from reinforcing strips laid horizontally in a frictional soil. This material he described as reinforced earth and he made first wall in 1968 in south of France.

Subsequently use of geotextiles was also invented in modern constructions, which was further promoted when synthetic polymer based materials arrived.

Fabric reinforced retaining walls have proved to be economical. Geotextiles were later used in the areas of separation, filtration and drainage.

Geotextiles materials can be divided into two categories—conventional and specials. Conventional geotextiles are products of textile industry and include nonwovens, woven, knitted and stretch bounded textiles. Special geotextiles are geosynthetics and are not textile products.

In 1980s, a special type of reinforcement in the form of an anchor was evolved. It was multi-anchor system which was developed in Japan by Fukoka (1980) for Japanese ministry of construction. The anchor is in the form of a rectangular steel plate.

The new retaining wall system developed in Austria is based on an elevated concrete facing and polymeric ties in the form of a closed loop (Figs. 1.11 & 1.12).

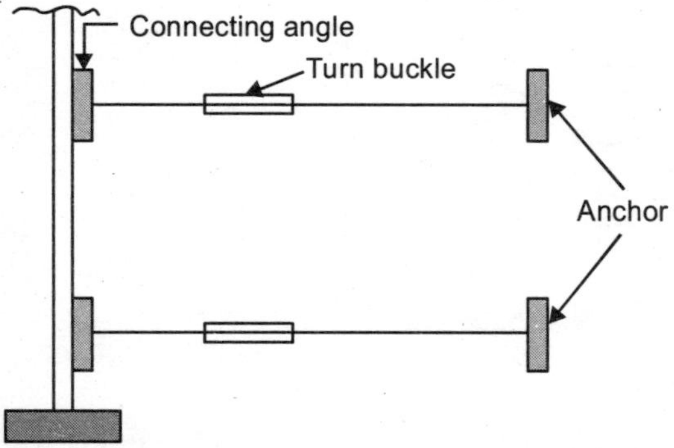

Fig. 1.11: Reinforcement form of an anchor

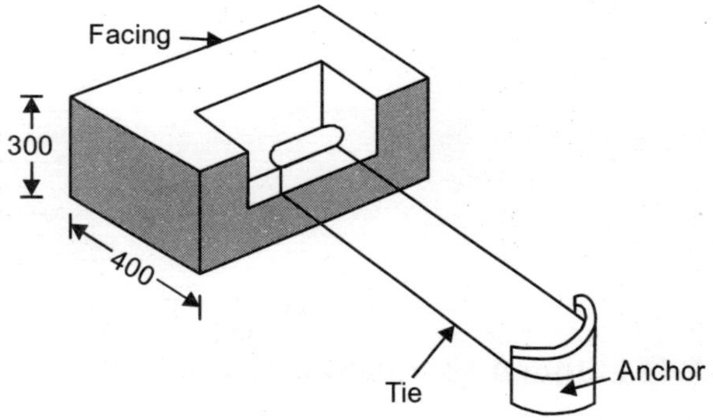

Fig. 1.12: Anchor ties with rectangular steel plate

In 1996, Jones et al., introduced another technique of soil reinforcement called tailed gabion, by which the stability of conventional gabion structures can be enhanced by addition of reinforcement (Fig. 1.13). Mittal (2012) has developed many such walls for a big housing project in Chandigarh (India).

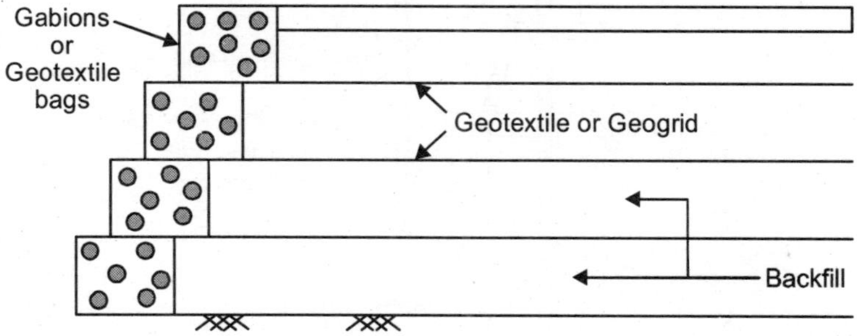

Fig. 1.13: Gabion wall with reinforcement

In 1981, the concept of stabilization of cuttings by reinforcing the earth in situ came. This was somewhat similar to those employed in ground anchor techniques. This is called soil nailing or lateral earth support system. Soil nailing system is defined in a separate chapter.

1.11 APPLICATION AREAS OF EARTH REINFORCEMENT (After Jones, 1996)

BRIDGEWORKS

1.11.1 Bridge abutment

Bridge abutment can be made using the concept of soil reinforcement.

Materials used: Reinforcement or prestressed concrete facing, strip, grid or anchor reinforcement, frictional or cohesive frictional fill. Technique is economical, hence may be used on poor sub soils. Speed of erection is high. It is able to accommodate compressive ground strains (Fig. 1.14).

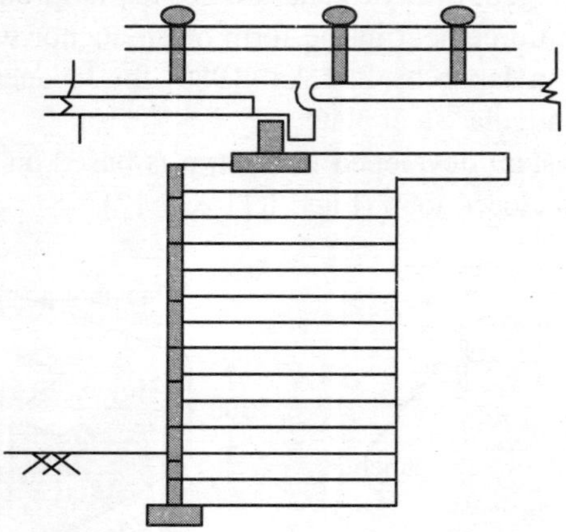

Fig. 1.14: Earth reinforcement of bridge abutments

1.11.2 Bridge abutment with piled bank seat

As evident from Fig. 1.15, the bridge abutment can be made with reinforced earth wall (retaining wall with attached reinforcing elements).

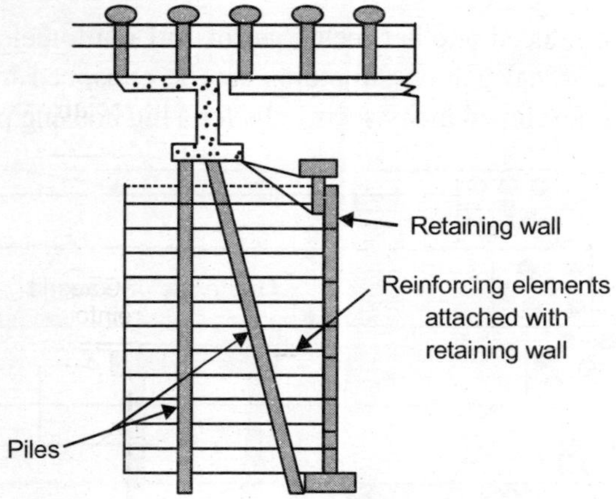

Fig. 1.15: Bridge abutments with reinforced earth wall

1.11.3 Sloping bridge abutment

Bridge abutment in sloping shape can also be made using reinforced earth concept.

Materials used: Masonry or precast concrete paving for facing, geotextiles, geogrid, strip or anchor reinforcement, frictional or $c - \Phi$ fill. (i.e. cohesive frictional soil). (Fig. 1.16)

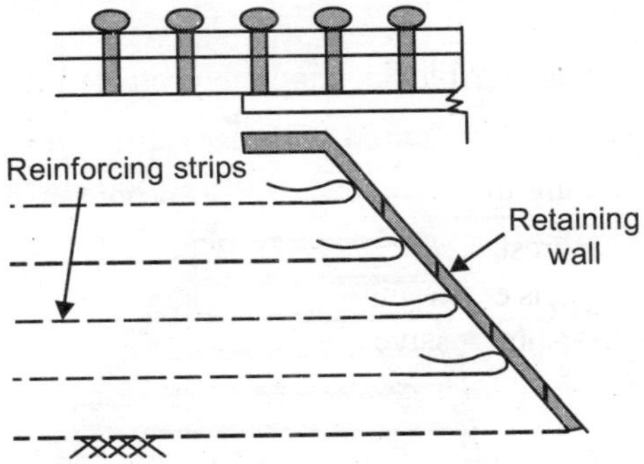

Fig. 1.16: Sloping bridge abutment

1.11.4 Geosynthetic reinforced bridge abutment

Materials used: Polyester geogrid reinforcement, sand bag temporary facing, full height mass concrete facing, cohesionless or cohesive fill.

Used in Japan to widen railway embankment or additional tracks. Even safe against earthquakes. (Fig.1.17)

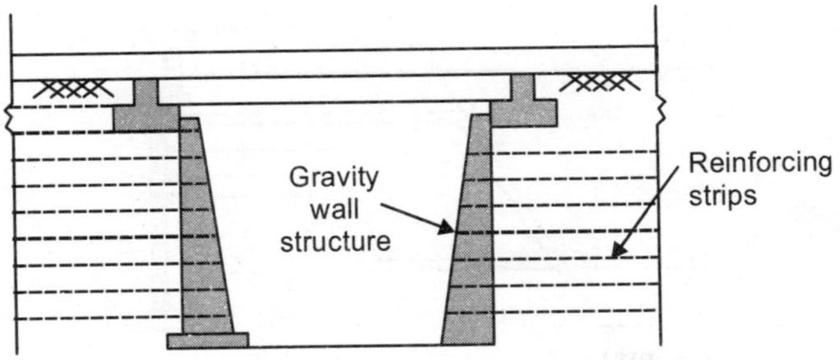

Fig. 1.17: Reinforced bridge abutment

1.11.5 Bridge approach piling

Materials used: High strength polymeric reinforcement and specially designed pile caps. (Fig. 1.18)

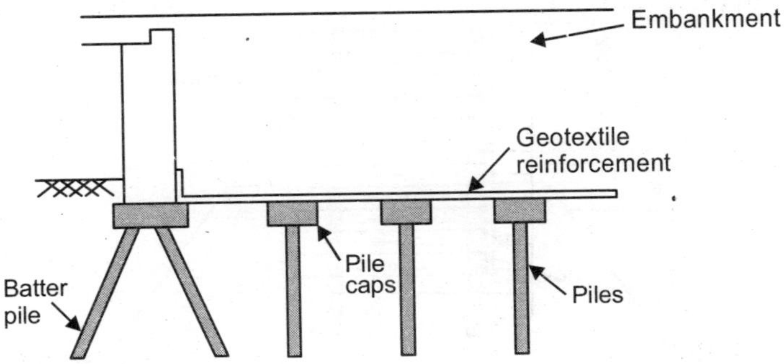

Fig 1.18: Bridge approach piling

Comments: Used to reduce the settlement of approach embankments for highways or railways when piles abutments are required. It provides a major reduction in cost over alternative methods.

1.12 DAMS APPLICATION

1.12.1 Earth fill dam

Materials used: RCC facing and anchor blocks concrete protected ties. (Fig. 1.19)

Comments: Special precautions are required with dam structures.

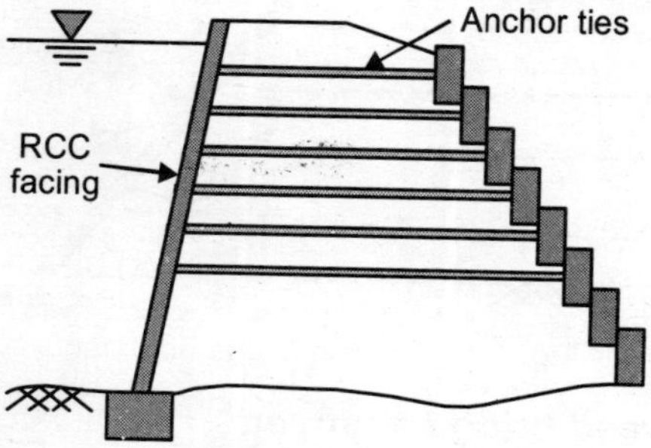

Fig. 1.19: Earth fill dam

1.12.2 Reinforced earth dam

Materials used: RCC facing, metal strip, grid or anchor reinforcement and selected frictional fill. (Fig. 1.20)

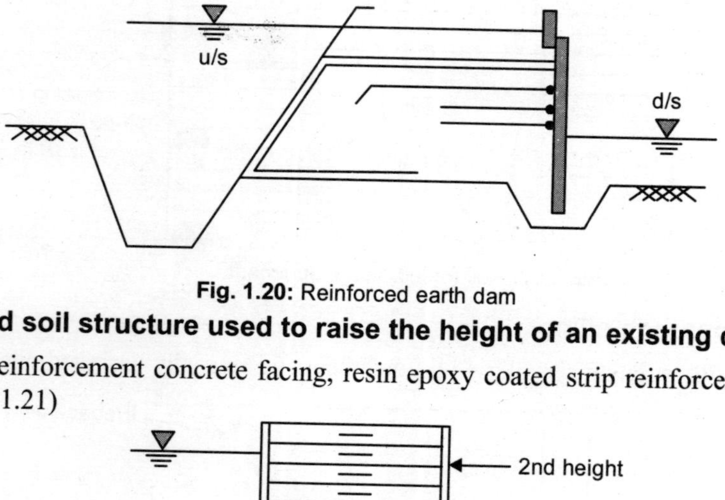

Fig. 1.20: Reinforced earth dam

1.12.3 Reinforced soil structure used to raise the height of an existing dam

Materials used: Reinforcement concrete facing, resin epoxy coated strip reinforcement and selected frictional fill. (Fig. 1.21)

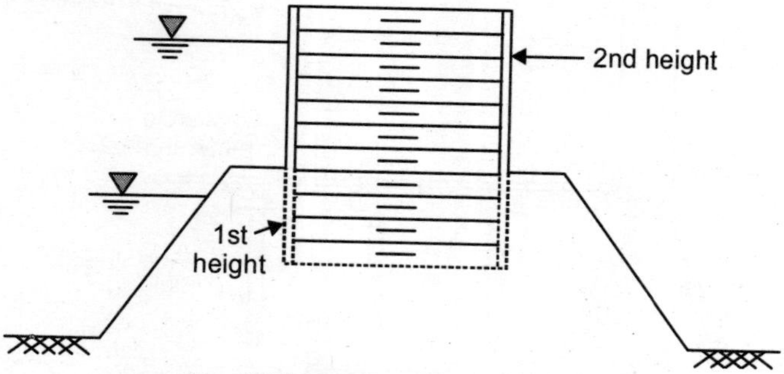

Fig. 1.21: Reinforced soil structure

1.13 APPLICATIONS IN EMBANKMENTS

1.13.1 Reinforced embankments

Materials used: Geotextiles or geogrid reinforcement and indigenous fill. (Fig. 1.22)

Comments: Reinforcement embankment may be undertaken for a variety of conditions, including steepening the side slopes to permit the use of marginal fill or to strengthen embankment.

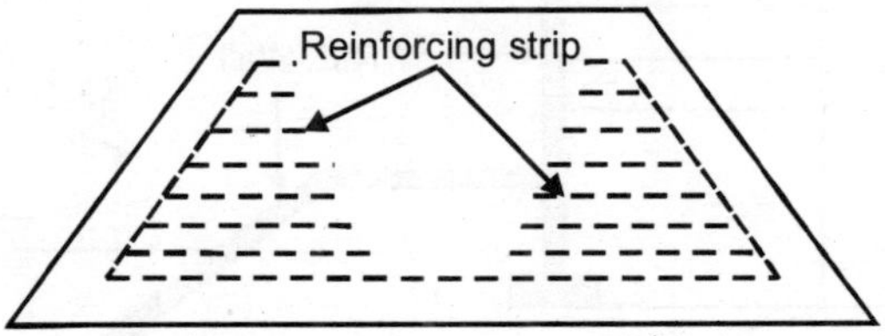

Fig. 1.22: Reinforced embankments

1.13.2 Geo-cell mattress used to increase embankment stability

Materials used: Geogrid materials for mattress filled with selected material. (Fig. 1.23)

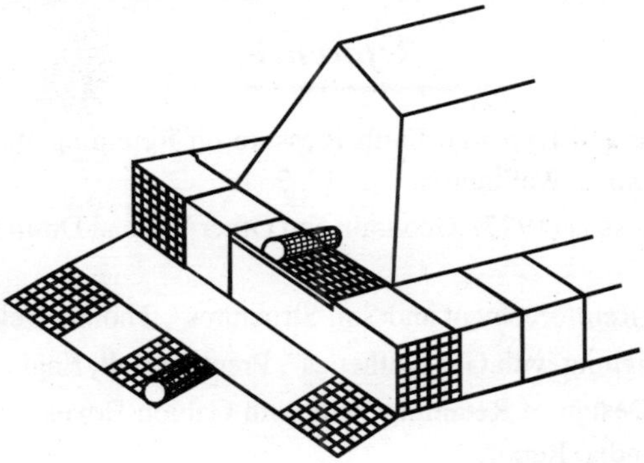

Fig. 1.23: Geocell enhances embankment stability

1.13.3 Application in sea wall or water way structures

Materials used: Special reinforcement concrete facing elements, together with strip or grid reinforcement and selected granular fill. (Fig. 1.24)

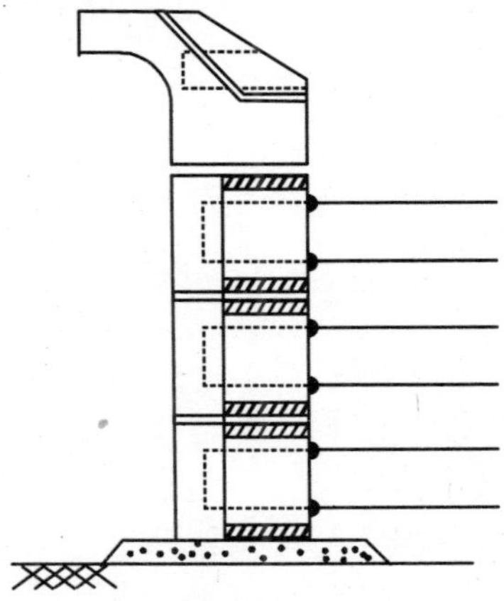

Fig. 1.24: Sea wall

1.13.4 Wall adjacent to river

Materials used: Reinforcement or pre-stressed concrete or timber facings together with grid or strip reinforcement and selected fill.

Comments: The earliest reinforcement soil quays built by Romans used timber for the facing and timber strips as reinforcement.

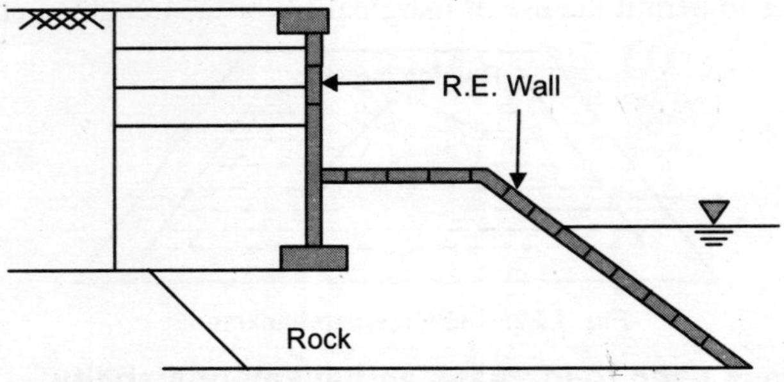

Fig. 1.25: Wall adjacent to river

References

1. Fukuoka, M. (1980). "Static and Dynamic Earth Pressure on Retaining Walls". *Proc. 3rd Australian–New Zealand Conf. on Geomechanics*, Wellington, 3, 3-37, 3–46.
2. Hansbo, S. and B.A. Tornstensson (1977). Geodrain and Other Vertical Drain Behaviour, *Proc. 9th INCSMFE*, Tokyo, Vol. 1, pp. 533–540.
3. Jones, CJFP (1996). " Earth Reinforcement and Soil Structures", Thomas Telford, ASCE Press.
4. Koerner, R.M. (1990). "Designing with Geosynthetics", Prentice Hall, Englewood Cliffs, New Jersey, USA.
5. Mittal, Satyendra (2012). "Design of Retaining Wall with Gabion Boxes for Housing Project in Panchkula, Chandigarh", IIT Roorkee (India) Report.
6. Pasley, C.W. (1822). "Experiments on Revetments", Vol. 2, Murray, London.
7. Raj, Purushothama, P. (2009). "Ground Improvement Technique", University Science Press, New Delhi.
8. Vidal, H. (1969). "The Principle of Reinforced Earth", Highway Research Record No. 282, USA.

Reinforced Earth Application in Foundations

CHAPTER 2

2.1 INTRODUCTION

Foundation is the integral part of any superstructure. Foundations are required to be intact dependable, firm and sound. Two types of foundations are well known that is shallow foundations and deep foundation. Sometimes shallow foundation have to be provided due to site constrains, limited budgets etc. Size of shallow foundation has to be in commensurate to the loads of the superstructure. There may be such situations where it may be difficult to provide that much space on ground as per design loads. The reason could be the neighbours' property, some water body, historical monuments etc. In such cases, if the size of foundation can be restrained suiting to site conditions but its bearing capacity be increased by use of reinforced element, the purpose gets solved. These days many reinforcing material made of polymers, textiles are available which have made the foundation designs quite innovative, economical and effective. The development of polymeric materials in the form of geosynthetics have brought major changes in the civil engineering profession. Geosythetics are available in a wide range of compositions appropriate to different applications and environments. Over the past three to four decades, civil engineers have shown an increasing interest in geosynthetics and in understanding their correct uses. Simultaneously, significant advances have been made regarding the use of geosythetics in civil engineering applications. Reinforced earth technology is gaining momentum worldwide. For reinforcing the earth, many materials can be used e.g., galvanized iron strips, aluminum strips, steel nails etc. Geosysthetics is also a very useful material which can be used successfully for reinforcing the ground. In this chapter, some useful tips are given by which it can be seen that how geosynthetics can be used on construction projects. Figures given in this chapter illustrate usefulness of these materials, which are self-explanatory.

A brief introduction of geosynthetic products is given below:

2.2 GEOGRIDS

Geogrid are integrally extruded polymer meshes. The meshes are made from select high strength engineering polymer (HDPE/PP) and are totally non-biodegradable and anti corrosive. They have fused or thermally welded joints having high dimensional stability and strength (Fig. 2.1). These grids are immune from naturally occurring soil alkalies, acids and mineral salts and are protected from UV radiations.

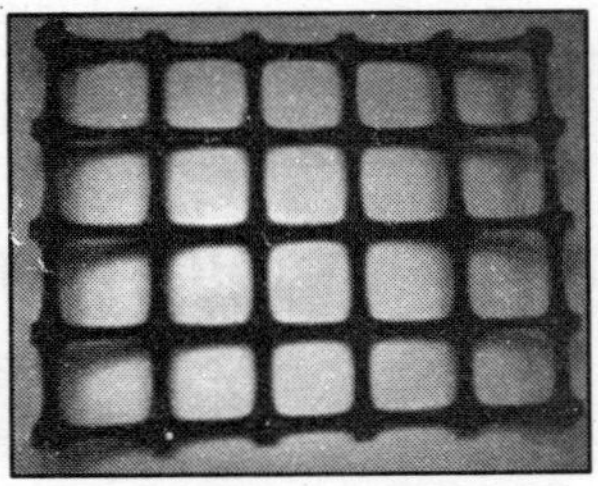

Fig. 2.1: Geogrid

They are capable of mobilizing high strength at relatively low operating/working strain of soil. These geogrids are flexible and identify closely to ground contours. These are very light weight and thus easy and fast to install.

2.3 APPLICATIONS FOR POLYMER REINFORCEMENT

Various applications for polymer reinforcements are explained in Table 2.1.

Table 2.1: Various applications for polymer reinforcement

1.	Embankment	Separator, reinforcement, tension membrane	Base reinforcement over soft ground is produced by either a mat, wedding or as a reinforced mattress, known as geocell.  Geocell
2.	Slopes	Reinforcement, erosion barrier, container	Internal reinforcement through placement of several layers. Can be applied to embankment or as a remedial measure to cuttings. Also bank protection.
3.	Retaining walls	Reinforcement, container	Vertical embankment reinforcement by several layers with an interlocking jigsaw facing or a soft wrap around facing.
4.	Foundations	Reinforcement, separator	To reinforce soil overlying culvert or cavities. To build over soft ground using geocells.
5.	Gabions	Separator, container	To form tubular or box containers for walls or slope protection.
6.	Coastal and inland waterway	Reinforcement, container, separator, erosion mat, screen	For breakwater structure, land reclamation, outfall protection, waterfront structure, wharves and revetments.
7.	Tunnels and sewer lining	Reinforcement, protector	Applied as Reinforce mentfor sprayed concrete.
8.	Roads	Reinforcement, separator, cushion	No reinforcement paved or unpaved road tracks.
9.	Structures	Reinforcement, protector	Roofing panels, structural reinforcement concrete.

2.4 IMPORTANT TERMS RELATED TO GEOTEXTILES

1. Tex – It is indicative of the linear density of the linear elements of geotextile. 1 Tex is equivalent to 106 kg/m. Tex is normally expressed as the weight in grams of 1000 m length of fibre.

2. Warp – The longitudinal yarns in a geotextile fabric structure.
3. Denier – The unit for linear density. It is expressed as weight in grams of 9000 m length of a fibre.
4. Weft – The yarns along the width of a geotextile fabric.
5. Creep – The resistance of geotextile to continued time dependent elongation, under the application of constant static stress.
6. Tensile strength – The strength of geotextile is basically a measure of load, the geotextile can support.
7. Fatigue strength – The ability of geotextile to withstand repetitive loading before undergoing failure.
8. Burst resistance – It is resistance of fabric to rupture, due to pressure applied normal to the plane of fabric.
9. Tear strength – The resistance of the fabric to propagate a tear. It ranges from 110 Newtons to 1300 Newtons.

2.5 PROPERTIES COMMON TO GEOTEXTILES

1. Good resistance to chemicals
2. Resistant to biological degradation
3. Thermal stability 0 to 120°C

2.6 AVERAGE PROPERTIES OF GEOTEXTILE

1. Material – 100% polypropylene
2. Specific gravity – 0.91
3. Weight/sq m – 206–303 gm/m
4. Thickness – 0.58–0.78 mm
5. Breaking strength
   - Warp way (kg) – 245–400 kg
   - Weft way (kg) – 181–333 kg
6. Tear strength
   - Warp way (kg) – 21–69 kg
   - Weft way (kg) – 18–71 kg
7. Water permeability (at 10 cm water head) – 4.2–55 litre/sec/metre
8. Pore size in microns – 60–243
   - Warp way (kg) – 32–45 kg
   - Weft way (kg) – 29–33 kg

2.7 USEFULNESS OF GEOGRID MATERIALS

2.7.1 Column foundation for embankment on weak subsoil [Fig. 2.2 (*a*) & (*b*)]

An embankment can be constructed even on weak soil by providing geogrid reinforcement as shown. It can be provided by providing layers or tubes.

Comments: Reinforcement webs and columns installed in situ.

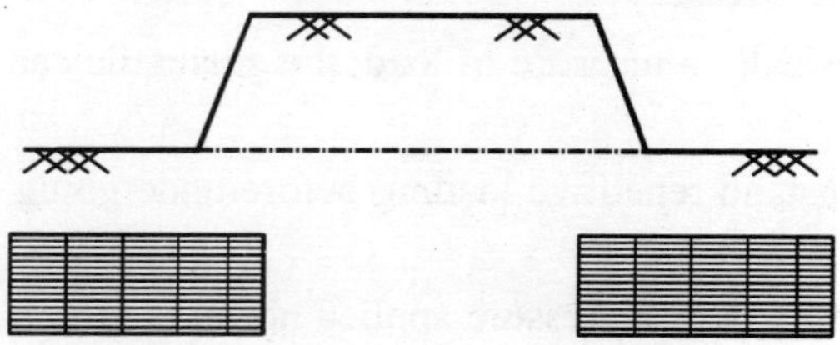

Fig. 2.2: (a) Geogrid reinforcement of sub-soil beneath embankment

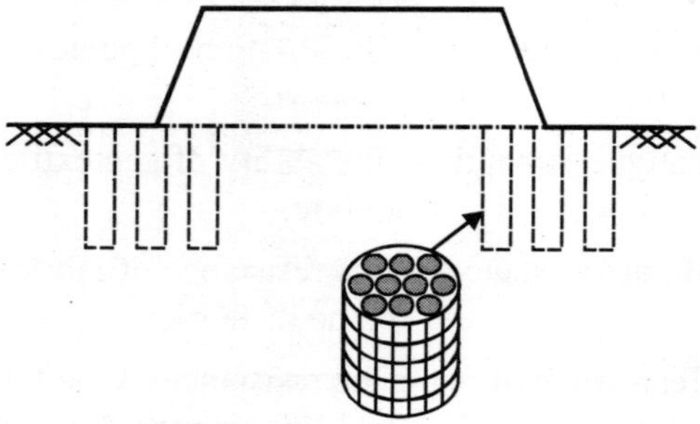

Fig. 2.2: (b) Stone columns formed from geogrid tubes

(*i*) **Excavations in urban conditions** (Fig. 2.3)

Vertical cuts can be strengthened in situ by providing soil nails (explained separately in a chapter) failing which, neighbouring soil could flow. This method also increases bearing capacity of soil. The facia material is spray concrete while steel nails are used as reinforcement.

Comment: Used in place of anchor systems.

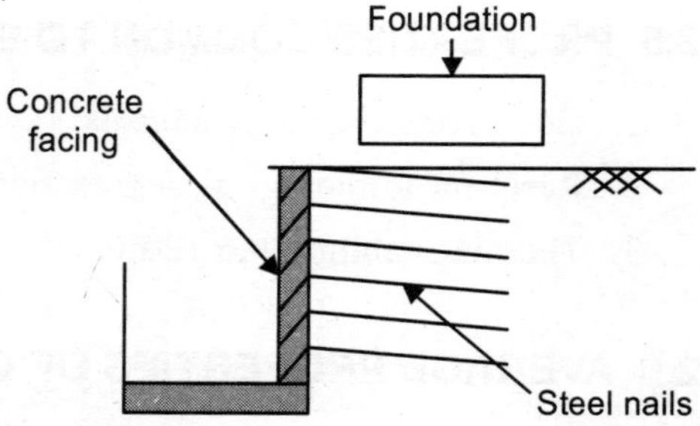

Fig. 2.3: Nails used for foundation

(*ii*) **Formation of cutting using soil nailing or the lateral earth support system** (Fig. 2.4)

Steel nails can be used to stitch landslide area also.

Materials used: Steel soil nails and spray concrete facing.

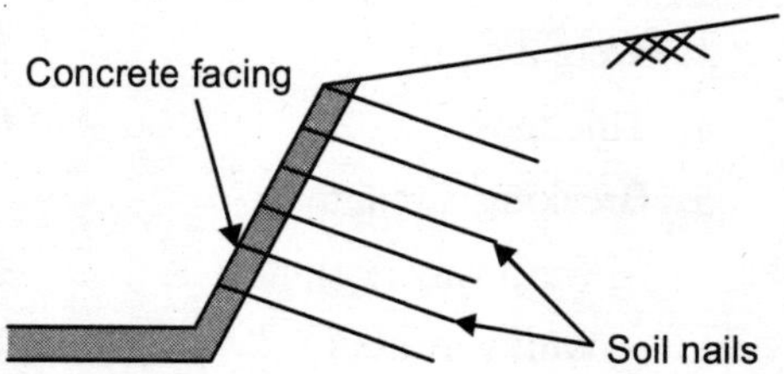

Fig. 2.4: Cutting formed using soil nailing

(*iii*) **Housing applications**

Reinforcement soil used to form terraced housing on sloping sites (Fig. 2.5)

Materials used: Concrete facing units and steel nail reinforcement in naturally existing ground.

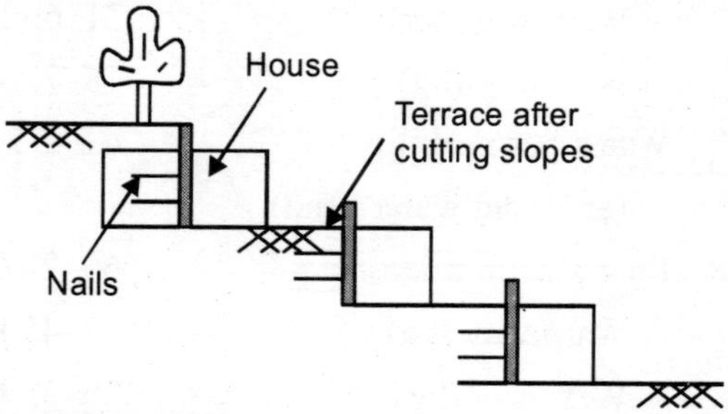

Fig. 2.5: House construction on terraces using soil nailing

(*iv*) **Industrial applications**

Underground tank: Underground tanks may be constructed as shown (Fig. 2.6) by stabilizing the side walls with the help of driven /grouted nails.

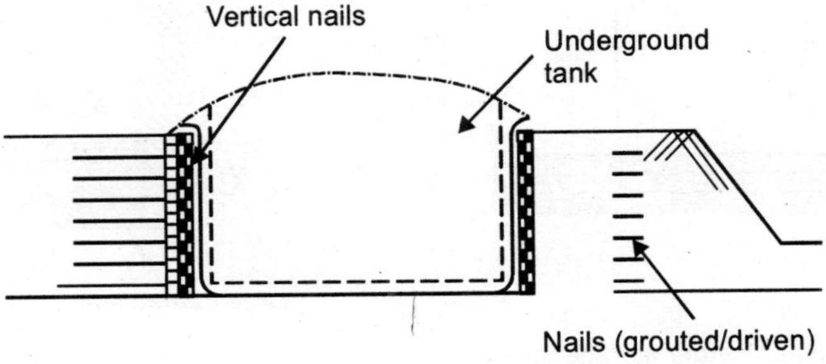

Fig. 2.6: Nails stabilising side cut

(*v*) **Tension membrane spanning voids or potential voids** (Fig. 2.7)

Materials used: High strength geogrid as reinforcing material and selected quality of fill, no facings are required.

Comments: Used to guard against sudden collapse or to retain serviceability following formation of void.

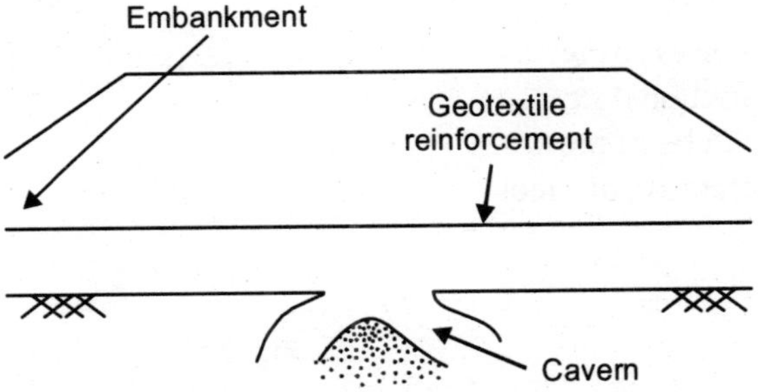

Fig. 2.7: Geotextile used to prevent against collapse

(*vi*) **Reinforced footings beneath foundations** (Fig. 2.8)

To increase bearing capacity of foundations, geogrid can be used as explained below. It ensures stability and reduce settlement.

Materials used: Strip or grid reinforced and frictional fill or cohesive frictional fill.

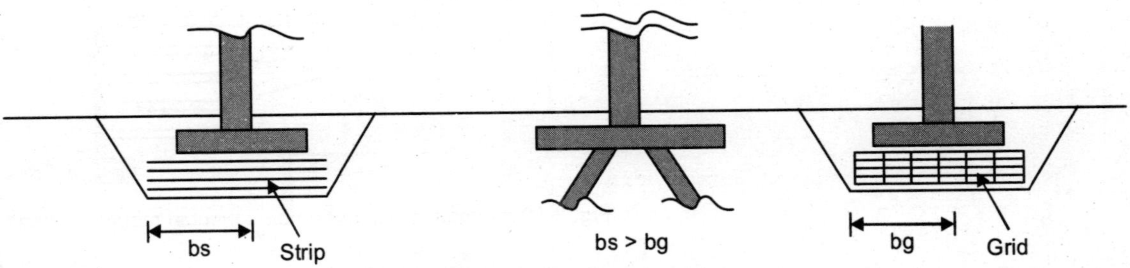

Fig. 2.8: Reinforcement under footings

(*vii*) **Reinforced foundations beneath storage tanks** (Fig. 2.9 and 2.10)

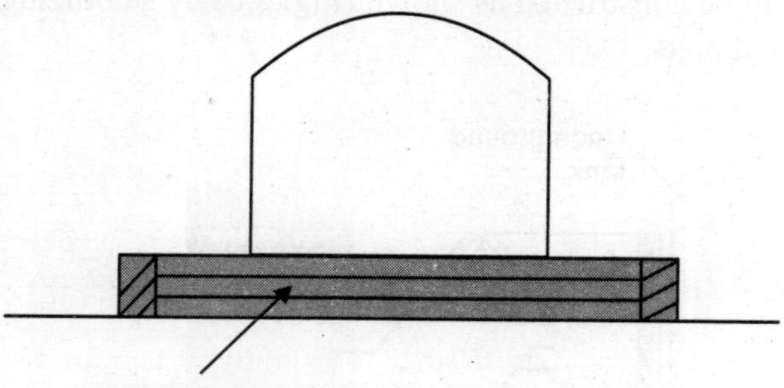

Fig. 2.9: Reinforcement under storage tanks

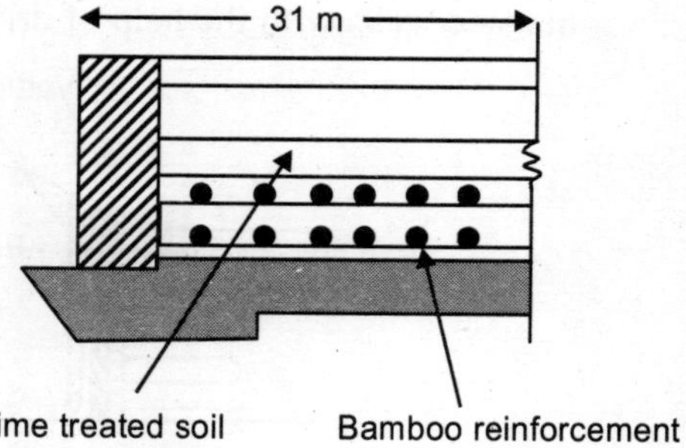

Fig. 2.10: Storage tank foundations

2.7.2 Application in highways

(*i*) **Reinforced earth wall** (Fig. 2.11)

This method is very useful for construction of roads in hilly areas, particularly for widening of road or new constructing of road.

Materials used: Geosynthetics reinforcement and frictional or c–ø soil backfill. Facia can be of precast reinforcement concrete or of steel facing, panels.

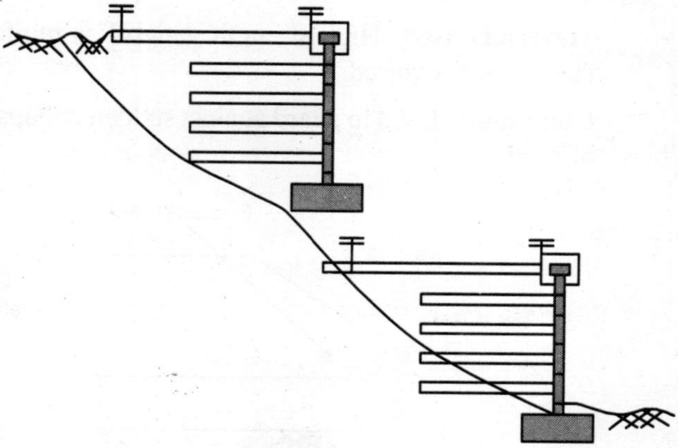

Fig. 2.11: Stepped highway structure

(*ii*) **Reinforcement soil support to embankment in mountainous region** (Fig. 2.12)

This method is very effective as a support to existing embankments in mountainous region to enhance the stability to embankment. This ensures stability of toe of slope.

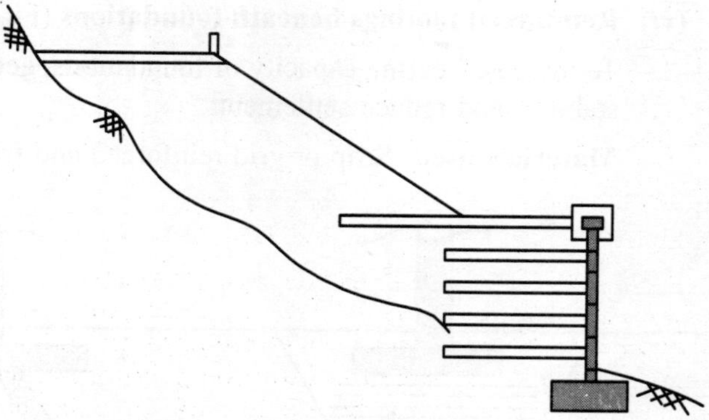

Fig. 2.12: Method to support embankment in mountain areas

(*iii*) Reinforcement embankment supporting highways [Fig 2.13 (*a*) & (*b*)]

In hilly area, the solutions are site specific. The embankments at such sites can be made by providing on facia the gabion walls and the backfill can be reinforced with geogrid materials while the geogrid shall be attached to the gabion box/wall.

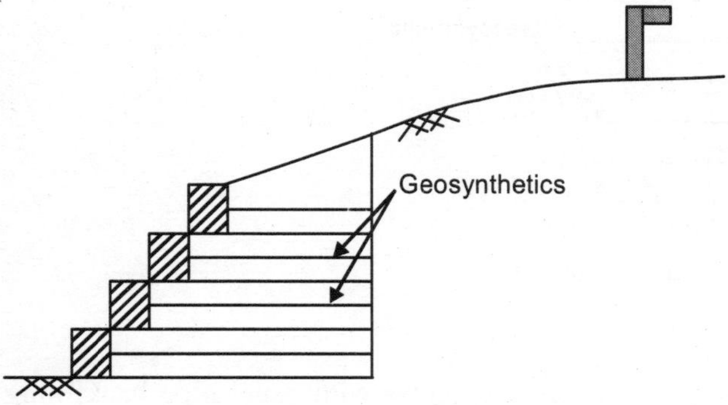

Fig. 2.13: (a) Stepped embankment or retaining wall

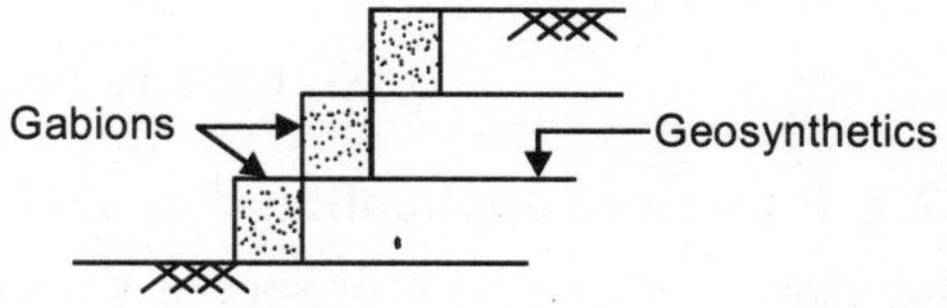

Fig. 2.13: (b) Gabion faced reinforced soil structure

(*iv*) Repair of embankment failure (Fig. 2.14 & 2.15)

The failed embankment where toe has yielded can be restored by providing a passive support. It can be done by dumping a soil mass ahead of failed toe of existing embankment and facia of this soil mass can be unused tyres of project sites.

Materials used: Waste car tyres and selected fill, the tyre may be tied together with steel hair pins or geotextile tape.

Fig. 2.14: Type/Geotextile composite

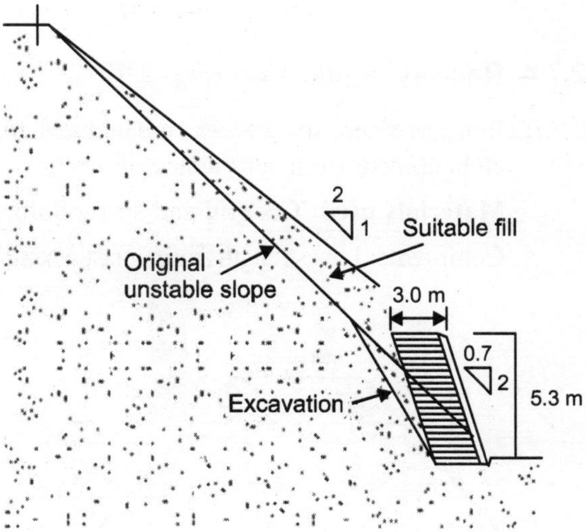

Fig. 2.15: Tyre wall repair of slope failure

(*v*) **Reinforced soil for repair of cutting failure** (Fig. 2.16)

Material used: Geogrid reinforcement used with failed material.

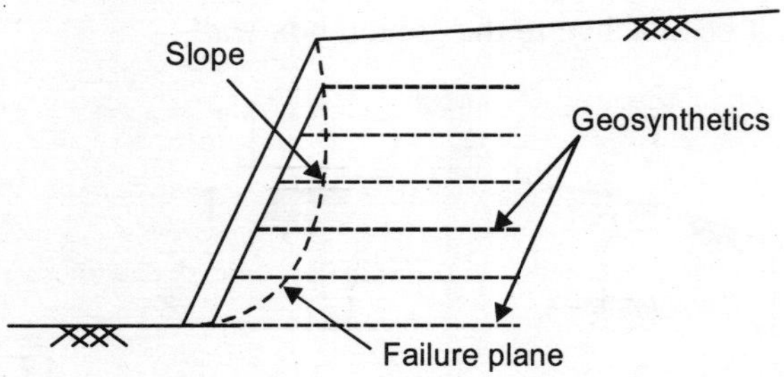

Fig. 2.16: Geosynthetics for repair of cut

2.7.3 Pipe work application (Fig. 2.17)

Materials used: Geogrid or special shaped inclusions acting together with usual pipe backfilling material.

Comments: Produces improved bedding conditions around pipes and increases lateral earth support.

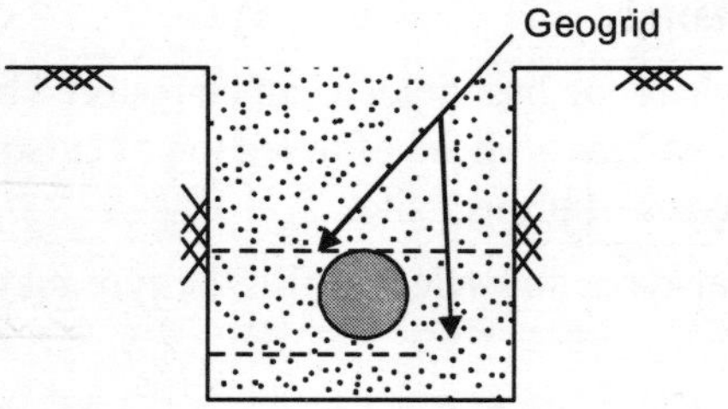

Fig. 2.17: Geogrid in pipe work application

2.7.4. Railway application (Fig. 2.18)

(*i*) Reinforcement of railway embankment to provide stability over poor sub soil and to protect embankment from subsidence, if any.

Materials used: Geogrid and strip reinforcement and indigenous fill.

Comments: Low strength thermoplastic netting has been found to be effective as a reinforcing material.

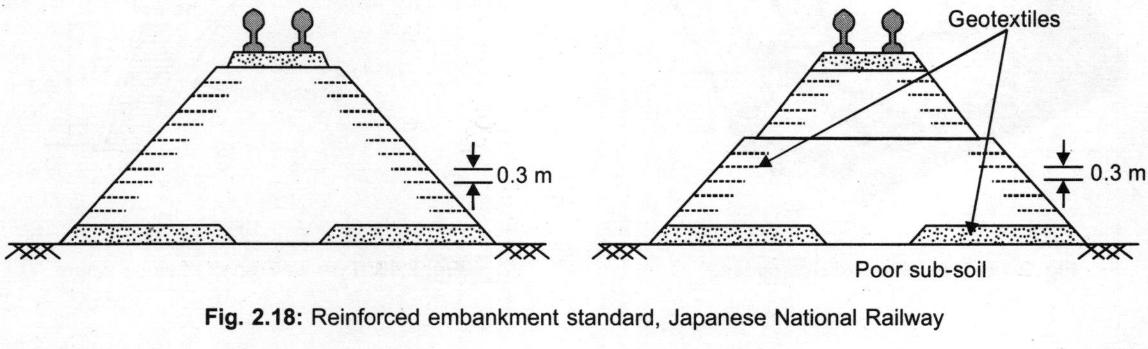

Fig. 2.18: Reinforced embankment standard, Japanese National Railway

(*ii*) **Railway formation rehabilitation** (Fig. 2.19)

Geogrid drastically reduces track attention, ballast puncture loss, mud pumping and differential settlement.

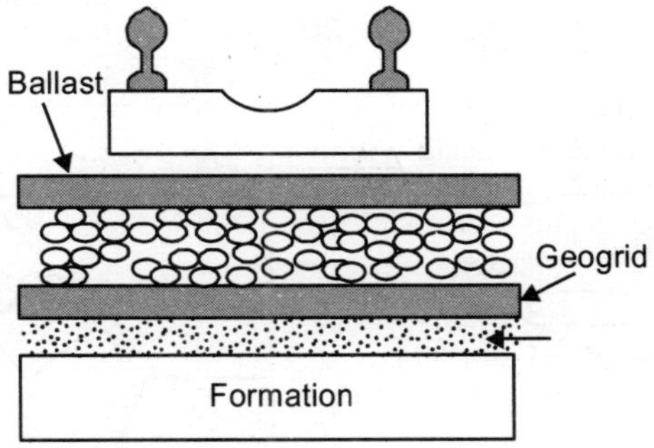

Fig. 2.19: Formation rehabilitation in railways

2.7.5 Road applications

(*i*) **Geotextiles can be used to improve performance of subgrade.** (Fig. 2.20)

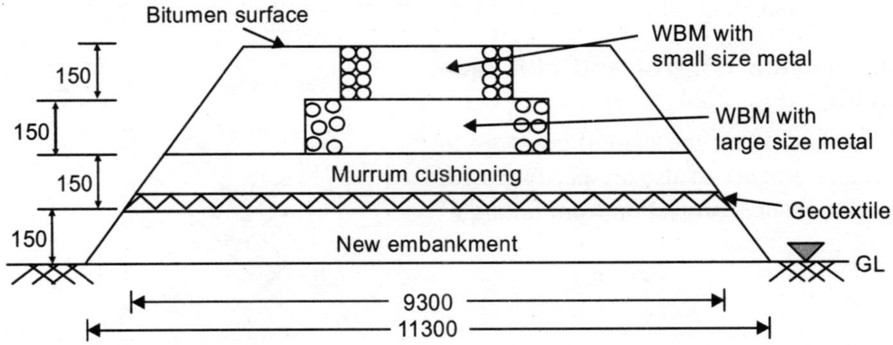

Fig. 2.20: A typical c/s of a road showing use of geotextile over new embankment

(*ii*) **Soil reinforcement and restraintment** (Fig. 2.21)

Geogrid is highly useful in ground improvement for roads, airfields and reduction or settlement.

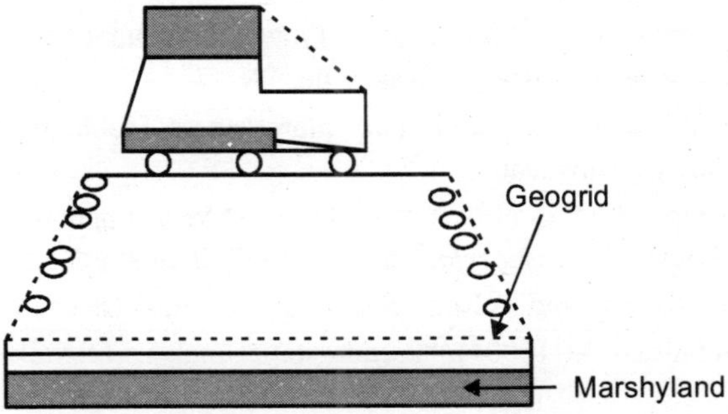

Fig. 2.21: Improving soil strength by using geogrid

2.7.6 Storage bunkers (Fig. 2.22)

Materials used: Sloping reinforcement concrete facing units or anchor and granular selected fill.

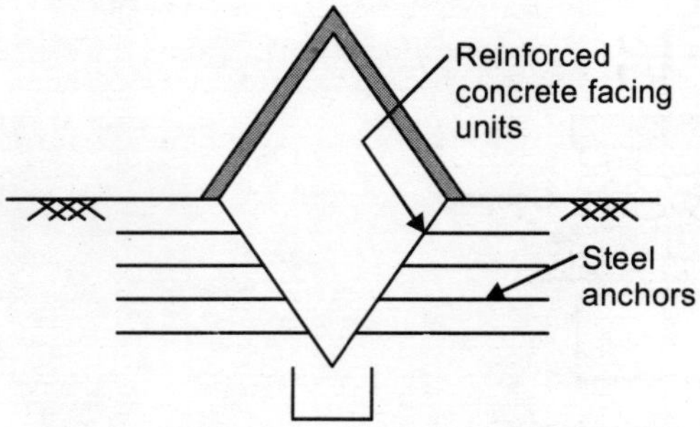

Fig. 2.22: Steel anchors used to create underground space

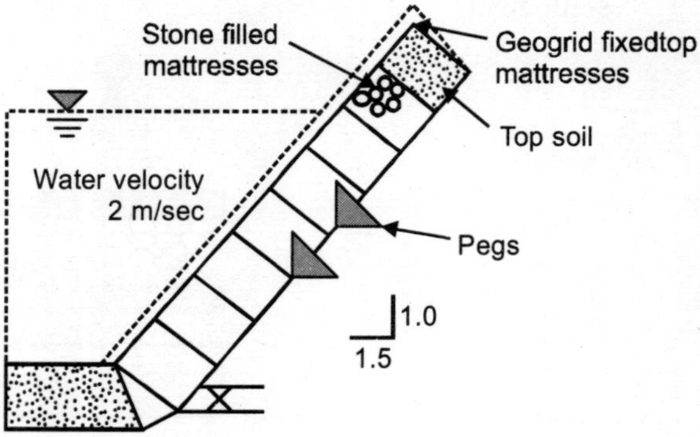

Fig. 2.23: Geogrid to control slope erosion

2.7.7 Canal protection works (Fig. 2.23)

Geogrid helps in dissipating pore water pressure and protects embankment from erosion.

2.7.8 Slope erosion control and cliff face protection (Fig. 2.24)

Geogrid can be laid along the natural contours to prevent soil displacement and also on speedy growth of vegetation and ensures protection from falling of loose stores.

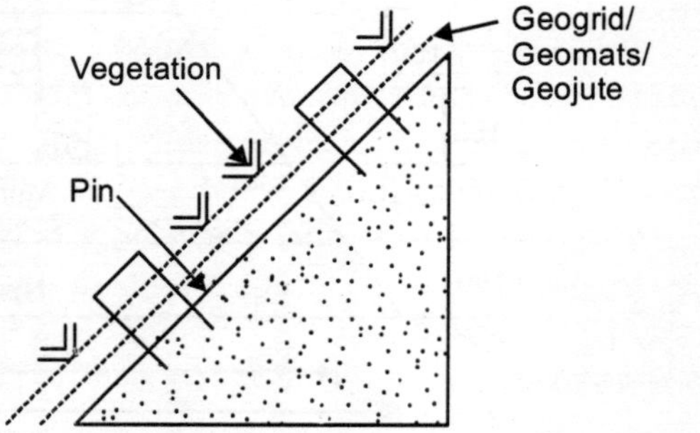

Fig. 2.24: Geogrid to protect the cliff from soil erosion

References

1. John, N.W.M. (1979). "Soil Reinforcement", *Civil Engineering*, pp. 47–53.
2. Lawrence, C.A. and Koerner, R.M. (May, 1988). "Flow Behaviour of Kinked Strip Drains", *Proc. Sym., Geosythetics for Soil Improvement.* "ASCE", Nashville, TN.
3. Mittal, S. (2012). "Recent Experiences of Ground Improvement Techniques" Keynote Lecture delivered in Indian Institute of Technology, Guwahati.
4. Ramana, G.V. and Venkatappa Rao, G. (1997). "Solid Waste Management—The Need for Developing Design and Construction Guidelines", *Proc. Geosythetics Asia*, 1997, Bangalore, I: IV.3-IV.14.
5. Saran, S. (2010). "Reinforced Soil and its Engineering Applications" (Second Edition), p. 423.
6. Saran, S., Mittal, S. and Gosavi, M. (2004). "Pseudo-static Analysis of Nailed Vertical Excavations", *Indian Geotechnical Journal.*
7. Venkatappa Rao, G. (1988). "Geosynthetics—A Revolution in Civil Engineering" *Journal of NESS*, Nepal 1:1:1–5.

Reinforced Earth Retaining Walls

CHAPTER 3

3.1 INTRODUCTION

The concept of soil reinforcement is very old and the technique of soil reinforcement in crude form has been practiced for construction of dwellings, foundation and embankments. But it is only recently that Vidal (1966) carried out a systematic and rational study of the beneficial effects of reinforcing the soil and demonstrated its application in retaining wall construction.

Reinforced earth retaining walls consists of a skin made up of pre-fabricated elements or units of metal, concrete or reinforced plastics to which strips or sheets of tensile stress resistant material are attached and embedded horizontally in the backfill. The reinforcing element interact with the frictional backfill and shear forces which develop along their surfaces, as the wall face or skin moves under lateral thrust, stabilises the backfill.

In analysing these walls, considerations must be given to both external and internal stability. In external stability, the safety of wall against sliding, overturning and bearing failure is checked in conventional manner. The assessment of internal stability involves checking the adequacy of each tension and pull-out failure.

In this chapter, a rational procedure of designing reinforced earth walls situated in seismic zones considering all the aspects mentioned above and taking into account both horizontal and vertical seismic forces has been presented.

On the basis of both analytical and experimental studies, two analyses namely—

(*i*) Tie back-wedge analysis and

(*ii*) Coherent gravity analysis have been suggested for checking the internal stability of reinforced earth wall.

The main differences between these two analyses originate from the basic assumptions made about the shape of the failure wedge, rotation of the wall facing and the lateral earth pressure acting within the reinforced soil.

The tie back-wedge analysis is more popular in use and the same is discussed here briefly.

3.2 TIE BACK-WEDGE ANALYSIS

The tie back-wedge analysis for the internal stability of reinforced soil structures considers the stability of individual reinforcement, resistance to horizontal sliding of the upper element of the wall and the stability of wedge within the reinforced fill. Lateral pressures are considered to be in active condition throughout height of structure. The 'at rest' pressure developed during the construction phase is assumed to be transitory. The potential failure plane is considered passing through the toe of wall and making an angle (45° + $\phi/2$) from the horizontal.

The tensile force (T_i) in a strip at depth h_i is calculated by

$$T_i = K_{aw} \cdot \sigma_{vi} \cdot S_H S_V \qquad ...(3.1)$$

If continuous sheet of reinforcement (geogrid/geotextile/mesh etc.) is used, the tensile force (T_i) per metre width in the reinforcement at depth h_i is given by

$$T_i = K_{aw} \cdot \sigma_{vi} \cdot S_V \qquad ...(3.2)$$

where,

K_{aw} = coefficients of active earth pressure (Rankine) of wall fill

σ_{vi} = vertical soil pressure at reinforcement

S_H = horizontal spacing of reinforcing strips

S_V = vertical spacing of reinforcement.

Factor of safety against pullout failure is determined by

$$F_p = \frac{2.w.f^* L_e \sigma_i}{T_i} \qquad ...(3.3)$$

where,

w = width of reinforcement (w = 1.0 for sheets)

f^* = apparent coefficient of friction

L_e = effective length of reinforcement = $L - (H - h_i) \tan\left(45° - \frac{\phi}{2}\right)$

In IIT Roorkee, an extensive model test program had been conducted (Khan, 1991). The same in brief is discussed below.

3.2.1 Model studies

A model reinforced earth retaining wall 1.0 m high with wooden skin panel was monitored. Uniformly distributed surcharge was placed on entire backfill. Few tests were conducted by placing the uniformly distributed surcharge beyond reinforced zone. Pressure cells were employed to measure earth pressure distribution on skin and on base of backfill, but they did not perform satisfactorily. Reinforcing strips were instrumented with staingauges to measure tension induced into them.

3.2.2 Soil and Reinforcement

Dry sand was used as backfill at 1.6 t/m^3 density. Three types of reinforcements (Geogrid, aluminium strips and wooden strips) were used for this study.

3.2.3 Instrumentation

Instrumentation of reinforcing strips of R-1 and R-2 was done by pasting strain gauges on both sides of strips that formed opposite arms of bridge to measure tensile force. The strips were calibrated them vertically and putting weights on hanger as shown in Fig. 3.1. Pressure cells were employed to measure earth pressure coming on skin panels, as shown in Fig. 3.2, and at the bottom of test tank vertical pressure distribution. Pressure cells were removed after few tests as they did not perform satisfactorily.

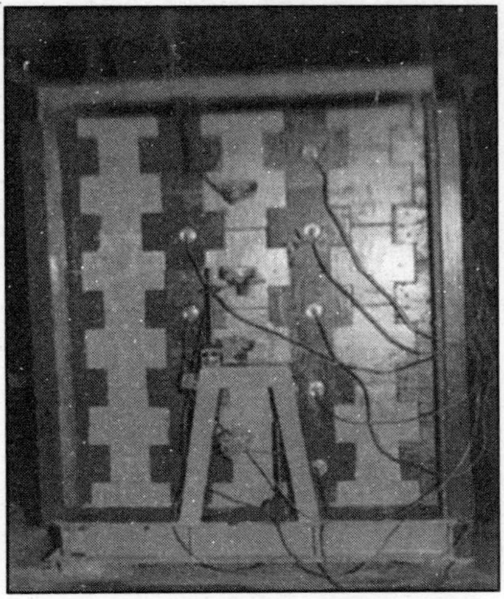

Fig. 3.1: Calibration of strips

Fig. 3.2: Close up of reinforced earth retaining wall showing pressure cells

3.2.4 Test set-up

A test tank was used with some modifications. The loading frame, rigid wall, clamps, screws to hold wall, screw jack and proving ring were removed. Side wall was covered with transparent sheet up to the front channel of the frame. Wooden panels were fabricated to use them as skin. A stand for fixing the dial gauges was also provided. Details of panel and tank are given in Fig. 3.3. Surcharge was applied in the form of sand bags, each weighing 40 kg put on a 4 mm thick steel sheet of 51 kg weight.

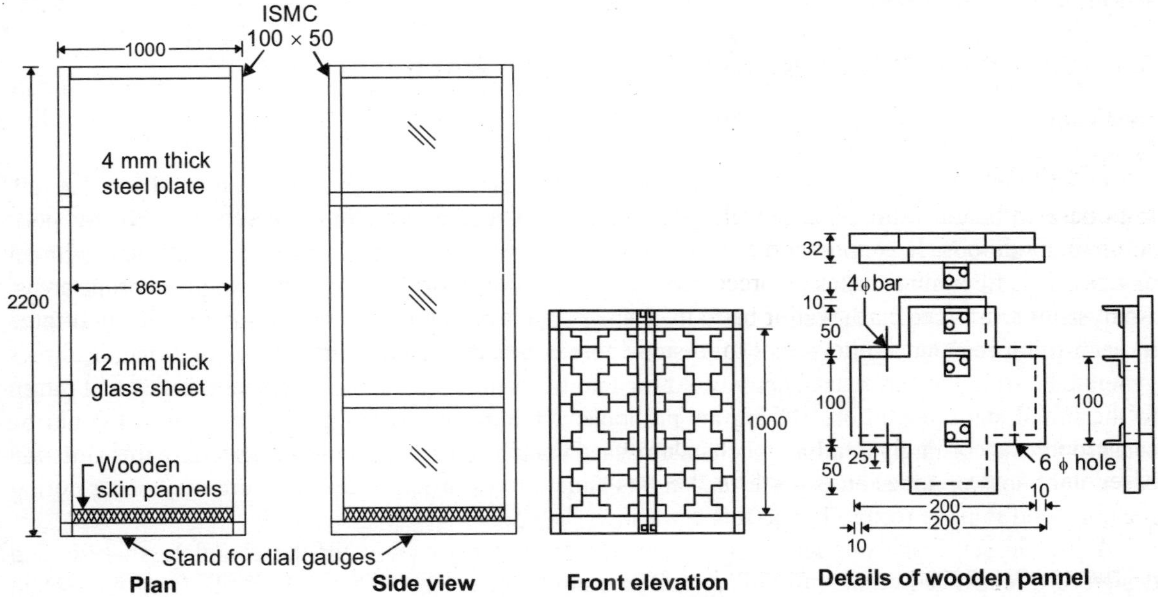

(All dimensions are in mm)

Fig. 3.3: Model tank with skin panel details

3.2.5 Test procedure

The procedure adopted for the construction of model wall was similar to that used in the field and described in the following steps:

(*i*) The first row of skin panels, consisting of 2 side panels, 3 half panels and 2 full panels was assembled in a line at the base of the tank and ensured plumb with a trisquare. Arrangements were made to keep the panels in straight alignments and vertical, and also to prevent flow of sand through sides. A vertical line was also to prevent flow of sand through sides. A vertical line was also marked on the perspex sheet side wall to see its deviation from the plumb during back-filling.

(*ii*) The backfilling of sand was done by using rainfall technique up to the level at which the first layer of reinforcement has to be laid.

(*iii*) After making the sand surface level, the reinforcing strips were laid on it and properly attached to the skin panels (Fig. 3.4). One central reinforcing strip was instrumented with strain gauges and the leads were carefully taken to the data acquisition system.

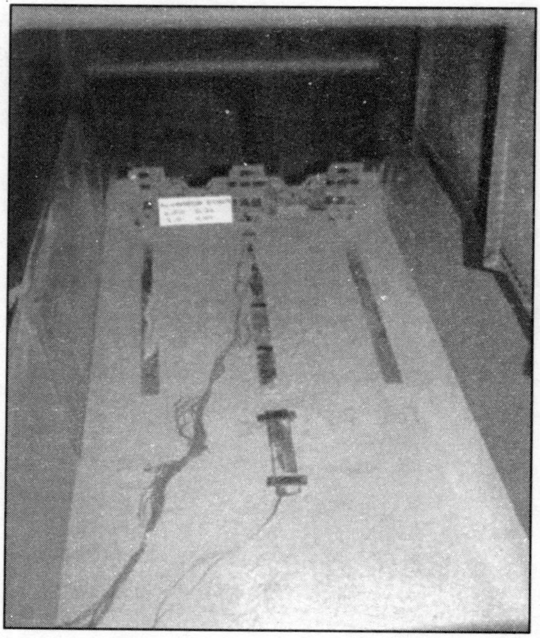

Fig. 3.4: Laying of reinforcement

The summary of tests performed in this study and the results thereof are described elsewhere (Khan, 1991).

3.3 SEISMIC DESIGN OF REINFORCED EARTH WALL

Beneficial effects of reinforcing the soil have been demonstrated by its large scale adoption in the live structures of varying nature.

Unlike rigid retaining walls the stress concentration near the toe is small because of the relatively high base to height ratio. This renders reinforced earth retaining walls eminently suitable for poor sub-soil conditions. It is appreciably flexible and can withstand large differential settlement without distress. This flexibility makes the reinforced earth structures more suitable for earthquake prone areas. Analysis of reinforced earth wall located in seismic area is based on the assumption that the total force in each reinforcement is the sum of the static forces before seismic event plus the dynamic forces generated during the earthquake activity. At the end of earthquake the reinforcement forces will return to the initial static condition. During an earthquake, failure by breakage of reinforcement could be catastrophic, accordingly a failure mechanism based upon lose of adhesion is preferred. In this situation based upon loss of adhesion is preferred. In this situation distortion of the structure may occur during seismic conditions is restored once this ceases.

A pseudo static analysis has been presented in following section for the complete analysis and design of a reinforced earth wall having uniform surcharge on the wall and backfill. Both forces had been taken into account. Findings of the model and prototype studies had been also incorporated in the analysis.

3.3.1 Analysis

Analysis of reinforced earth retaining wall being presented is based upon prediction of dynamic earth pressure resulting from the ground motion. The analysis covers the following modes of failure.

(*a*) External stability
- (*i*) Sliding failure
- (*ii*) Overturning failure
- (*iii*) Tilting or bearing failure

(*b*) Internal stability
- (*i*) Tension failure
- (*ii*) Pull-out failure

A reinforced earth wall is considered of width L and height H as shown in Fig. 3.5. The parameters are defined below:

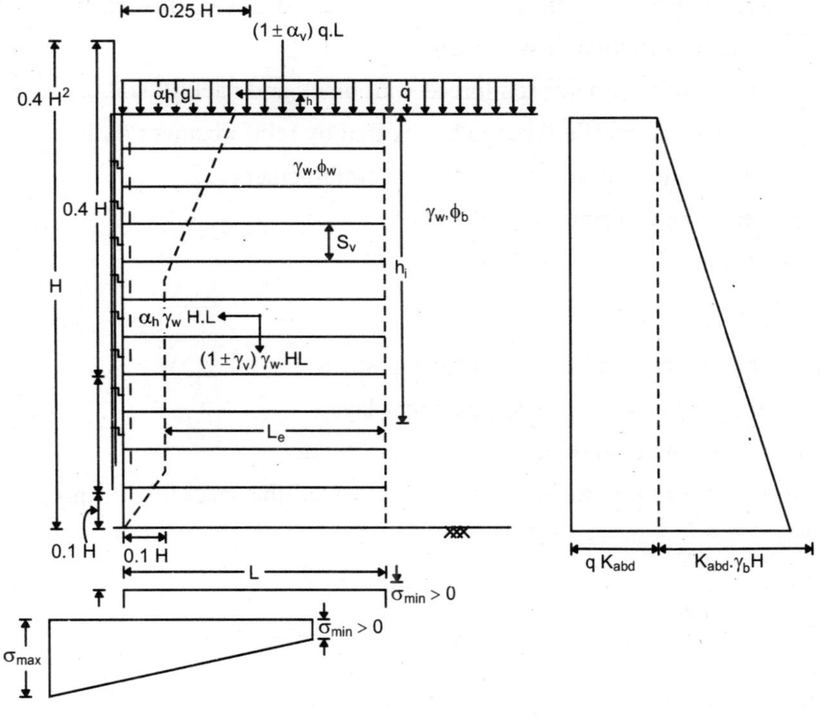

Fig. 3.5: Reinforced earth wall

H = overall height of the reinforced wall

h_i = height of reinforced earthwall/fill above level of the *i*th layer of reinforcement

L = length of reinforcement, at right angles in plan to the face of the wall

L_e = Length of reinforcement beyond the potential failure plane, at right angles in plan to the face of wall

γ_w = unit weight of wallfill

γ_b = unit weight of backfill

q = intensity of uniformly distributed surcharge

h = height of C.G. of the surcharge q from backfill surface

ϕ_w = angle of internal friction of the wallfill

ϕ_b = angle of internal friction of the backfill

K_{awd} = dynamic coefficient of active earth pressure

K_{abd} = dynamic coefficient of active earth pressure

σ_{max} = ground bearing pressure under front edge of wall trapezoidal distribution of pressure

σ_{min} = ground bearing pressure under back edge of the wall assuming trapezoidal distribution of pressure

μ = coefficient of friction between the soil (at base) reinforced earthwall

f^* = apparent coefficient of friction between soil (wall fill) reinforcement

δ_w = angle of friction of wall facing

δ_b = angle of friction of reinforced wall block and backfill soil

T_i = maximum tensile force to be resisted by reinforcement ith layer

T_{allow} = permissible tensile force for the reinforcement

α_h = horizontal seismic coefficient

α_v = vertical seismic coefficient

$= \dfrac{\alpha_h}{2.0}$

S_H = horizontal spacing of reinforcing strips

S_V = vertical spacing of reinforcement layers

w = width of reinforcing strip

In developing the non-dimensional charts, it is assumed that backfill soil properties are same as those of wall i.e.,

$$\gamma_b = \gamma_w = \gamma$$

$$\phi_b = \phi_w = \phi$$

$$\delta_b = \delta_w = \delta$$

$$K_{awd} = K_{awd} = K_{ad}$$

(a) External Stability

(i) Sliding failure

$$\text{Factor of safety, } F_s = \frac{\text{Resisting force}}{\text{Sliding force}}$$

$$F_s = \frac{\mu\ (\gamma_{w\,HL+qL})\ (1 \pm \alpha_v)}{\frac{1}{2}K_{abd}\gamma_b H^2 + K_{abd} qH + \alpha_h \alpha_w HL = \alpha_h qL}$$

$$\frac{L}{H} = \frac{F_s.K_{ad}\ (0.5 + \frac{q}{\gamma_H})}{\left(1 + \frac{q}{\gamma_H}\right)[\mu\ (1 \pm \alpha_v)\ -\ F_s \alpha_h]}$$

where,

$$K_{ad} = \frac{\cos^2(\phi - \psi) \times (1 \pm \alpha_v)}{\cos\psi \cos(\delta + \psi)} \times \frac{1}{\left[1 + \sqrt{\frac{\sin(\phi+\delta)\sin(\phi+\psi)}{\cos(\delta+\psi)}}\right]^2} \quad ...(3.5)$$

$$\psi = \tan^{-1}\left(\frac{\alpha_h}{1 \pm \alpha_v}\right) \quad ...(3.6)$$

$$\alpha_h = \frac{\alpha_h}{2} \quad ...(3.7)$$

Factor of safety against sliding lies between 1.5 and 2.0.

(ii) Overturning failure

Factor of safety against overturning F_0

$$= \frac{\text{Stablilisation moment}}{\text{Overturning moment}}$$

or,

$$F_0 = \frac{(1 \pm \alpha_v)\left[\frac{\gamma_w HL^2}{2} + \frac{qL^2}{2}\right]}{\frac{K_{abd}\gamma_b H^3}{6} + K_{abd}\frac{qH^2}{2} + \alpha_h \frac{\gamma_w H^2 L}{2} + \alpha_h\, qL\,(H+h)} \quad ...(3.8)$$

Equation (3.8) can be made non-dimensional by dividing numerator and denominator by $\frac{1}{2}\gamma H^2$ and can be written as

$$(1 \pm \alpha_v)\left(1 + \frac{q}{\gamma H}\right)\left(\frac{L}{H}\right)^2 - \alpha_h F_0\left[1 + \frac{2q}{\gamma H}\left(1 + \frac{h}{H}\right)\right]\left(\frac{L}{H}\right) - F_0 K_a\left(\frac{1}{3} + \frac{q}{\gamma H}\right) = 0$$

$$\frac{L}{H} = \frac{-b + \sqrt{b^2 - 4ac}}{2a} \quad ...(3.9)$$

where,

$$a = (1 \pm \alpha_v)\left(1 + \frac{q}{\gamma H}\right) \quad ...(3.10)$$

$$b = -\alpha_h F_0\left[1 + \frac{2q}{\gamma H}\left(1 + \frac{h}{H}\right)\right] \quad ...(3.11)$$

$$c = -F_0 K_a\left(\frac{1}{3} + \frac{q}{\gamma H}\right) \quad ...(3.12)$$

(iii) Tiltling or bearing failure

Maximum pressure, σ_{max} at base (toe) is given by

$$\sigma_{max} = (1 \pm \alpha_v)(\gamma_w H + q) + \frac{\frac{1}{6}K_{abd}\gamma_b H^3 + \frac{1}{2}K_{abd}qH^2 + \alpha_h \gamma_w \frac{LH^2}{2} + qL\alpha_h(H+h)}{\frac{L^2}{6}} \quad ...(3.13)$$

$$\frac{\sigma_{max}}{\gamma H} = (1 \pm \alpha_v)\left(1+\frac{q}{\gamma H}\right)+\left[K_{ad}+\left(\frac{H}{L}\right)^2\right]\left(1+\frac{3q}{\gamma H}\right)+3\alpha_h\frac{H}{L}\cdot\left[1+\frac{2q}{\gamma H}\left(1+\frac{h}{H}\right)\right] \quad ...(3.14)$$

Similarly the minimum pressure σ_{min} at base (heel) is given by

$$\frac{\sigma_{min}}{\gamma H} = (1 \pm \alpha_v)\left(1+\frac{q}{\gamma H}\right)-\left[K_{ad}+\left(\frac{H}{L}\right)^2\right]\left(1+\frac{3q}{\gamma H}\right)+3\alpha_h\frac{H}{L}\cdot\left[1+\frac{2q}{\gamma H}\left(1+\frac{h}{H}\right)\right] \quad ...(3.15)$$

3.4 INTERNAL STABILITY

Analysis of internal stability is based on the findings of the experimental results on reinforced earth wall and frictional characteristics of soil-reinforcement interface. Basically these are:

(*i*) Tension force in reinforcing strips develop corresponding to active earth pressure distribution.

(*ii*) Rupture wedge in reinforced earth wall develops as shown in Fig. 3.6.

(*iii*) The coefficient of friction f^* between rough reinforcing strips and soil is given by

$$f^* = [1 - 0.625\,(\alpha - 2)]\,[1 - 0.35\,(L - 1)]\,\mu \quad ...(3.16)$$

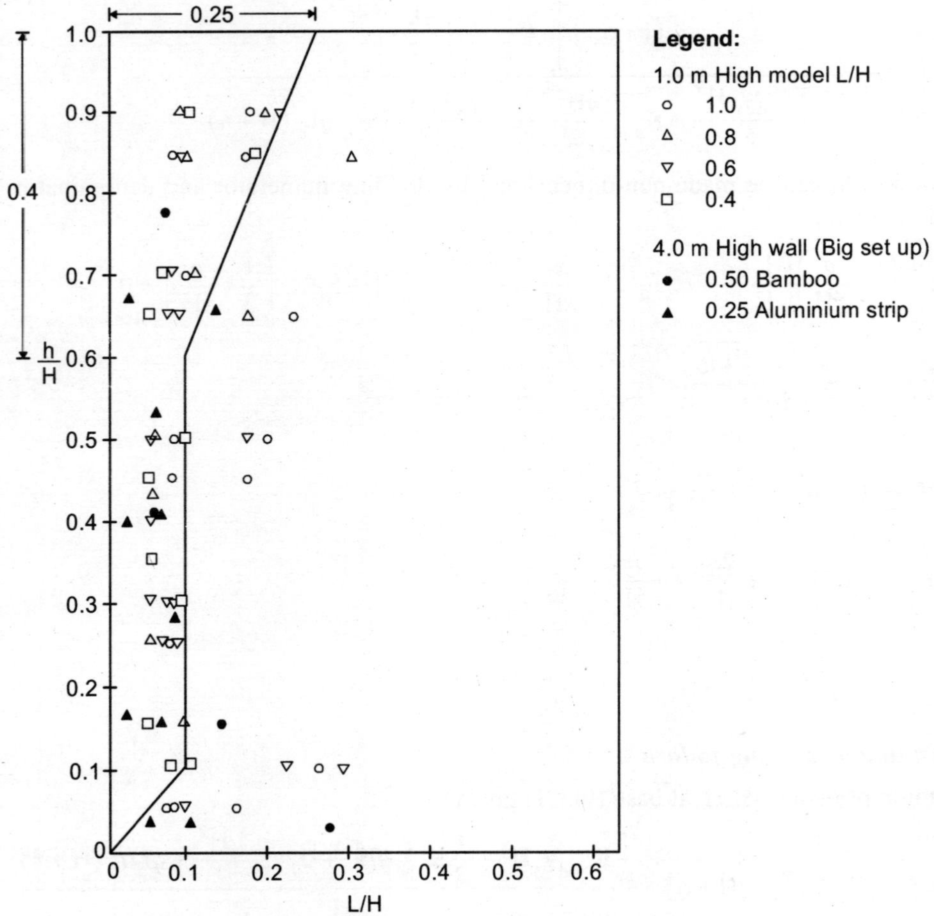

Fig. 3.6: Location of maximum tension in strips

3.4.1 Tension failure

The maximum tension (T_i) in reinforcement strip can be calculated by eq. (3.16).

$$T_i = K_{ad} (q + \gamma H_1 (1 \pm \alpha_v) S_H S_V \quad ...(3.17)$$

When continuous sheet of reinforcement like geogrids or geotextiles is to be used, the maximum tension to be resisted by per unit width of reinforcement may be determined by eq. (3.18).

$$T_i = K_{ad} (q + \gamma h_i (1 + \alpha_v) S_V \quad ...(3.18)$$

If uniform spacings of reinforcement are adopted for entire height of wall the maximum tension will be induced in bottommost layer. T_i must always be less than T_{allow}.

3.4.2 Pull-out failure

Rupture wedge is considered as shown in Fig. 3.6.

$$L_e = \frac{T_i F_p}{2w\sigma_i \, f^*} \quad ...(3.19)$$

where,

F_p = factor of safety against pull-out

σ_i = normal stress at ith strip level $(1 \pm \alpha_v)(q + \gamma H_i)$

$$L_e = \frac{K_{ad} S_H S_V F_p}{2w \, f^*} \quad ...(3.20)$$

or

$$\frac{L_e}{H} = \frac{K_{ad}}{2D_p} F_p \quad ...(3.21)$$

$$D_p = \frac{wf^* H}{S_H S_V}$$

is a non-dimensional factor which takes into account the reinforcement characteristics.

When $h_i < 0.4$ H, then L_e from Fig. 3.6.

$$L_e = L - 0.1H - (0.4\,H - h_i) \frac{0.25H - 0.1H}{0.4H} \quad ...(3.22)$$

Now eq. (3.20) becomes

$$\frac{L}{H} = 0.25 - 0.375 + \frac{h_i}{H} + \frac{K_{ad} F_p}{2D_p} \quad ...(3.23)$$

Value of L/H will be maximum when $h_i = (S_V/2)$ i.e., the upper most reinforcement layer will give us the critical value of L/H. Ignoring the small quantity $0.375 \times (h_i/H)$, we get

$$\frac{L}{H} = 0.25 + \frac{K_{ad} F_p}{2D_p} \quad ...(3.24)$$

3.5 NON-DIMENSIONAL DESIGN CHARTS

Calculations were made using eqs. (3.4, 3.9, 3.14, 3.15 and 3.23) for uniformly distributed surcharge ($q/\gamma H$ = 0.0, 0.5 and 1.0 and h/H = 0 and 0.2). However, the presentation of charts has been limited to q/H = 0.0. A brief description is as under.

3.5.1 Sliding failure

The non-dimensional charts between L/H and ϕ for different values of μ (0.4, 0.5 and 0.6) have been prepared for F_s = 1.5 and 2.0 and α_h = 0.0, 0.05, 0.10 and 0.15. Computations were done using eq. (3.4) for both positive and negative values of α_v. It was found that the computed values of L/H were higher when α_v is taken negative, the same are given in Fig. 3.7.

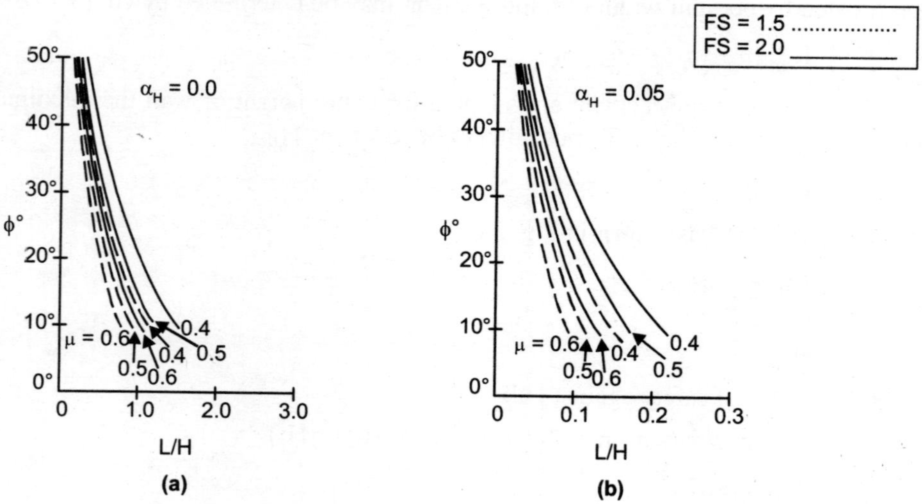

Fig. 3.7: (A) Design chart for sliding failure ($q/\gamma H = 0.0$)

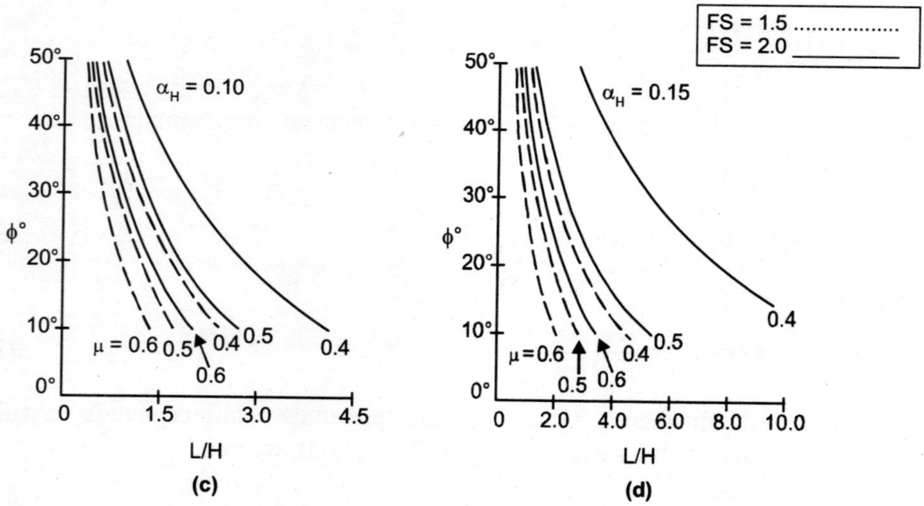

Fig. 3.7: (B) Design chart for overturning failure

3.5.2 Overturning failure

Value of factor of safety against overturning also lies between 1.5–2.0. Using eq. (3.9), non-dimensional charts, as shown in Fig. 3.8, have been prepared to give the value of L/H for the given value of ϕ, F_0 and α_h. In this case also, negative values of α_v gave the higher value of L/H.

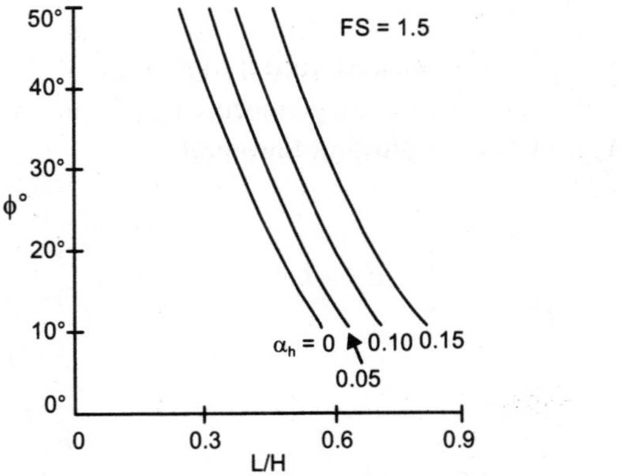

Fig. 3.8: (a) Design chart for sliding failure ($q/\gamma H = 0.0$)

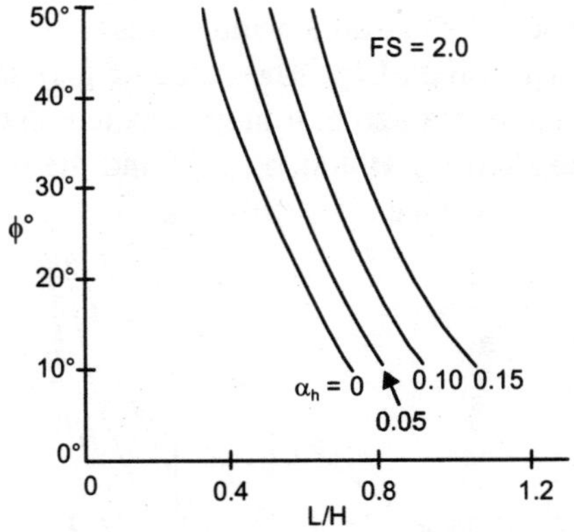

Fig. 3.8: (b) Design chart for overturning failure

3.5.3 Tilting or bearing failure

Using eq. 3.14 non-dimensional charts have been prepared (Fig. 3.9) to evaluate the value of L/H for given values of ϕ and σ_{max}/H. σ_{max} may be taken equal to the allowable soil pressure, σ

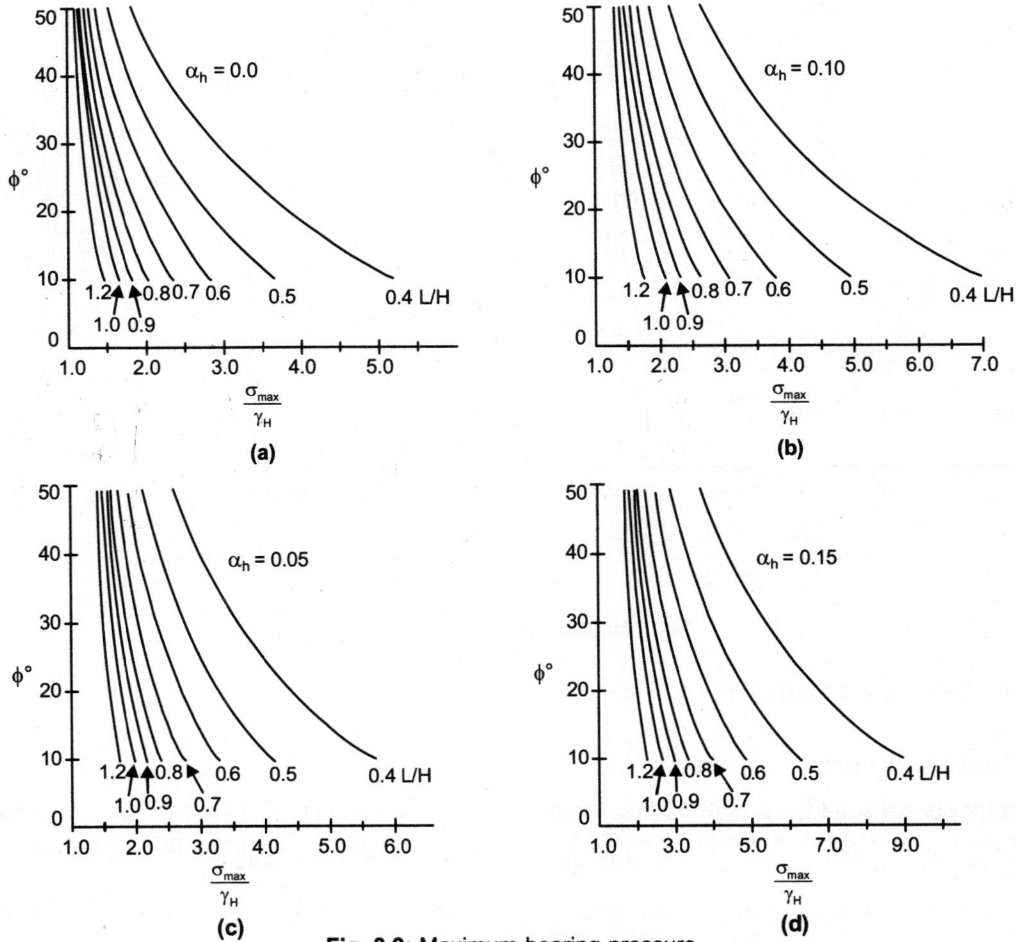

Fig. 3.9: Maximum bearing pressure

(bearing capacity). Non-dimensional charts between σ_{min} /H and ϕ are prepared with using eq. 3.15 and are shown in Fig. 3.10. The value of L/H for given value of ϕ should be such that the value of σ_{min} /H is greater than zero. It may be noted that positive values of α_v gave (in eq. 3.14 and 3.15) the higher values of σ_{max} /H and σ_{min} /H and the same are used for plotting these charts.

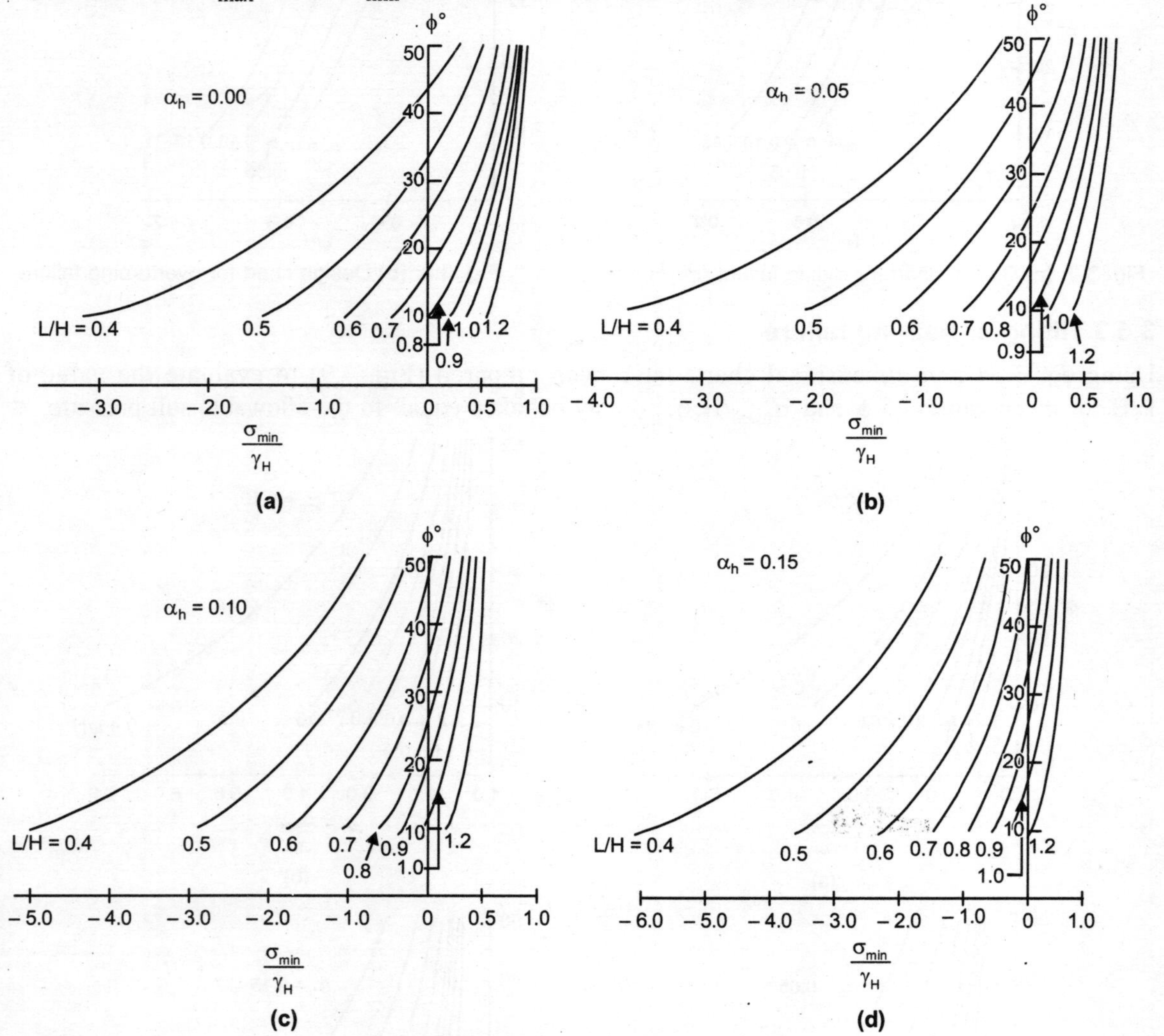

Fig. 3.10: Minimum bearing pressure

3.6 INTERNAL STABILITY

3.6.1 Tension failure

The maximum tension (T_i) in reinforcement can be calculated using eq. (3.17) or eq. (3.18), as the case may be. Value of K_{ab} can obtained by eq. (3.5). Positive value of α_v will give the maximum value of tension.

If uniform spacings of reinforcement are adopted for maximum tension will be induced in bottommost layer. T_i must always be less than T_{allow}.

3.6.2 Pull-out failure

Figure 3.11 gives the plots of calculations based on eq. (3.23) for D_p = 0.5, 1.0, 2.0, 4.0, and 8.0. Positive values of α_v give higher values of K_{ab} and the same are included in charts shown in Fig. 3.11.

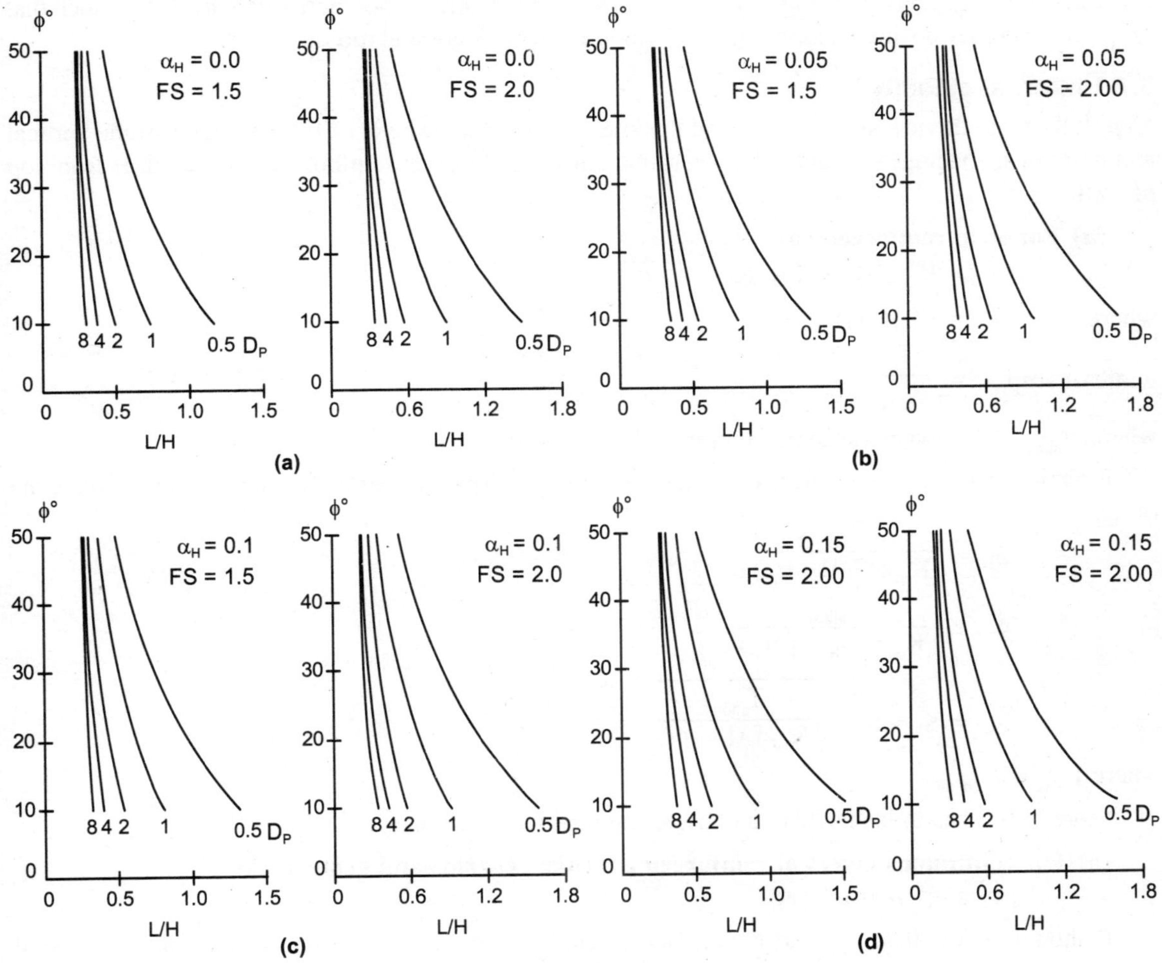

Fig. 3.11: ϕ *vs* L/H for different D_p values

3.6.3 Design procedure

Detailed design procedure and one example to illustrate the use of design charts is presented under this section.

Design steps

Given data:

Height of wall, H; Density of soil, γ; Allowable soil pressure, σ; Angle of internal friction, ϕ; Friction coefficient at base, μ; Vertical seismic coefficient, α_v; and Horizontal seismic coefficient, α_h.

Step 1: Select suitable reinforcing material. Find its permissible tensile T_{allow} and coefficient of friction f^*.

3.7 EXTERNAL STABILITY

Step 2: Choose suitable factor of safety and obtain L/H from non-dimensional charts (Figs. 3.7–3.9) for sliding, over-turning and bearing failure. Pick-up maximum of the above three L/H ratio. For this L/H ratio, find $\sigma_{min}/\gamma H$ from Fig. 3.10. If it is positive then O.K., if not increase value of L/H such that $\sigma_{min}/\gamma H$ becomes positive. Adopt this L/H value for further computations.

3.7.1 Internal stability

Step 3: Keeping in view size of panel and thickness of compacted layer of fill material, assume vertical and horizontal spacings (S_V and S_H). Determine tension induced in reinforcement at depth H from top of wall.

(*a*) For strip reinforcement

$$T_i = K_{ad}\sigma_i(1+\alpha_v)S_H.S_V$$

where $\sigma_i = (\gamma H + q)$

width of strip, $w = \dfrac{T_i}{t_{allow}}$

where, t_{allow} is allowable tension strength per unit width of strip.

If choice of size of strip is restricted. Assume one of the spacings either S_H and S_V and calculate the other.

$$S_H = \frac{T_{allow}}{K_{ad}\sigma_i S_V(1+\alpha_v)}$$

or

$$S_V = \frac{T_{allow}}{K_{ad}\sigma_i . S_H(1+\alpha_v)}$$

or

$$S_H = S_V = S = \sqrt{\frac{T_{allow}}{K_{ad}\sigma_i(1+\alpha_v)}}$$

where $T_{allow} = w.t_{allow}$.

Value of K_{ad} has to be calculated considering positive value of α_v.

(*b*) For continuous layers of reinforcement (like geogrids and geotextiles)

$$T_i = K_{ad}\sigma_i(1+\alpha_v)S_V$$

T_i should be less than T_{allow}, if not replace T_i and T_{allow} in the above equation and find the vertical spacing.

$$S_V = \frac{T_{allow}}{K_{ad}\sigma_i(1+\alpha_v)}$$

If the spacing of reinforcement is to be varied with the height of wall, replace H by h_i and repeat step 3 (*a*) or (*b*).

Divide the wall height in sections of equal spacings and design skin panels accordingly.

Step 4: Pullout failure check.

(a) Find friction coefficient f^*

(*i*) Find f^* from pull out test data, if possible

(*ii*) If not, find friction angle by sliding shear test. Multiply it by some suitable factor, *c*, which can take care of length of reinforcement and overburden pressure on it. Find f^*,

$$f^* = c.\tan\delta$$

(b) Compute D_p factor

$$D_p = \frac{f^* w.H}{S_H.S_V} \text{ for strips}$$

$$D_p = \frac{f^* H}{S_V} \text{ for sheets}$$

(*c*) Choose suitable factor of safety against pull out. Find L/H from Fig. 3.11.

(*d*) Compare this net L/H ratio with the L/H ratio computed under step 2.

(*e*) Adopt maximum of these two L/H ratios for construction.

Note: (1) Loss of strength with time should be considered according to the designed life of the wall. Highway Research Record No. 242 (Vidal 1969) suggests following equation for metal loss expressed in millimetres, $y = 0.035 \times - 0.09$, where x is the number of years. This equation is based on survey carried out in America on highway culverts made of galvanised steel.

(2) All possible potential slip surfaces should be investigated using prevailing theories.

EXAMPLE 3.1: Given data:

$\alpha_h = 0.05$, H = 5.0 m, $\gamma = 1.6$ t/m³, $\phi = 30°$, $\mu = 0.6$, $\sigma = 15$ t/m², F.S. = 2.0 against all type of failures.

Step 1: Galvanised strips 2 .0 mm thick; permissible tensile force = 300 kg/cm width friction coefficient $f^* = 0.30$

Step 2:

From Fig. 3.7 (*b*) for $\phi = 30°$ and $\mu = 0.6$; $\frac{L}{H} = 0.65$

From Fig. 3.8 (*b*) for $\phi = 30°$ and $\alpha_h = 0.05$; $\frac{L}{H} = 0.53$

From Fig. 3.9 for $\phi = 30°$ for $\frac{\sigma_{max}}{\gamma H}$ and $\mu = 0.6$; $\frac{L}{H} = 0.74$; Maximum $\frac{L}{H} = 0.75$

From Fig. 3.10 (*b*) for $\phi = 30°$ and $\frac{L}{H} = 0.75$, $\frac{\sigma_{max}}{\gamma H} = 0.19$ it is positive hence O.K.

Step 3: Let

S_H = 20 cm = 0.2 m

S_V = 25 cm = 0.25 m

(*a*) $T_{bottom} = K_{ad}\sigma_i(1+\alpha_v)S_H.S_V$

$= 0.337 \times 1.6 \times 5 \times (1 + 0.025) \times 0.2 \times 0.25 = 0.138$ ton

= 138 kg (say 140 kg)

The metal loss due to corrosion for 40 years of life span will be

$y = 0.035 \times - 0.09$ (as per Highway Research Record No. 242; Vidal, 1969)

where x is the number of years.

$y = (0.035 \times 40 - 0.09)$ mm

= 1.31 mm (say 1.50 mm)

Now available thickness of strip = 2.0 – 1.5 = 0.5 mm

and strength of strip per cm width $= \frac{300}{2} \times 0.5 = 75$

T_{allow} = 75 kg

Width of strip $w = \dfrac{T_i}{t_{allow}} = \dfrac{140}{75} = 1.87$ cm

Provide 2.0 cm wide strip.

Step 4:

(*a*) $f^* = 0.30$ (from Step 1)

(*b*) $$D_p = \frac{f^* w.H}{S_H.S_V} = \frac{0.3 \times 0.02 \times 5.0}{0.20 \times 0.25} = 0.6$$

(*c*) Factor of safety = 2.0 given value of D_p does not change with height of wall.

From Fig. 3.11, for $D_p = 0.6$ and $\phi = 30°$

$\frac{L}{H}$ comes out to be 0.85 (say 0.9)

Provide

Length of G.I. strip = 4.5 m
Width of G.I. strip = 0.02 m
Thickness of G.I. strip = 0.002 m
Horizontal spacing = 0.20 m
Vertical spacing = 0.25 m

3.8 ADVANTAGES AND DISADVANTAGES OF REINFORCED EARTH WALL

3.8.1 Advantages of mechanically stabilised earth (MSE) walls

MSE walls have many advantages compared with conventional reinforced concrete and concrete gravity retaining walls. MSE walls:

- Use simple and rapid construction procedures and do not require large construction equipment.
- Do not require experienced craftsmen with special skills for construction.
- Require less site preparation than other alternatives.
- Needless space in front of the structure for construction operations.
- Reduce right-of-way acquisition.
- Do not need rigid, unyielding foundation support because MSB structures are tolerant to deformations.
- Are cost effective.
- Are technically feasible to heights in excess of 25 m (80 ft).

The relatively small quantities of manufactured materials required, rapid construction, and competition among the developers of different proprietary systems has resulted in a cost reduction relative to traditional types of retaining walls. MSE walls are likely to be more economical than other wall systems for walls higher than about 3 m (10 ft) or where special foundations would be required for a conventional wall.

One of the greatest advantages of MSE walls is their flexibility and capability to absorb deformations due to poor subsoil conditions in the foundations. Also, based on observations in seismically active zones, these structures have demonstrated a higher resistance to seismic loading than have rigid concrete structures.

Precast concrete facing elements for MSE walls can be made with various shapes and textures (with little extra cost) for aesthetic considerations. Masonry units, timber, and gabions also can be used with advantage to blend in the environment.

3.8.2 Advantages of reinforced soil slopes (RSS)

The economic advantages of constructing a safe, steeper RSS than would normally be possible are the resulting material and rights-of-way saving. It also may be possible to decrease the quality of materials required for construction. For example, in repair of landslides it is possible to reuse the slide debris rather than to import higher quality backfill.

Right-of-way savings can be a substantial benefit, especially for road widening projects in urban areas where acquiring new right-of-way is always expensive and in some cases unobtainable. RSS also provide an economical alternative to retaining walls. In some cases, reinforced slopes can be constructed at about one-half the cost of MSEW structures.

The use of vegetated-faced reinforced soil slopes that can be landscaped to blend with natural environments may also provide an aesthetic advantage over retaining wall type structures. However, there are some potential maintenance issues they must be addressed such as mowing grass-faced steep slopes, however, these can be satisfactorily handled in design.

In terms of performance, due to inherent conservatism in the design of RSS. They are actually safer than flatter slopes designed at the same factor of safety. As a result, there is a lower-risk of long-term stability problems developing in the slopes. Such problems often occur in compacted fill slopes that have been constructed to low factors of safety and/or with marginal materials (e.g. deleterious soils such as shale, fine grained low cohesive silts, plastic soils, etc.). The reinforcement may also facilitate strength gains in the soil over time from soil aging and through improved drainage, further improving long-term performance.

3.8.3 Disadvantages

The following general disadvantages may be associated with all soil reinforced structures:

Require a relatively large space behind the wall or outward face to obtain enough wall width for internal and external stability.

3.8.3.1 MSEW require select granular fill. (At sites where there is a lack of granular soils, the cost of importing suitable fill material may render the system uneconomical). Requirements for RSS are typically less restrictive.

3.8.3.2 Suitable design criteria are required to address corrosion of steel reinforcing elements, deterioration of certain types of exposed facing elements such as geosynthetics by ultra violet rays, and potential degradation of polymer reinforcement in the ground.

3.8.3.3 Since design and construction practice of all reinforced systems are still evolving specifications and contracting practices have not been fully standardised, especially for RSS.

3.8.3.4 The design of soil-reinforced systems often requires a shared design responsibility between material suppliers and owners and greater input from agencies geotechnical specialists in a domain often dominated by structural engineers.

3.9 RELATIVE COSTS

Site specific costs of a soil-reinforced structure are a function of many factors, including cut-fill requirements, wall/slope size and type, *in situ* soil type, available backfill materials, facing finish, temporary or permanent application. It has been found that MSB walls with precast concrete facings are usually less expensive than reinforced concrete retaining walls for heights greater than about 3 m (10 ft) and average foundation conditions. Modular block walls (MBW) are competitive with concrete walls at heights of less than 4.5 m (15 ft).

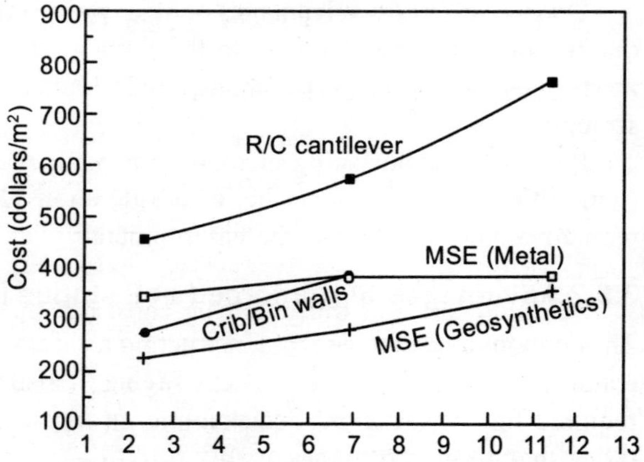

Fig. 3.12: Cost comparison for retaining walls

In general, the use of MSB walls results in savings on the order of 25–50 per cent and possibly more in comparison with a conventional reinforced concrete retaining structure, especially when the latter is supported on a deep foundation system (poor foundation condition). A substantial savings is obtained by elimination of the deep foundations, which is usually possible because reinforced soil structures can accommodate relatively large total and differential settlements. Other cost saving features include ease of construction and speed of construction. A comparison of wall material and erection costs for several reinforced soil retaining walls and other retaining wall systems, based on a survey of state and federal transportation agencies, is shown in Fig 3.12. Typical total costs for MSB walls range from \$200–\$400 per m^2 (\$19–\$37 per ft^2) of face, generally as function of height, size of project and cost of select fill.

The actual cost of a specific MSEW structure will depend on the cost of each of its principal components. For segmental precast concrete faced structures, typical relative costs are:

- Erection of panels and contractors profit — 20–30 per cent of total cost
- Reinforcing materials — 20–30 per cent of total cost
- Facing system — 25–30 per cent of total cost
- Backfill materials including placement — 35–40 per cent of total cost

where the fill is a select granular fill from an offsite borrow source.

The additional cost for panel architectural finish treatment ranges from \$5–\$15 per m^2 (\$0.50–\$1.50 per ft^2) depending on the complexity of the finish. Traffic barrier costs average \$550 per linear meter (\$170 per linear foot). In addition, consideration must be given to the cost of excavation which may be somewhat greater than for other systems. MBW faced walls at heights less than 4.5 m (15 ft) are typically less expensive by 10 per cent or more.

The economy of using RSS must be assessed on a case-by-case basis, where use is not dictated by space constraints. For such cases, an appropriate benefit to cost ratio in analysis should be carried out to see if the steeper slope with the reinforcement is justified economically over the alternative flatter slope with its increased right-of-way and materials costs, etc. It should be kept in mind that guardrails or traffic barriers are often necessary for steeper embankment slopes and additional costs such as erosion control systems for slope face protection must be considered.

With respect to economy, the factors to consider are as follows:

- Cut or fill earthwork quantities.
- Size of slope area.

- Average height of slope area.
- Angle of slope.
- Cost of non select *vs* select backfills.
- Temporary and permanent erosion protection requirements.
- Cost and availability of right-of-way needed.
- Complicated horizontal and vertical alignment changes.
- Need for temporary excavation support systems.
- Maintenance of traffic during construction.
- Aesthetics.
- Requirements for guardrails and traffic barriers.

The actual bid cost of a specific RSS structure depends on the cost of each of its principal components. Based on limited data, typical relative costs are:

- Reinforcement — 45–65 per cent of total cost
- Backfill — 30–45 per cent of total cost
- Face treatment — 5–10 per cent of total cost

High RSS structures have relatively higher reinforcement and lower backfill costs. Recent bid prices suggest costs ranging from \$1 10/m² to \$260/m² (\$10/ft² to \$24/ft²) as a function of height.

For applications in the 10–15 m (30–50 ft) height range bid costs of about \$170/m² (\$16/ft²) have been reported. These prices do not include safety features and drainage details.

Fig. 3.13 provides a rapid, first-order assessment of cost items for comparing a flatter unreinforced slope with a steeper reinforced slope.

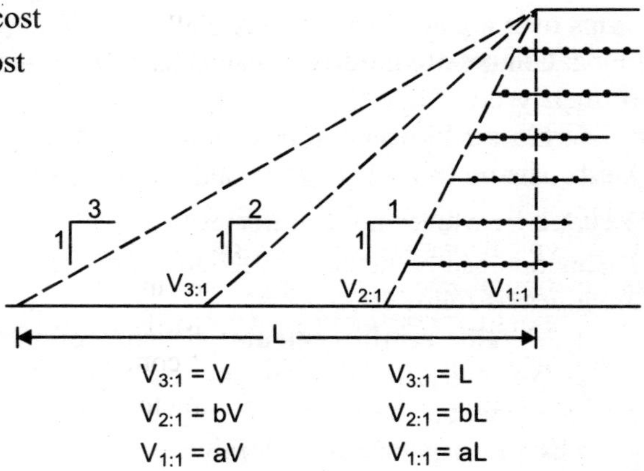

Fig. 3.13: Cost evaluation of reinforced soil slopes

Cost:

$$3H:1V = V_{SOIL} + L_{LAND} + \text{Guardrail*} + \text{Hydroseeding}$$

$$2H:1V = bV_{SOIL} + bL_{LAND} + \text{Guardrail*} + \text{Erosion Control} + \text{High Maintenance}$$

$$1H:1V = aV_{SOIL} + aL_{LAND} + \text{Reinforcement} + \text{Guardrail} + \text{Erosion Control}$$

3.10 DESCRIPTION OF MSE/RSS SYSTEMS

3.10.1 Systems differentiation

Since the expiration of the fundamental process and concrete facing panel patents obtained by the Reinforced Earth Co. for MSEW systems and structure, the engineering community has adopted a generic term Mechanically Stabilised Earth to describe this type of retaining wall construction.

Trademarks, such as Reinforced Earth®, Retained Earth, Genesis, etc. describe systems with some present or past proprietary features or unique components marketed by nationwide commercial suppliers.

*Include guardrail or traffic barrier cost if required.

Other trademark names appear yearly to differentiate systems marketed by competing commercial entities that may include proprietary or novel components or for special applications.

A system for either MSEW or RSS structures is defined as a complete supplied package that includes design, specifications and all prefabricated materials of construction necessary for the complete construction of a soil reinforced structure. Often technical assistance during the planning and construction phase is also included. Components marketed by commercial entities for integration by the owner in a coherent system are not classified as systems.

3.10.2 Types of systems

MSE/RSS systems can be described by the reinforcement geometry, stress transfer mechanism, reinforcement material, extensibility of the reinforcement material, and the type of facing and connections.

3.10.2.1 Reinforcement geometry

Three types of reinforcement geometry can be considered:

1. **Linear unidirectional:** Strips, including smooth or ribbed steel strips, or coated geosynthetic strips over a load-carrying fibre.

2. **Composite unidirectional:** Grids or bar mats characterised by grid spacing greater than 150 mm (6 inches).

3. **Planar bidirectional:** Continuous sheets of geosynthetics, welded wire mesh, and woven wire mesh. The mesh is characterised by element spacing of less than 150 mm (6 inches).

3.10.2.2 Reinforcement material

Distinction can be made between the characteristics of metallic and nonmetallic reinforcements. They are given as follows:

1. **Metallic reinforcements:** Typically of mild steel is usually galvanised or may be epoxy coated.

2. **Nonmetallic reinforcements:** Generally polymeric materials consisting of polypropylene, polyethylene, or polyester.

The performance and durability considerations for these two classes of reinforcement van-considerably and are detailed in the companion corrosion/degradation document.

3. **Reinforcement Extensibility:** The two classes of extensibility are given as follows:

1. ***Inextensible*:** The deformation of the reinforcement at failure is much less than the deformability of the soil.

2. ***Extensible*:** The deformation of the reinforcement at failure is comparable to or even greater than the deformability of the soil.

3.10.3 Facing systems

The types of facing elements used in the different MSE systems control their aesthetics because they are the only visible parts of the completed structure. A wide range of finishes and colours can be provided in the facing. In addition, the facing provides protection against backfill sloughing and erosion, and provides in certain cases drainage paths. The type of racing influences settlement tolerances. Major facing types are:

3.10.3.1 Segmental precast concrete panels

Segmental precast concrete panels summarised and illustrated in Fig. 3.14. The precast concrete panels have a minimum thickness of 140 mm (5.6 inches) and are of a cruciform, square, rectangular, diamond, or hexagonal geometry. Temperature and tensile reinforcement are required but will vary with the size of the panel. Vertically adjacent units are usually connected with shear pins.

Fig. 3.14: Segmental precast concrete panels

3.10.3.2 Dry cast modular block wall (MBW) units

These are relatively small, squat concrete units that have been specially designed and manufactured for retaining wall applications. The mass of these units commonly ranges from 15–50 kg (30–110 lbs), with units of 35–50 kg (75–110 lbs) routine *y* used for highway projects. Unit heights typically range from 100–200 mm (4–8 inches) for the various manufacturers. Exposed face length usually varies from 200–450 mm (8–18 inches). Nominal width (dimension perpendicular to the wall face) of units typically ranges between 200 and 600 mm (8 and 24 inches). Units may be manufactured solid or with cores. Full height cores are filled with aggregate during erection. Units are normally dry-stacked (i.e., without mortar) and in a running bond configuration. Vertically adjacent units may be connected with shear pins. They are referred to by trademarked names such as Keystone, Versa-Lok, Allan etc. They are illustrated in Fig. 3.15.

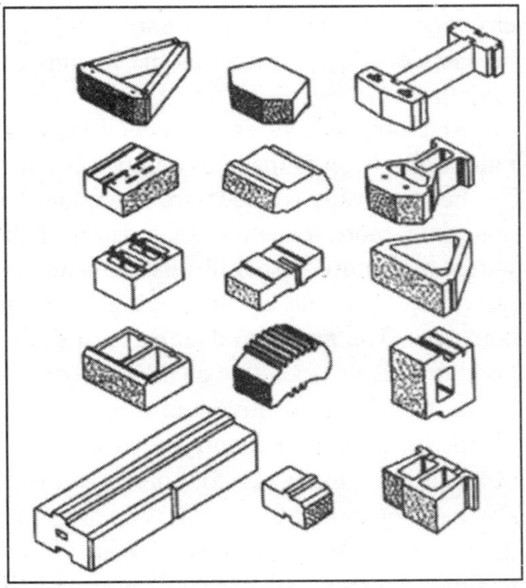

Fig. 3.15: Examples of commercially available MBW units (from NCMA design manual for segmental retaining walls)

3.10.3.3 Metallic facings

The original reinforced earth* system had facing elements of galvanised steel sheet formed into half cylinders. Although precast concrete panels are now commonly used in reinforced earth walls, metallic facings may be appropriate in structures where difficult access or difficult handling requires lighter facing elements.

3.10.3.4 Welded wire grids

Wire grid can be bent up at the front of the wall to form the wall face. This type effacing is used in the Hilfiker, Tensar and Reinforced Earth wire retaining wall systems.

3.10.3.5 Gabion facing

Gabions (rock-filled wire baskets can be used as facing with reinforcing elements consisting of welded wire mesh, welded bar-mats, geogrids, geotextiles or the double-twisted woven mesh placed between or connected to the gabion baskets.

3.10.3.6 Geosynthetic facing

Various types of geotextile reinforcement are looped around at the facing to form the exposed face of the retaining wall. These faces are susceptible to ultraviolet light degradation, vandalism (e.g., target practice) and damage due to fire. Alternately, a geosynthetic grid used for soil reinforcement can be looped around to form the face of the completed retaining structure in a similar manner to welded wire mesh and fabric facing. Vegetal ion can grow through the grid structure and can provide both ultraviolet light protection for the geogrid and a pleasing appearance.

3.10.3.7 Post construction facings

For wrapped faced walls, the facing-whether geotextile, geogrid, or wire mesh- can be attached after construct ion of the wall by shotcreting. guniting, cast-in-place concrete or attaching prefabricated facing panels made of concrete, wood, or other materials. This multi-staging facing approach adds cost but is advantageous where significant settlement is anticipated.

Precast elements can be cast in several shapes and provided with facing textures to match environmental requirements and blend aesthetically into the environment. Retaining structures using precast concrete elements as the facings can h.ive surface finishes similar to any reinforced concrete structure.

Retaining structures with metal facings have the disadvantage of shorter life because of corrosion, unless provision is made to compensate for it.

Facings using welded wire or gabions have the disadvantages of an uneven surface, exposed backfill materials, more tendency for erosion of the retained soil, possible shorter life from corrosion of the wires, and more susceptibility to vandalism. These disadvantages can. of course, be countered by providing shotcrete or by hanging facing panels on the exposed face and compensating for possible corrosion. The greatest advantages of such facings are low cost, ease of installation, design flexibility, good drainage (depending on the type of backfill) that provides increased stability, and possible treatment of the face for vegetative and other architectural effects. The facing can easily be adapted and well-blended with natural country environment. These facings, as well as geosynthetic wrapped facings, are especially advantageous for construction of temporary or other structures with a short-term design life. Dry cast segmental block MBW facings may raise some concerns as to durability in aggressive freeze-thaw environments when produced with water absorption capacity significantly higher than that of wet-cast concrete. Historical data provide little insight as their usage history is less than two decades. Further, because the cement is not completely hydrated during the dry cast process, (as is often evidenced by efflorescence on the surface of units), a highly alkaline regime may establish itself at or near the face area, and may limit the use of some geosynthetic products as reinforcements. Freeze-thaw durability is enhanced for products produced at higher compressive strengths and low water absorption ratios. The current specifications in Chapter 8 have been developed to address this issue.

The outward faces of slopes in RSS structures are usually vegetated if 1:1 or flatter. The vegetation requirements vary by geographic and climatic conditions and are therefore, project specific.

3.11 TYPES OF REINFORCEMENT

Most, although not all systems using precast concrete panels use steel reinforcements which are typically galvanized but may be epoxy coated. Two types of steel reinforcements are in current use:

Steel strips: The currently commercially available strips are ribbed top and bottom 50 mm (2 inches) wide and 4 mm (5/32 inch) thick. Smooth strips 60–I20 mm (2d–$4^{3/4}$ inch) wide, 3–4 mm (c–5/32 inch) thick have been used.

Steel grids: Welded wire grid using 2–6 W 7.5–W24 longitudinal wire spaced at either 150 or 200 mm (6 or 8 inches). The transverse wire may vary from WL1 to W20 and are spaced based on design requirements from 230–600 mm (9–24 inches). Welded steel wire mesh spaced at 50 by 50 mm (2 by 2 inch) of thinner wire has been used in conjunction with a welded wire facing. Some MBW systems use steel grids with 2 longitudinal wires.

Most MBW systems use geosynthetic reinforcement, principally geogrids. The following types are widely used and available:

High density polyethylene (HDPE) geogrid: These are of uniaxial manufacture and are available in up to 6 styles differing in strength.

PVC coated polyester (PET) geogrid: Available from a number of manufacturers. They are characterised by bundled high tenacity PET fibres in the longitudinal load carrying direction. For longevity the PET is supplied as a high molecular weight fibre and is further characterised by a low carboxyl end group number.

Geotextiles: High strength geotextiles can be used principally in connection with reinforced soil slope (RSS) construction. Both polyester (PET) and polypropylene (PP) geotextiles have been used.

3.11.1 Reinforced backfill materials

3.11.1.1 MSEW structures

MSB walls require high quality backfill for durability, good drainage, constructability and good soil reinforcement interaction which can be obtained from well graded, granular materials. Many MSB systems depend on friction between the reinforcing elements and the soil. In such cases, a material with high friction characteristics is specified and required. Some systems rely on passive pressure on reinforcing elements, and in those cases, the quality of backfill is still critical. These performance requirements generally eliminate soils with high clay contents.

From a reinforcement capacity point of view, lower quality backfills could be used for MSBW structures; however, a high quality granular backfill has the advantages of being free draining, providing better durability for metallic reinforcement, and requiring less reinforcement. There are also significant handling, placement and compaction advantages in using granular soils. These include an increased rate of wall erection and improved maintenance of wall alignment tolerances.

3.11.1.2 RSS structures

Reinforced soil slopes are normally not constructed with rigid facing elements. Slopes constructed with a flexible face can thus readily tolerate minor distortions that could result from settlement, freezing and thawing, or wet-drying of the backfill. As a result, any soil meeting the requirements for embankment construction could be used in a reinforced slope system. However, a higher quality material offers less durability concerns for the reinforcement, and is easier to handle, place and compact, which speeds up construction.

3.11.1.3 Miscellaneous materials of construction

Walls using precast concrete panels require bearing pads in their horizontal joints that provide some compressibility and movement between panels and preclude concrete to concrete contact. These materials are neoprene, SBR rubber or HDPE.

All joints are covered with a polypropylene (PP) geotextile strip to prevent the migration of fines from the backfill. The compressibility of the horizontal joint material should be a function of the wall height. Walls with heights greater than 15 m (50 ft) may require thicker or more compressible joints to accommodate the larger vertical loads due to the weight of panels in the lower third of the structure.

3.11.1.4 Construction of MSEW systems with precast facings

The construction of MSEW systems with a precast facing is carried out as follows:

1. **Preparation of subgrade:** This step involves removal of unsuitable materials from the area to be occupied by the retaining structure. All organic matter, vegetation, slide debris and other unstable materials should be stripped off and the subgrade compacted. In unstable foundation areas, ground improvement methods, such as dynamic compaction, stone columns, wick drains or other foundation stabilisation/improvement methods would be constructed prior to wall erection.

2. **Placement of a levelling pad for the erection of the facing elements:** This generally unreinforced concrete pad is often only 300 mm (1 ft) wide and 150 mm (6 inches) thick and is used for MSEW construction only, where concrete panels are subsequently erected. A gravel pad has been often substituted for MBW construction.

The purpose of this pad is to serve as a guide for facing panel erection and is not intended as a structural foundation support.

3. **Erection of the first row of facing panels on the prepared levelling pad:** Facings may consist of either precast concrete panels, metal facing panels, or dry cast modular blocks.

The first row effacing panels may be full or half-height panels, depending upon the type of facing used. The first tier of panels must be shored up to maintain stability and alignment. For construction with modular dry-cast blocks, full sized blocks are used throughout with no shoring.

The erection of facing panels and placement of the soil backfill proceed simultaneously.

4. **Placement and compaction of backfill on the subgrade to the level of the first layer of reinforcement and its compaction:** The fill should be compacted to the specified density, usually 95–100 per cent of AASHT3T–99 maximum density and within the specified range of optimum moisture content. Compaction moisture contents dry of optimum are recommended.

A key to good performance is consistent placement and compaction. Wall fill lift thickness must be controlled based on specification requirements and vertical distribution of reinforcement elements. The uniform loose lift thickness of the reinforced backfill should not exceed 300 mm (12 inches). Reinforced backfill should be dumped into or parallel to the rear and middle of the reinforcement and bladed toward the front face. Random fill placement behind the reinforced volume should proceed simultaneously.

5. **Placement of the first layer of reinforcing elements on the backfill:** The reinforcements are placed and connected to the facing panels, when the compacted fill has been brought up to the level of the connection they are generally placed perpendicular to back of the facing panels.

6. **Placement of the backfill over the reinforcing elements to the level of the next reinforcement layer and compaction of the backfill:** The previously outlined steps are repeated for each successive layer.

7. **Construction of traffic barriers and copings:** This final construction sequence is undertaken after the final panels have been placed, and the backfill has been completed to its final grade.

8. **Construction of MSE systems with flexible facings:** Construction of flexible-faced MSE walls, where the reinforcing material also serves as facing material, is similar to that for walls with precast facing elements. For flexible facing types such as welded wire mesh, geotextiles, geogrids or gabions, the erection of the first level facing element requires only a level grade. A concrete footing or levelling pad is not usually required unless precast elements are to be attached to the system after construction.

Construction proceeds as outlined for segmental facings with the following exceptions.

9. **Placement of first reinforcing layer:** Reinforcement with anisotropic strength properties (i.e., many geosynthetics) should be placed with the principal strength direction perpendicular to face of

structure. It is often convenient to unroll the reinforcement with the roll or machine direction parallel to the face. If this is done then the cross machine tensile strength must be greater than the design tension requirements.

Secure reinforcement with retaining pins to prevent movement during reinforced till placement. Overlap adjacent sheets a minimum of 150 mm (6 inches) along the edges perpendicular to the face. Alternatively, with geogrid or wire mesh reinforcement, the edges may be butted and clipped or tied together.

3.12 CONSTRUCTION METHODOLOGY

In the actual construction, geosynthetic (geotextile/geogrid)—reinforced soil retaining walls have continued to demonstrate excellent performance characteristics and exhibit many advantages over conventional retaining walls. To achieve better performance, the following points must be considered on site-specific basis.

1. Any unsuitable foundation soils should be replaced with compacted granular backfill material.
2. The geosynthetic layer should be installed with its principal strength (warp strength) direction perpendicular to the wall face.
3. With the geosynthetic layer, it is necessary to take care not to tear it in the direction parallel to the wall face because a partial tear of this type will reduce the amount of tensile force carried out by geotextile layer.
4. The overlap along the edges of the geosynthetic layer should generally exceed 200 mm. If there is possibility of large foundation settlements, then sewn or other suitable joints may be recommended between adjacent geosynthetic layers.
5. There should not be any wrinkles or slack in the geosynthetic layer as they can result in differential movement.
6. Granular backfill soil should generally be compacted to at least 95% of the standard proctor maximum dry unit weight. Compacted lift thickness should vary from 200–300 mm. Efforts should be made to compact uniformly to avoid differentials settlement.
7. Backfill soil should be compacted, taking care not to get the compactor very close to lateral displacement of the wall face. It is, therefore, recommended to use light weight hand-vibratory compactors within 1 m of the wall face.
8. Wraparound geosynthetic facing can be constructed using a temporary formwork as shown in Fig. 3.16. The lap length should be generally not less than 1 m.

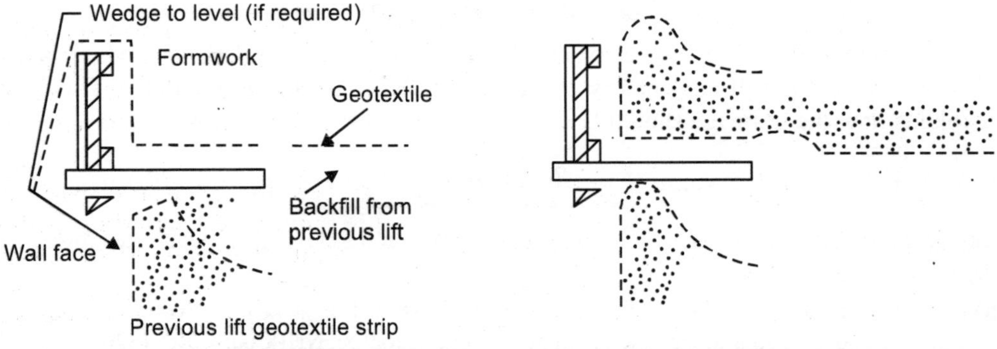

Fig. 3.16: Lift construction sequence for geotextile reinforced soil walls

9. Geogrid-reinforced retaining walls can be constructed with geotextile filters near the face. Major construction steps are shown in Fig. 3.17.
10. In the case of facing made with segmental or modular concrete blocks (MCBs), full-height precast concrete panels, welded wire panels, gabion baskets, or treated timber panels, should be placed prior to placing the backfill and must be carefully checked as per the design guidelines because the success of a geosynthetic-reinforced retaining wall is highly dependent on the facing connections.
11. It is necessary to have tight construction specifications and quality inspection to ensure that the wall face is constructed properly; otherwise an unattractive wall face, or a wall face failure, could result.

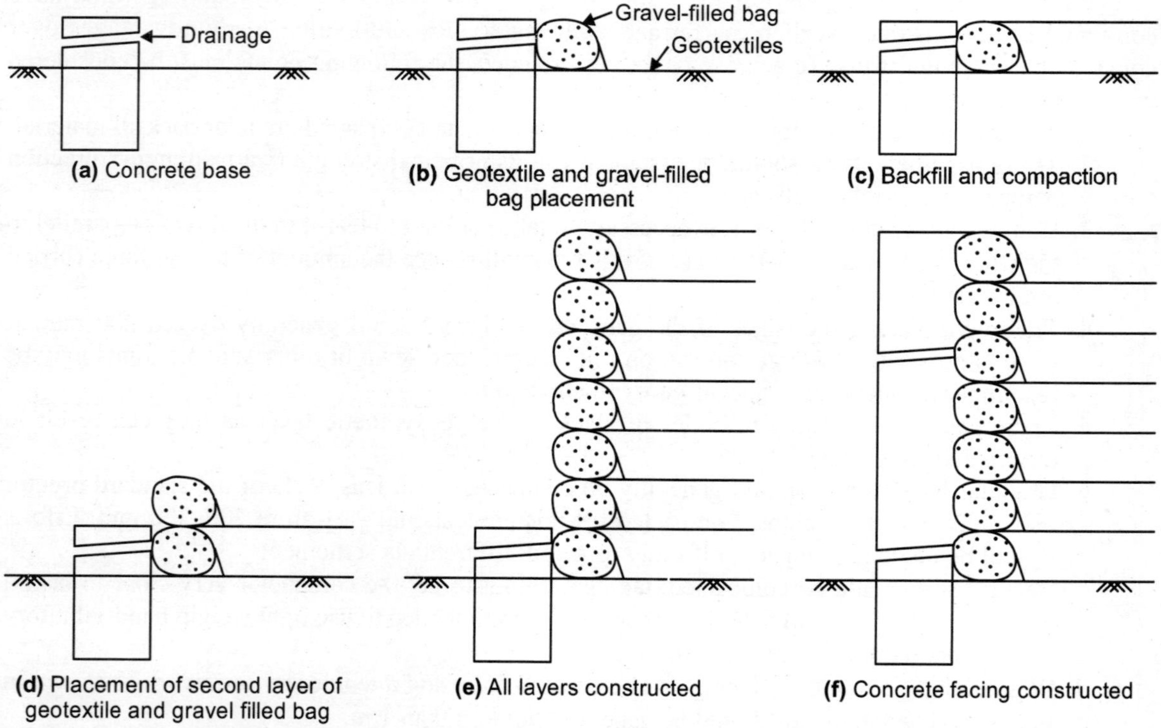

Fig. 3.17: Standard staged construction procedure for a Geosynthetic reinforced soil retaining wall (GRS-RW) (after Tatsuoka et al., 1997).

In Japan, simple methodology was also developed for construction of R.E. wall, which is illustrated in Fig. 3.17 below and is self-explanatory constructed (after Tatsuoka et al., 1997). This figure also clearly explains how the drainage should be ensured in such walls.

References

1. Khan, Iqraz Nabi (1991). "A Study of Reinforced Earth Wall and Retaining Wall with Reinforced Backfill", Ph.D. Thesis, IIT Roorkee.
2. Mechanically Stabilised Earth Walls and Reinforced Soil Slopes Design & Construction Guidelines (2001). U.S. Department of Transportation Federal Highway Administration, NHI Course No. 132042.
3. Vidal, H. (1969). English Translation of Text Entitled "La terre Armee" as published in the Annales de I' Institut Technique du Batiment et des Travaux Publics, Series, Materials (38) – No. 259–260, July/August.

Use of Geosynthetics on Roads and Pavements

CHAPTER 4

"American roads are not good because America is rich
But America is rich because American roads are good" **– John. F. Kennedy**

4.1 INTRODUCTION

In a world where dependence is constantly on rise, the processes involved in constructing these roads are becoming increasingly significant. In developed countries, mostly roads, pavements, airstrips, etc. are being constructed with new materials and technology. In India also, geotextiles, jute, geomembranes are now a days being used for expressways, national highways and even for city roads.

In 1970s there was worldwide demand for reduction of immense repair costs of aged bituminous road surface. This resulted in development of new materials.

Alligator cracking is common mode of failure in asphalt surfaces. When an asphalt road is re-surfaced, cracks in the old asphalt layer eventually are reflected or transferred into the new asphalt surface. The main causes of tertiary cracking which is the reflection of secondary cracking into the new bituminous layer are shown in Fig. 4.1.

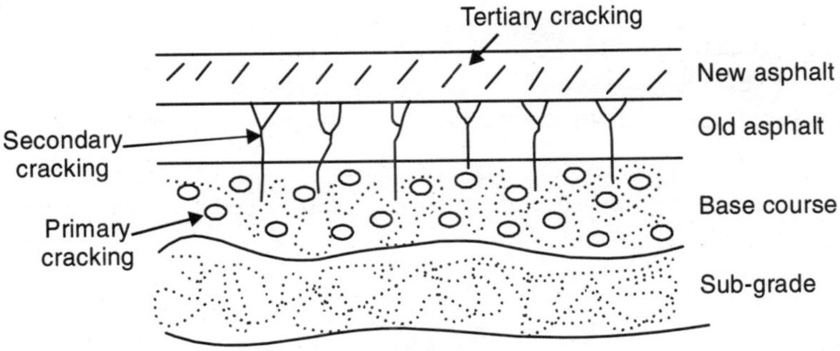

Fig. 4.1

4.2 USE OF GEOSYNTHETICS IN OVERLAYS (OVER EXISTING PAVEMENT)

A geotextile installed between old and new bituminous surface helps prevent the reflection of cracking into the new asphalt surface. This is accomplished through several mechanism namely prevention of moisture penetration reduction of pavement fatigue, retardation of reflective cracks and reduction of flexural fatigue (Fig. 4.2).

The flexural fatigue (N_f) of the bituminous concrete surface is a function of traffic load and the load bearing capacity of soil.

The flexural fatigue is also a parameter that can be considerably influenced through the use of a geotextile. The reduction of the flexural fatigue extending the life expectancy of the road surface can be expressed through the fabric effectiveness factor (FEF).

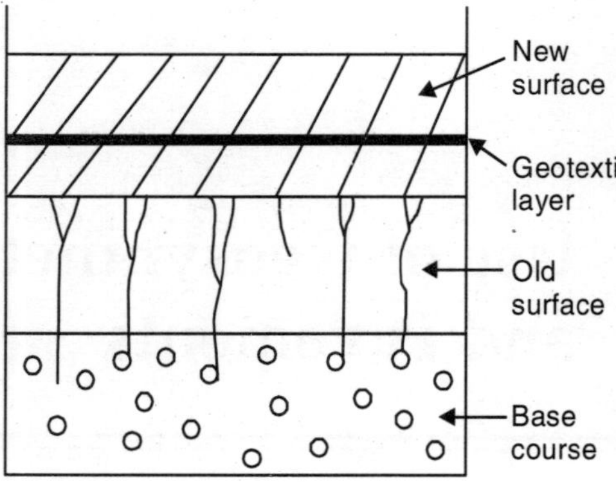

Fig. 4.2

Thus, $$\text{FEF} = \frac{N_{fg}}{N_f} \qquad ...(4.1)$$

The FEF value is obtained through the flexural fatigue ratio (N_{fg}/N_f) between geotextile reinforcement and non reinforcement asphalt concrete determined through testing. In evaluating geotextile efficiency the following factors must be considered:

- Road geometry
- Road surface flexibility
- Position of geotextile
- Type of liquid bitumen
- Geotextile type
- Stress level
- Surface course thickness

4.2.1 Unpaved roads

The use of reinforcement (particularly geotextiles) for unpaved aggregate roads is a major application area. Giroud et al. (1994) have developed most advanced analytical method for the same. They have demonstrated that there is difference in required. Thickness of stone base despite considering the cost of geogrid.

For unreinforcement case, Giroud (1994) have adapted corps of engineers empirical formula which includes the number of vehicles passages. For geogrid reinforcement case, new concepts are developed, which include three beneficial mechanism attribute to inclusion of geogrid.

Methodology

1. An increase in soil sub-grade strength from non reinforced case to the reinforcement case is indicated in following equations:

$$P_e = \pi C_{uN} + \gamma h_0 \qquad ...(4.2)$$

$$P_{lim} = (\pi + 2) C_{uN} + \gamma h \qquad ...(4.3)$$

where, P_e = bearing capacity pressure based on elastic limit (non reinforcement case)

P_{lim} = bearing capacity pressure based on plastic limit (reinforcement case)

C_{uN} = undrained soil strength at the Nth vehicle passage

γ = unit weight of aggregate

h_0 = aggregate thickness without reinforcement

h = aggregate thickness with reinforcement

2. An improved load distribution to the soil sub-grade due to load spreading which is quantified
3. On the basis of pyramidal geometric shape (as shown in Fig. 4.3).

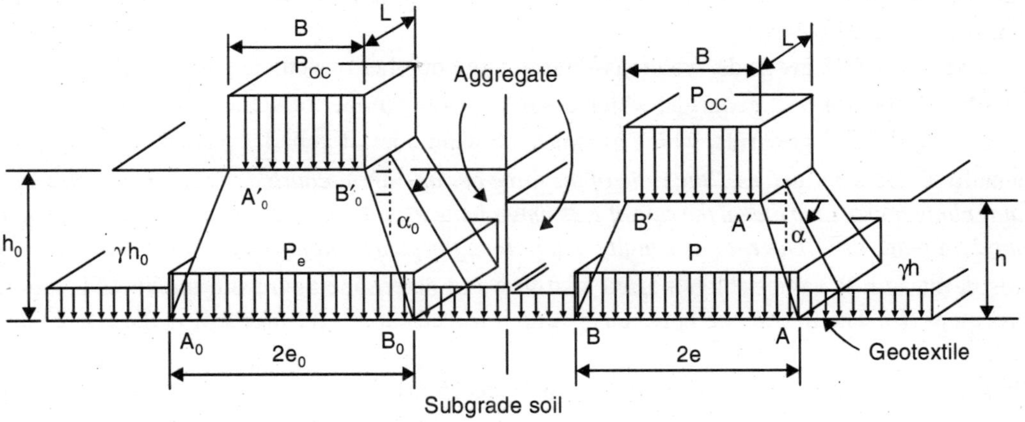

(a) Case without geotextile **(b)** Case with geotextile

Fig. 4.3: Load distribution by aggregate layer (after Giroud and Noiray, 1994)

Here α_0 = angle for unreinforcement case (≈ 26°), α = angle for reinforced case (may be as high as 60°). The ratio of reinforcement to non reinforcement situations is expressed as a ratio of $\tan \alpha/\tan \alpha_0$ which is ≥ 1.

4. A tensioned membrane effect which is a function of tensile modulus and elongation of the geogrid and the deformed surface of sub-grade soil.

By taking the combined effect of the above mentioned geogrid reinforcement mechanism and comparing it to the unreinforcement case, Giroud et al. (1994) have developed the design charts, which we shown in Fig. 4.4.

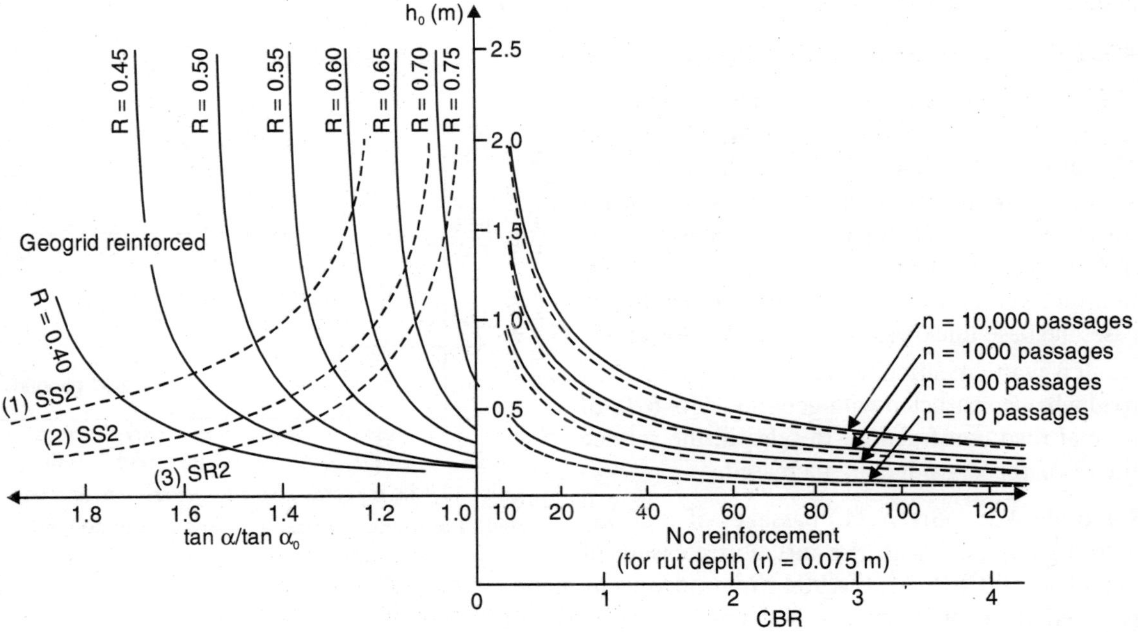

Fig. 4.4

Here, on the RHS of graph, for a standard axle load of 80 kN (now officially it is 102 kN in India) and any number of vehicle passes from 10 to 10,000, a thickness of non reinforced stone base (h_0) can be obtained knowing the soil sub-grade strength. This value is then carried to the left side of figure, where it is intersected with either:

- Curve 1, for SS2 geogrids, which assumes a large number of vehicle passes (N>1000)
- Curve 2, also for SS2 geogrids, which assumes a low number of vehicles passes.
- Curve 3, for SR2 geogrids, which assumes a low number of vehicles passes.

(Incidentally, CBR is defined as, *"the ratio of the force per unit area required to penetrate a soil mass with a circular plunger of 50 mm dia. at the rate of 1.25 mm/minute to that required for corresponding penetrations of a standard material." The ratio is usually determined for penetrations of 2.5 mm & 5 mm.*)

This defines an R value which is used as follows to determine aggregate thickness using geogrid reinforcement (h). The difference between h_0 (unreinforced) and h (reinforced) is the aggregate saved (Δh).

Thus, $$h = R \cdot h_0 \text{ (for } r < 0.15 \text{ m)} \quad ...(4.4)$$

Example 4.1: Given a soil sub-grade with CBR strength as 1.0 with 1000 vehicles passes with a rut depth = 0.075 m. What is the required aggregate depth without a Geogrid. Also compute the aggregate depth with SS2 grade geogrids, and the difference in aggregate thickness between 2 cases.

Solution: Refer design charts, the unreinforced case gives

$$h_0 = 0.60 \text{ m}$$

For geogrid-reinforcement case, curve 2 gives

$$R = 0.50$$

$$H = Rh_0$$

$$= 0.5 \times 0.60 = 0.30 \text{ m}$$

and, aggregate saved is $\Delta h = 0.6 - 0.3 = 0.30$ m.

4.2.2 Base thickness with geogrid

The base thickness with geogrid can be computed from following graph (Fig. 4.5) also, which is a typical graph for Tensar SS1 geogrid. The x-axis shows non-reinforced base thickness in mm (or inches, which is mentioned in parenthesis) while the y-axis shows the equivalent base thickness for reinforced case. It clearly shows that in reinforced case, the base thickness is considerably lower.

It is worth mentioning here that overall cost of road when constructed with geogrid is 50–67% of normal road construction. Besides it, the service life of road is considerably increased.

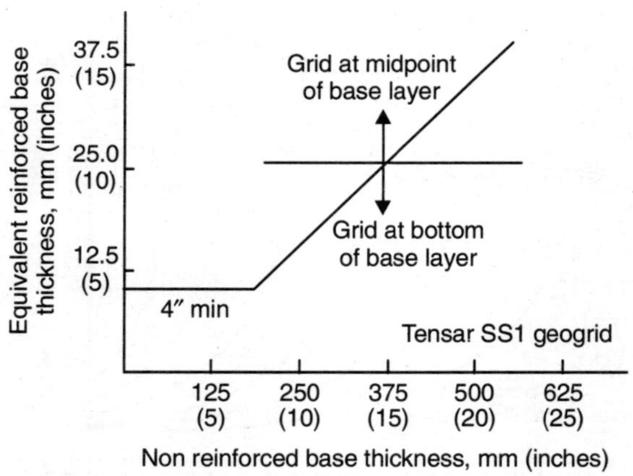

Fig. 4.5: Reinforced base thickness with geogrid

Example 4.2: Given 340 passages of a 80 kN axle load vehicle with tire inflation pressure of 480 kPa, soil CBR = 1, E = 90 kN/m (modulus of fabric), and on allowable rut depth = 0.3, what is the required aggregate thickness of an unpaved road. Refer following graph (Fig. 4.6).

Solution: Fig. 4.6 gives $h_0' = 0.35$ (i.e., thickness when no geogrid is used) for N = 340, CBR = 1. It also gives $\Delta h = 0.15$ m for E = 90 kN/m and CBR = 1 (i.e., the reduction in aggregate thickness for use of a fabric reinforcement layer. The difference i.e., 0.20 m is the required aggregate depth when fabric is used.

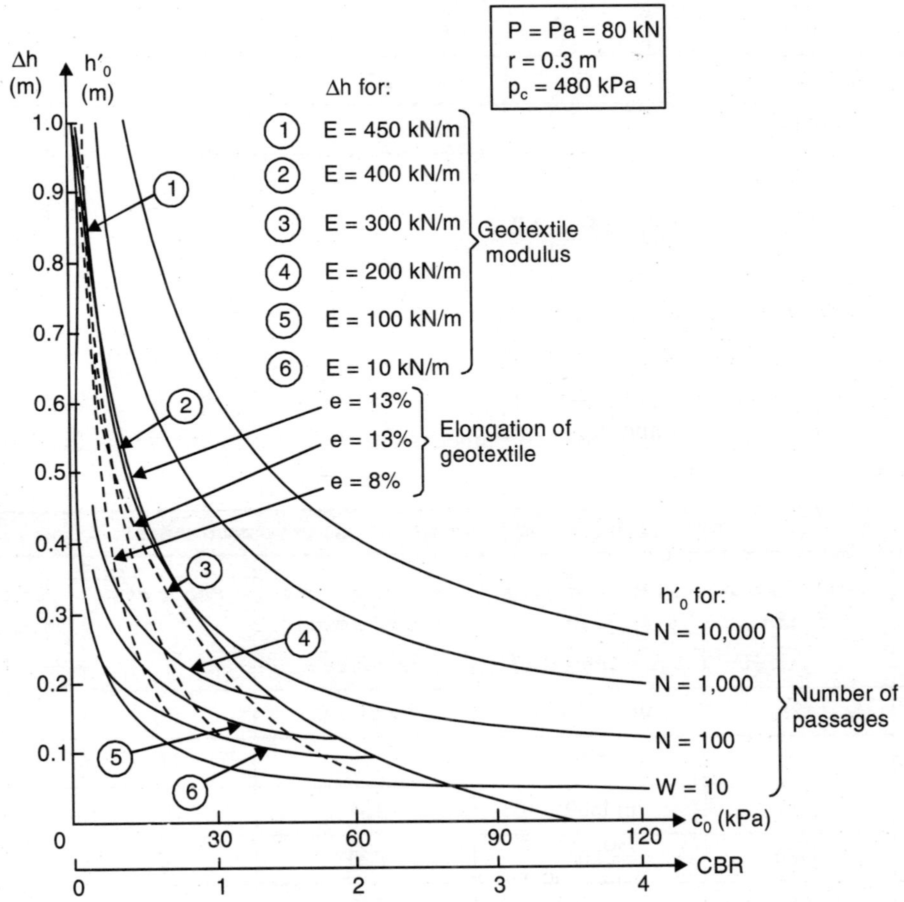

Fig. 4.6: Reducing aggregate thickness with a geotextile (after Giroud and Noiray). Chart is related to a highway truck with standard axle load

Example 4.3: Given 1000 passages of a single axle load (9 tonnes) vehicle with a tire inflation pressure of 492 kPa on a soil with CBR = 2, geotextile modulus E = 0.00197 t/m and an allowable rut depth = 0.28 m (*i*) Plot the response curve from Fig. 4.7. (*ii*) Determine the aggregate savings and (*iii*) Conduct an economic study based on the distance the actual field work is done from the quarry and place of manufacturer of geotextile as given in following table. Assume that the stone weighs about 2.3 t/m^3.

Distance (Km)	*Aggregate cost (₹ / ton)*	*Geotextile cost ₹ / m²*
≤ 10	600	60
10–25	810	63
25–50	1160	65
50–100	1680	70
100–200	2500	75

Solution:

(*i*) The required curve is shown in Fig. 4.7

(*ii*) At CBR = 2,

$$h_0' \text{ (without geotextile)} = 0.3 \text{ m}$$

$$h' \text{ (with geotextile)} = 0.2 \text{ m}$$

$$\Delta h \text{ (saving in stone thickness)} = 0.1 \text{ m}$$

(*iii*) Presuming that 1.9 t/m^2 – *m* stone is consumed at site. It results in following (Table.4.1) which is self-explanatory.

Table 4.1: Study of road estimate with use of geosynthetics

Distance	*Cost of stone ₹ / ton*	*Stone laying cost ₹ / m² (unreinforced)*	*Saving in stone cost when reinforced*	*Fabric cost ₹ / m²*	*Overall savings in construction ₹ / m²*
(1)	(2)	(3) = 1.9 × (2)	(4) = (3) × 0.1	(5)	(6) = (4) – (5)
< 10	600	1140	114	60	54
10–25	810	1539	154	63	91
25–50	1160	2204	220	65	155
50–100	1680	3192	319	70	249
100–200	2500	4750	475	75	400

Giroud & Noiray Procedure:

$$p = 9 \text{ t}$$

$$p_c = 49 \text{ t/m}^2$$

$$r = 0.28 \text{ m}$$

$$N = 1000 \text{ passes}$$

The column (6) of above table clearly shows that there is considerable saving in cost of construction.

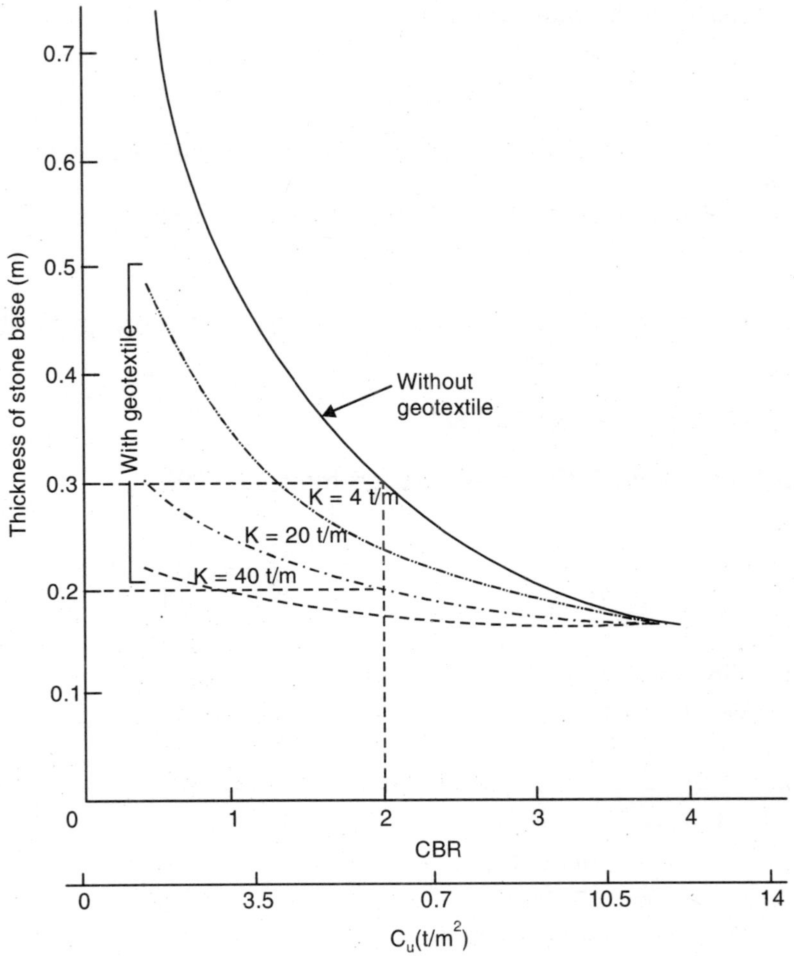

Fig. 4.7: CBR *vs.* thickness of stone base with and without geotextiles

Author designed a road for a new township coming up near Chandigarh by DLF (India) Ltd. The soil and other data for the site were as follows (Table 4.2).

Table 4.2: Soil data

Test/Location No.	***CBR values at 97% of maximum dry density***	
	Soaked	***Unsoaked***
CBR–1	5.0	14.5
CBR–2	5.3	8.9
CBR–3	4.8	10.3
CBR–4	2.8	9.8
CBR–5	3.1	8.2

For the soaked condition, the lowest value of CBR was 2.8, hence it was adapted for design of new road in township.

(A) Traffic data

The clients, M/s DLF desired that road be designed for class A-A loading only. Following data is taken for design purpose:

(*i*) Commercial vehicle per day (CVPD) = 10 (for weight more than 3 tonnes, i.e., of the order of 9-10 tonnes)

(*ii*) MSa = 10

On the basis of traffic data and clients requirements, it was decided to design flexible pavement for the site.

4.3 DESIGN OF FLEXIBLE PAVEMENT

4.3.1 As per AASHTO flexible pavement design method, 1993

Input Parameters

Reliability level, R	—	90%
Normal derivate, Z_R	—	– 1.282
Standard deviation, S_0	—	0.45
Initial serviceability index, p_i	—	4.2
Terminal serviceability index, p_t	—	2
Required traffic, $W_{18,\ Req}$	—	10 MSa
Sub-grade soil, CBR	—	3.0%

Details of Pavement Layers

Thickness of bituminous concrete, $D_{1,\ BC}$	—	40 mm
Thickness of dense bituminous concrete, $D_{1,\ DBM}$	—	75 mm
Thickness of granular base/WBM/WMM, D_2	—	250 mm
Thickness of granular sub-base, D_3	—	250 mm
Layer coefficient of bituminous concrete, $a_{1.BC}$	—	0.45
Layer coefficient of dense bituminous concrete, a_{1DBM}	—	0.45
Layer coefficient of granular base/WBM/WMM, a_2 (min. CBR of WBM/WMM = 60%)	—	0.15
Layer coefficient of granular sub-base, a_3 (min. CBR of granular sub-base = 20%)	—	0.12
Drainage factor for granular base, M_2	—	1.3
Drainage factor for granular sub-base, M_3	—	1.3

Design of Flexible Pavement as per AASHTO, 1993

$$\log_{10}(W_{18}) = Z_R S_0 + 9.36 \log_{10}(SN+1) - 0.20 + \frac{\log_{10}\left[\dfrac{p_i - p_t}{p_i - 15}\right]}{0.4 + \dfrac{1094}{(SN+1)^{5.19}}} + 2.32 \log_{10} M_R - 8.07 \quad ...(4.5)$$

where M_R = sub-grade resilient modules

SN = structural number of pavement

For unreinforced pavement section

$$SN = a_1D_1 + a_2D_2m_2 + a_3D_3m_3$$

Calculation

Structural number of unreinforced pavement, SN = 5.49

Design traffic satisfied by the unreinforced pavement, $W_{18,\ Des}$ = 10.15 MSa

Design Traffic > Required Traffic, Hence Safe

Hence, Total thickness of flexible pavement = 40 + 75 + 250 + 250
= 615 mm

4.3.2 As per IRC: 37–2001

For the given CBR data and traffic data, total thickness of flexible pavement shall be 580 mm as per following details (Table 4.3). However, this thickness is for 5 years life span of flexible pavement.

Table 4.3: Design of flexible pavement (CBR = 3%)

Sl. No.	*Item*	*Symbol*	*Thickness (mm)*
1	Sub-base	GSB	230
2	Base	WBM/WMM	250
3	Binder	BM	75
4	Wearing course	SDBC	25
	Total		580 mm

Legend: GSB – Granular Sub base, WBM – Water Bound Macadam, WMM – Wex Mix Macadam, BM – Bituminous Macadam, SDBC – Semi Dense Bituminous Concrete.

If life span of road is taken as 10 years, the corresponding thickness of flexible pavement shall be 630 mm.

The design thickness of pavement submitted by DLF to IIT was 615 mm (Fig. 4.7).

The CBR value of sub-grade is only 3. Therefore, pavement thickness is coming of the order of 2 feets. More thickness of pavement clearly means more expenditure.

4.4 DESIGN OF REINFORCED FLEXIBLE PAVEMENT

These days many ground improvement techniques are available. Use of geosynthetics is gaining popularity for road construction. Laying of geosynthetics layers on sub-grade level not only enhances the CBR value significantly but also minimises the maintenance cost. Hence design of reinforcement pavement is recommended. It is stated that CBR is going to be increased to minimum 6% by use of geosynthetics.

4.5 AS PER AASHTO FLEXIBLE PAVEMENT DESIGN METHOD, 1993

Input Parameters

Reliability level, R	—	90%
Normal derivate, Z_R	—	–1.282
Standard deviation, S_0	—	0.45
Initial serviceability index, p_i	—	4.2
Terminal serviceability index, p_t	—	2
Required traffic, $W_{18, Req}$	—	10 MSa
Sub-grade soil, CBR	—	3.0%
Reinforced in granular base (Yes/No)	—	No
Type of geogrid in granular base	—	Not used
Reinforced in granular sub-base (Yes/No)	—	Yes
Type of geogrid in granular sub-base	—	BX 3030 (Biaxial geogrid with tensile strength as 30 kN/m) in both direction with aperture of 34 mm× 34 mm.

Type of geotextile: 55 kN/m tensile strength in longitudinal direction and 42 kN/m in lateral direction.

Details of Pavement Layers

Thickness of bituminous concrete, $D_{1.BC}$	—	40 mm
Thickness of dense bituminous concrete, D_{1mDBM}	—	50 mm
Thickness of granular base/WBM/WMM, D_2	—	200 mm
Thickness of granular sub-base, D_3	—	235 mm
Layer coefficient of bituminous concrete, $a_{1.BC}$	—	0.45
Layer coefficient of dense bituminous concrete, $a_{1.DBM}$	—	0.45
Layer coefficient of granular base/WBM/WMM, a_2 (min. CBR of WBM/WMM = 60%)	—	0.15
Layer coefficient of granular sub-base, a_3 (min. CBR of granular sub-base = 20%)	—	0.12
Drainage factor for granular base, m_2	—	1.3
Drainage factor for granular sub-base, m_3	—	1.3

4.5.1 As per IRC: 37–2001

By use of geogrid, CBR value is bound to increase. Even if geogrid of extremely low strength is taken CBR value increase by 40% hence presuming that CBR increases to a conservative value of 4.2, (though CBR will not be less than 6%) the thickness of reinforced pavement shall be as follows (Table 4.4):

Table 4.4: Design of reinforced flexible pavement (for increased value of CBR = 4.2)

Sl. No.	*Item*	*Symbol*	*Thickness (mm)*
1	Sub-base	GSB	230
2	Base	WBM/WMM	150
3	Binder	BM	75
4	Wearing course	SDBC	25
	Total		480 mm

The above value is indicative only.

Note: This thickness is minimum for a life of 5 years. Strengthening layers may be required after 5 years as per actual site condition. After 5 years a strengthening layers of minimum of 50 mm BM + 25 mm SDBC may be laid.

4.5.2 Design by Giroud & Noiray's approach

Referring to the curves as per Fig. 4.6, the pavement thickness is given as below:

(*a*) For CBR = 3, pavement thickness
= 350 mm (for unreinforced)

(*b*) For CBR = 4.3, pavement thickness (for a particular tensile strength and elongation of reinforced)
= 350 – 150 = 200 mm (for reinforced, section with enhanced CBR)

The computations from Giroud & Noiray's approach are only indicative. However, it clearly shows that for reinforced section, the pavement thickness is bound to reduce.

4.6 MANDAL AND MHAISKAR APPROACH

The thickness of base course can be computed by following equation:

$$c = \frac{P}{\left\{2\pi\left[\sqrt{\frac{P}{P_c}} + 2h_0 \tan\alpha\right]\cdot\left[\sqrt{\frac{P}{2P_c}} + 2h_0 \tan\alpha\right]\right\}} \qquad ...(4.6)$$

where, h_0 = base course thickness

α = angle of load dispersion (= 26° for unreinforced pavement and is equal to 52° for reinforced pavement)

P = axle load (10 t)

P_c = type inflation pressure (62 t/m^2)

c = cohesion (t/m^2)

Taking c = 1.5 t/m^2, and substituting above values in above equation:

$$1.5 = \frac{10}{\left\{2\pi\left[\sqrt{\frac{10}{62}} + 2h_0 \tan 26\right]\left[\sqrt{\frac{10}{2\times 62}} + 2h_0 \tan 26\right]\right\}}$$

$$1.5 = \frac{10}{\{2 \times 3.14\}[0.4 + 40][0.28 + h_0]}$$

or, $$(0.4 + h_0)(0.28 + h_0) = 1.06$$

$$h_0 = \frac{-0.68 \pm \sqrt{(0.68)^2 + 4 \times 0.95}}{2} = 0.69 \text{ m (for } \alpha = 26°)$$

For reinforced pavement (i.e., α = 52°)

$$1.5 = \frac{10}{\{6.28[0.4 + 2.56\,h_0][0.28 + 2.56h_0]\}}$$

$$(0.4 + 2.56\,h_0)(0.28 + 2.56\,h_0) = 1.06$$

$$h_0 = \frac{-1.74 \pm \sqrt{(1.74)^2 + 4 \times 6.55 \times 0.95}}{2 \times 6.55}$$

$$= 0.27 \text{ m.}$$

Hence, it is clearly evident that reinforced section has much lesser base course thickness.

4.7 SUGGESTED DESIGN

Based on all the discussions above, soil proposed present at site, type of complex M/s DLF are developing and quality of people going to live there, it is recommended to provide a road which is maintenance free, robust, durable and sustainable. Modern techniques are available now to construct such roads. The design of unreinforced pavement as proposed by M/s DLF is shown in Fig. 4.8 and design of reinforced pavement is shown in Fig. 4.9.

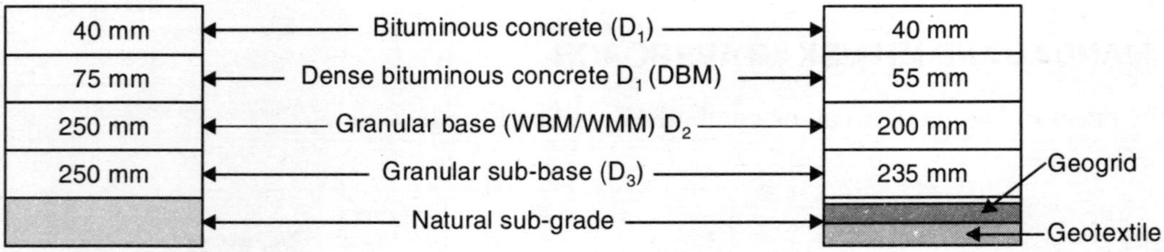

Fig. 4.8: Unreinforced section (Total thickness = 615 mm)

Fig. 4.9: Reinforced section (Total thickness = 525 mm)

4.8 COST ANALYSIS

The cost analysis has been done based on cost of material available today in market for geogrid, geotextile and other construction material. The costing has been done on per square metre basis. Following table 4.5 shows the cost analysis:

Table 4.5: Cost analysis of flexible pavement (unreinforced and reinforced) for class A-A loading

RATE ANALYSIS

Project: DLF Project at New Chandigarh

Summary of Pavement Design

Details	***Pavement***	
	Unreinforced	***Reinforced***
CBR (%)	3.0	4.3
Traffic. (MSa)	10	10
Pavement Details		
BC, mm	40	40
DBM, mm	75	50
WBM, mm	250	200
GSB, mm	250	235
Total	615	525

Sl. No.	*Item*	*Unit*	*Rate/Unit, ₹*	*Unreinforced section*		*Reinforced section*		
				Thk. (m)	*Cost/Sqm, ₹*	*Thk. (m)*	*Cost/Sqm, ₹*	
1	BC	cu.m	7091	0.04	283.64	0.04	283.64	
2	DBM	cu.m	5627	0.075	422.025	0.05	281.35	
3	WBM	cu.m	1021	0.25	255.25	0.200	178.6	
4	GSB	cu.m	962	0.25	240.5	0.235	221.00	
5	Biaxial Geogird, BX 3030	sq.m	102	-	-	1 layer	102	
6	Geotextile, GWF 52-240	sq.m	68.5	-	-	1 layer	68.5	
			Total cost/sq. m		**1201.415**			**1135.00**
						Total savings	**5.52 (%)**	

Saving = ₹ 66.32 per sq. m

As stated above cost analysis is on per sq. m basis. The clients may work out the total saving in construction cost themselves by simply working out total area of roads.

4.9 USE OF JUTE GEOTEXTILES IN ROAD CONSTRUCTION

In India, Jute Geotextile (JGT) has been used on following roads successfully:

1. Kakinada Port Trust Road, Andhra Pradesh
2. Munshirhat Rajpur Road, West Bengal
3. Andulia-Boyratala Road, West Bengal
4. Stabilisation of Slope of Mine-spoil, Uttarakhand
5. Slope Protection of Flood Embankment, Beluakhai, Orissa
6. Slope Stabilisation of Bridge Approach, West Bengal
7. Slope Stabilisation for Rail Track, Madhusudanpur, West Bengal
8. Hill Slope Stabilisation at TLDP Project (NHPC), Rambi, West Bengal
9. Hill Slope Stabilisation at Sonapur Land Area, Meghalaya

Typical applications of JGT are shown in Fig. 4.10.

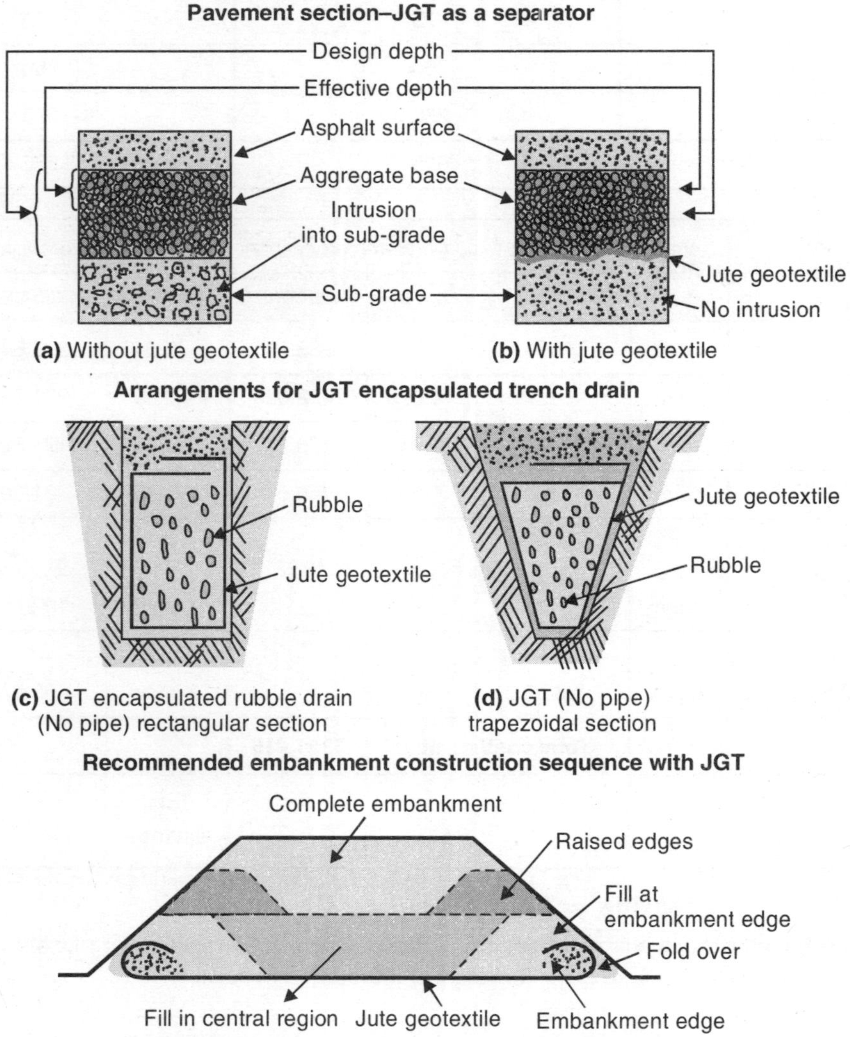

Fig. 4.10: Various applications of jute geotextile

Jute geotextiles applied in road construction at some sites are illustrated in Fig. 4.11 (a, b, c)

Fig. 4.11: (a) Application of JGT on soft soil

Fig. 4.11: (b) Road embankment on the JGT

Fig. 4.11: (c) Application of JGT for construction of road side drains

4.10 TYPES OF JUTE GEOTEXTILE USUALLY APPLIED IN ROAD CONSTRUCTION

Typical specifications

1. **Woven type as separation, soil retainer and filter:** Untreated weight–760 g/m^2, thickness–2 mm, width 76 cm, porometry–150 micron, permittivity–20 litres/m^2/sec under 10 cm head;

strength (warp × weft)–20 × 20 kN/m, strain at break (warp × weft)–10% × 10%, packing-bales (Fig. 4.12).

Fig. 4.12: Jute geotextile

2. **Non-woven type as filter and for drainage:** Weight–1000 g/m^2; thickness–8 mm, width–150 cm; co-efficient of permittivity 3.4 × 10^{-3} to 10^{-4} m/s; of transitivity 10–20 m/s; strength (warp × weft)—7 × 3 kN/m; strain at break (warp × weft)—25% × 30%; Packing–rolls.

4.11 RBI GRADE-81, PAVEMENT MATERIAL SOIL STABILISER

RBI grade-81 is recently developed unique and innovative product that has been developed for stabilisation of soils. It is inorganic, hydration activated powder-based stabiliser that reacts with soil particles to create layers that are interconnected through a complex inter-particle framework. All types of soils e.g., clay, silt, sand and gravel can be stabilised with this product. It can be used from rural roads to runways. Some of the test results obtained by various institutes and laboratory are shown in following Table 4.6.

Table 4.6: Comparative test results of RBI Grade-81 with different types of soils

Untreated soil (UTS) treated soil (TS with RBI)	*Type of soil*	*CBR %*					*UCS, kg/sq cm, (all results are with 4% Stabilizer) * 6%*				
		CRRI	*IIT Kharagpur*	*IIT Roorkee*	*IIT Chennai*	*RASTA*	*CRRI*	*IIT Kharagpur*	*IIT Roorkee*	*IIT Chennai*	*RASTA*
UTS	Clay	4	3	-	2	5	5	0		0	3
TS with RBI	Clay	25	78	-	22	42	15	19		14	14
UTS	Silt	8	-	13	-	10	4	-		-	16
TS with RBI	Silt	198	-	114	-	32	32	-		-	39
UTS	Sand	22	-	-	-	14	0.5	-		-	5
TS with RBI	Sand	53	-	-	-	143	7	-		-	23
UTS	Gravel	30	-	90	-	34	3	5		-	2.1
TS with RBI	Gravel	197	-	361	-	180	26	35		-	19*

UTS = Untreated Soil, TS with RBI =Treated Soil with RBI.

Untreated (UT) treated (T)	*Type of soil*	*PI (Plasticity Index) (%)*					*FSI (Free Swell Index) (%)*				
		Sample 1	*Sample 2*	*Sample 3*	*Sample 4*	*Sample 5*	*Sample 1*	*Sample 2*	*Sample 3*	*Sample 4*	*Sample 5*
UT	Clay	18	11.5	29	24.2	46.4	16.6	20	33	61.5	53.8
T with RBI	Clay	11.7	5.2	8.8	13.3	21.2	3.2	4.3	13	23.1	21.2
UT	Silt	10.3	11.6	10.3	5.6	23	5	16.6	0	9.1	8.3
T with RBI	Silt	5.4	6	2.3	1	10.6	0	0	0	0.8	2.1

In India, largest manufacturers and suppliers of Jute geotextiles are:

1. Indian Jute Industries' Research Association (IJIRA), Website: www.ijira.org.
2. Jute Manufactures Development Council (JMDC), Website: www.jute.com
3. Indian Jute Mill (IJMA), E-mail: ijira@vsnl.com

Mittal & Kannan (2012) conducted various tests on clayey soils as well as on cohesionless soils and found that RBI-81 performs very well for roads construction.

4.12 CONSTRUCTION METHODOLOGY

4.12.1 Requirements and Specification

Prior to the construction of the stabilised layer it should be ensured that underlying layer or foundation meets all specified requirements.

The underlying layer should be investigated to establish whether there is any damage, wet spots or other defects. Any defects to the layer should be rectified before the stabilised layer is constructed where the stabilised layer is constructed on the floor of a pavement excavation or on the top of an existing pavement layer i.e., where the underlying layer has not been reworked or reconstructed, the floor of the excavation or the top of the existing pavement layer should first be watered and the compaction of the layer should be verified.

4.12.2 Layer preparation

The material to be stabilised should be placed, in the case of an existing pavement layer (Fig. 4.13), scarified to the full depth specified, broken down, watered if necessary and mixed to achieve a homogenous layer. Any oversize material (>1/3 the thickness of the layer to be stabilised) should be removed.

Fig. 4.13: Layer preparation

4.12.3 Application of RBI Grade-81

Dosage determination should be performed prior to product application. After the layer of soil or stabilised material has been prepared the product should be spread uniformly over the full area of the layer by means of an approved type of mechanical spreader or by hand. Material that has been exposed to the open air for a period of 4 hours or more should not be used (Fig. 4.14).

If application is administered by hand, bags of the product should be accurately spaced at equal intervals along the section to be stabilised so that the specified rate of application can be achieved. The stabilising agent should be spread as evenly as possible over the entire surface. Do not apply RBI Grade-81 in high wind conditions, as determined by the on-site engineer or project manager, are such that the stabiliser becomes airborne and is wasted in such a manner as to become hazardous to traffic, workmen, adjacent workers, adjacent property, or results in adverse impact upon the public.

Fig. 4.14: Application of RBI Grade-81

4.12.4 Dry-mixing RBI Grade-81

Immediately after the product has been spread it should be mixed with the soil/material to the full depth of treatment. Special attention should be taken not to disturb the compacted layer underneath and especially not to mix the stabilising agent in below the desired depth. Mixing should continue for as long as necessary and repeated as often as required to ensure homogeneity and thorough mixing of the soil/ material with RBI Grade-81 over the full area of the application site.

(a) Spreading of RBI Grade-81 on soil

(b) Mixing of RBI Grade-81 with soil

Fig. 4.15: Dry-mixing of RBI Grade-81

Mixing should be done using a rotator (soil recycler or agricultural), grader or equivalent equipment over successive passes of the layer. Specific attention should be made to ensure that the mixing device does not cycle RBI Grade-81 and relocate the majority of the powder on the bottom of the layer to be constructed.

4.12.5 Watering

The water supply and watering equipment should be adequate to ensure that all the water required is added and mixed with the soil/material being treated within the prescribed period to enable compaction to be completed (Fig. 4.16).

Fig. 4.16: Watering to enable compaction

4.12.6 Uniform mixing of the product and water

It is paramount to the success of the stabilised layer. The number of passes required for RBI Grade-81 stabilisation to obtain adequate mixing is based on the type of soil and experience of the contractor.

Immediately after RBI Grade-81 has been properly mixed into the layer, the pre-determined optimum moisture content (OMC) should be administered. Each application or increment of water should be well mixed with the material so as to avoid concentration of water near the surface or in pockets within the layer (Fig. 4.17).

Fig. 4.17: Layer stabilising

Particular care should be taken to ensure satisfactory moisture distribution over the full depth, width and length of the section being stabilised and to prevent any portion of the work from getting excessively wet after the stabilising agent has been added. Ensure that the moisture content of the mixture is not below the OMC or more than 2 per cent above the OMC and less than the quantity that causes the roadbed to become unstable during compaction and finishing. If any portion of the work becomes too wet after the stabilising agent has been added and before the mixture has been compacted, as during rain or downpour, the application shall be rejected. These portions should be allowed to dry out to the required moisture content and then be scarified, re-stabilised, re-compacted and finished off (Fig. 4.18).

Fig. 4.18: Mixing of product

4.12.7 Compaction

Compaction of the material in the pavement layer should be carried out within 4 hours. In colder climates, a cool soil may slow the setting of RBI Grade-81 and this should be taken into consideration in the design stage (Fig. 4.19).

Fig. 4.19: Shaping the road during compaction

The type of compaction equipment to be used and the amount of rolling to be done should be such as to ensure that specified densities are obtained without damage being done to lower layer structures. Selecting the right compaction equipment is typically carried out by the contractor. The compacted layers should be adequately shaped (design specific) to prevent standing water from scouring the completed work. The process of shaping should be completed by an experienced grader operator and should be facilitated by shaping the road to reflect the required camber without excessive removal from or addition to the thickness of the layer.

Fig. 4.20: Compaction in progress

Compaction should be carried out in a series of continuous operations covering the full width of the layer concerned. The length of any section of a layer being compacted shall, wherever possible, not be more than can be properly compacted with the available equipment. During compaction, the layer should be maintained to the required shape and cross-section, and holes, ruts and laminations should be removed. Loss of moisture from evaporation should be corrected by further light applications of water over the surface. The minimum compaction required shall be 98% of Mod AASHTO density for the stabilised layer and the sub-base or immediate underlying layers. During final compaction, field density determinations should be done to determine accurately the applied compaction effort and to ensure the minimum compaction requirement has been obtained. Density testing should be carried out within 24-hours of compaction having been completed.

Fig. 4.21: Density testing during compaction

4.12.8 Usages

It improves the *in situ* soil spoiled pavements (lower CBR & high plasticity). Thus allows the use of the *in situ* soil to meet the required CBR, plasticity, etc. in construction of new pavements. Thus the cost of purchasing commercial quarried aggregates and the transport cost to site in the delivery of this commercial quarried material is dispensed with. Furthermore the amount of heavy vehicles in the transport of these commercial quarried materials are reduced which in itself saves maintenance costs of these vehicles as well as the damage to the environment. Add to this the discomfort of residents along these routes or roads under construction due to excessive dustiness and vehicle accidents

The improvement of the sub-grade *in situ* material thus reducing the number of selected sub-grade layers to be constructed thus reducing the overall pavement design layers. In order to demonstrate benefit 1, let's say the pavement layers have deteriorated over time (+15 yrs) due to excessive traffic loading etc. from Figs. 4.22 to 4.23.

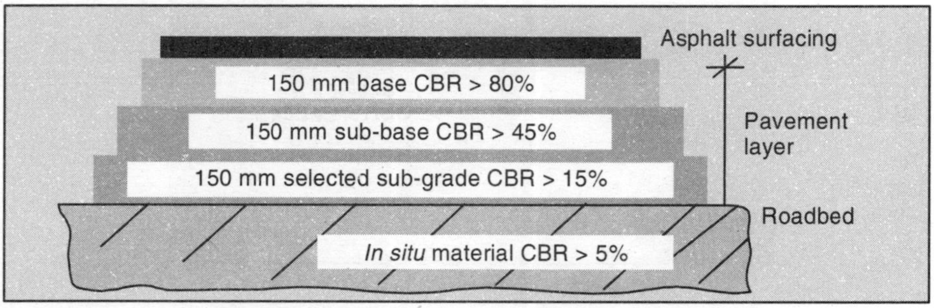

Fig. 4.22: Normal conventional pavement layers

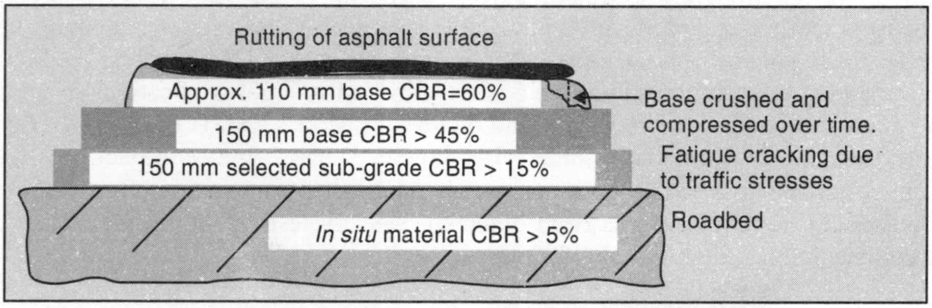

Fig. 4.23: Pavement structure after 15 years

The following remedial actions can be implemented when stabilising with RBI Grade-81, see Figs. 4.24 (*i*) and Fig. 4.24 (*ii*), with the dosage predetermined to the requirement.

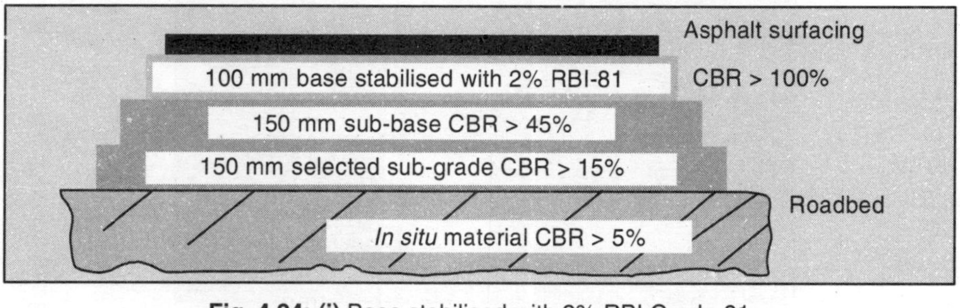

Fig. 4.24: (i) Base stabilised with 2% RBI Grade-81

In Fig. 4.24 (*i*) the base material has been improved to its original strength and the road should last at least another 5–10 years.

In Fig. 4.24 (*ii*) the base together with the sub-base material is stabilised with 4% RBI Grade-81 creating layer strength in excess of 200% (CBR) thus a layer of 250 mm thick is applicable here representing the base and sub-base layers. The sub-base is stabilised with the base material for in this instance the sub-base strength has also changed due to age, i.e., decomposition of the materials which make-up the sub-base material, reducing the CBR from >45% to a CBR = 30%. Therefore to save time and cost (construction) the layers were constructed as one (250 mm) and not as two individual layers which in itself is a benefit.

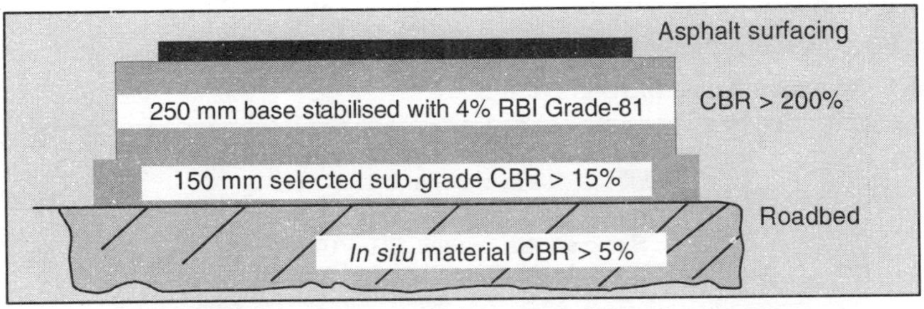

Fig. 4.24: (ii) Base/Sub-base stabilised with 4% RBI Grade-81

Benefit 2, improvement of the *in situ* roadbed material, is illustrated in Fig. 4.25 (*i*) and Fig. 4.25 (*ii*) where different CBR strengths of the *in situ* are used.

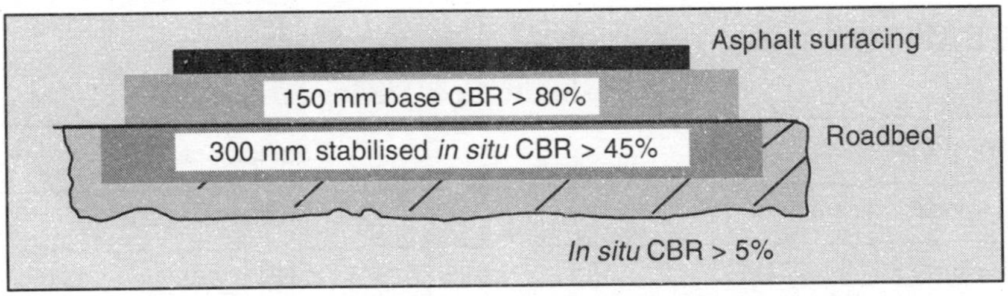

Fig. 4.25: (i) *In situ* material stabilised with RBI Grade-81 to CBR > 20%

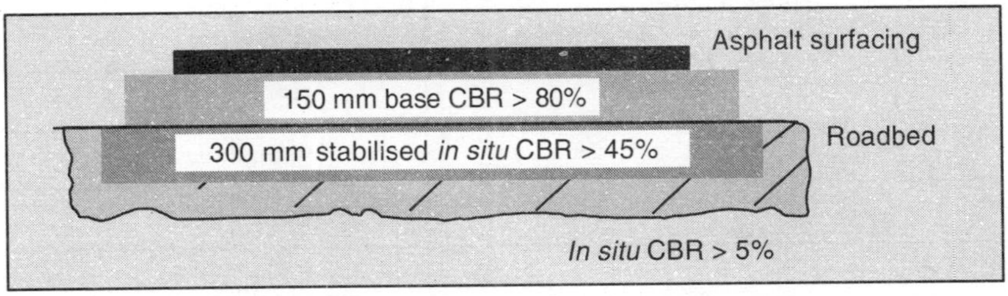

Fig. 4.25: (ii) *In situ* material stabilised with RBI Grade-81 to CBR > 20%

In Fig. 4.25 (*i*) 150 mm of the *in situ* material is stabilised with RBI Grade-81 increasing the CBR to above 20% reducing the need of the selected sub-grade layer which CBR specification is > 15%.

In Fig. 4.25 (*ii*) 300 mm of the *in situ* material was stabilised with RBI Grade-81 where the CBR has been increased from 5% to >45% reducing the need of the selected sub-grade layer as well as the sub-base layer.

Benefit 3 is the stabilisation of natural *in situ* material as a wearing course layer for farm access roads (Fig. 4.26) where only light traffic is involved and existing gravel road surfaces as a wearing course layer (Fig. 4.27).

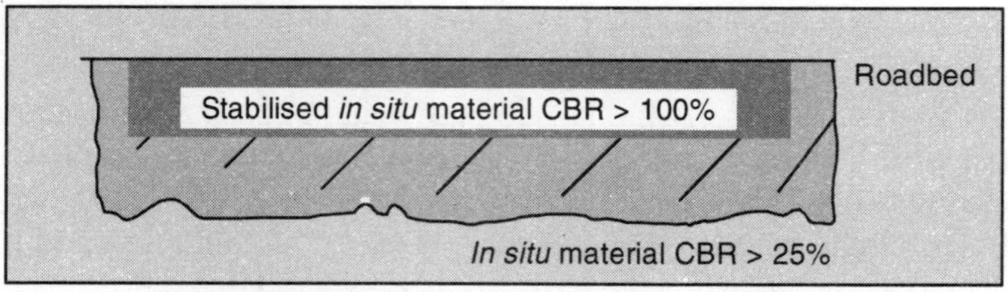

Fig. 4.26: Farm access (track) road

(*i*) Reduce dustiness
(*ii*) Improve its bearing strength and
(*iii*) Reduce the maintenance of re-gravelling.

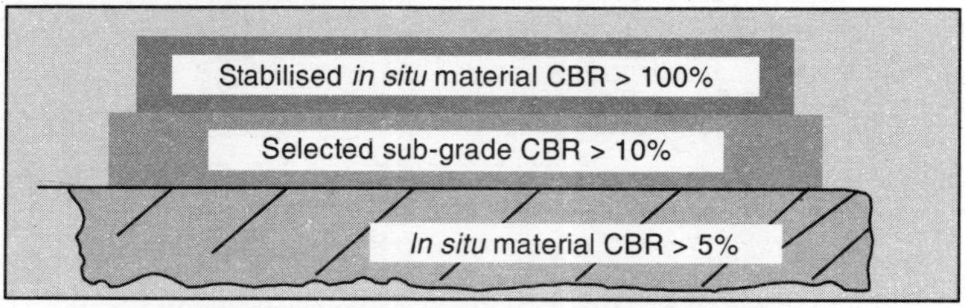

Fig. 4.27: Existing gravel road

4.13 COST SAVINGS

Approximate Cost Savings

0	Equipments cost	Equipments cost
10	Engineering cost	Engineering cost
20	Clearing and Grubbing	Clearing and Grubbing
30	Material transport cost	Material transport cost
		BC
		DBM
		Stone dust + RBI 81
40	BC	Soil + RBI 81
50	DBM	Soil + RBI 81
	BM	Soil + RBI 81
60	WMM	Soil + RBI 81
70	GSB	
80		Saving
90	Earthwork	
100		

4.14 PROJECTS UNDERTAKEN BY USE OF RBI-GRADE-81 IN INDIA

1. Road in Jammu
2. Road near Siri Fort, Delhi
3. Road in Barijangs, Orissa
4. Road in Leh
5. Road in Thirupur, Tamilnadu
6. Road in Srinagar

International projects

1. Road in Dominican Republic
2. Cycle path in north of Italy
3. Construction yard in Brazil

4.15 USE OF GEOTEXTILE IN ROAD CONSTRUCTION (Experiences of M/s DuPont Typar® SF)

Depending on the different applications, the main functions of the geotextile are separation, filtration, reinforcement, protection to stabilisation. For most applications a combination several functions required. A further requirement is its resistance to damage during installation.

The purpose of this section is to provide a basic, technical understanding of these functions and requirements with respect to geotextiles and to the different mechanisms within each function. It should aid in the selection of the appropriate geotextile for a specific purpose, a difficult task, because the interaction between interrelated factors, such as mechanical and hydraulic properties, cogging, structure, time and degradation etc. is very complex.

4.15.1 Separation

Separation is defined as: *"The preventing from intermixing of adjacent dissimilar soils and/or fill materials by the use of a geotextile or a geotextile-related product."*

The main application areas of a geotextile used as a separator are in road and railway projects. The use of the geotextile preserves and improves the integrity and function of the different material. Two mechanisms occur when an aggregate base is laid over a soft sub-soil and a vertical load is applied.

First of all the geotextile prevents the loss of aggregate into the soft sub-grade (Fig. 4.28). An engineering adage describes this very well: "10 kilos of stone placed on 10 kilos of mud results in 20 kilos of mud". The geotextile confines the aggregate base and thus a higher degree of compaction can be obtained with subsequent higher bearing capacity. Secondly the contamination of the aggregate base with soil of the sub-grade is prevented and thus the reduction of the bearing capacity avoided. The migration of fine soil particles into the clean aggregate occurs especially under dynamic stress and is called 'pumping effect'. These fines act as a lubricant between the coarse aggregate grams and can so substantially reduce the shear strength of the aggregate (Fig. 4.28).

In its separation function a geotextile can:

- Prevent the reduction of load bearing capacity caused by the mixing of fine-grained sub-grade with the aggregate base.
- Increase the bearing capacity by preventing the loss of aggregate into soft sub-grade and increases the degree of compaction.

- Reduce the deterioration of roads through frost heave effects.
- Eradicate the need to remove weak sub-grade.
- Maintain the drainage capacity of the aggregate base.
- Prevent migration of fine particles especially under dynamic loads.

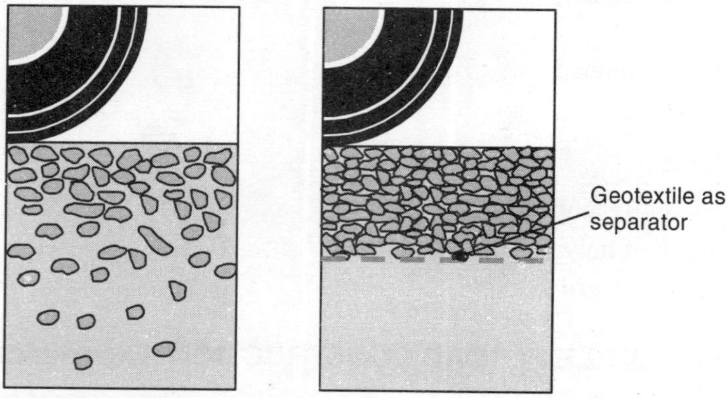

Fig. 4.28: Left: Without geotextile–loss of aggregate into soft sub-grade, **Right:** With geotextile–no loss of aggregate, better compaction

4.15.2 Stabilisation and Reinforcement

In many applications a geotextile fulfills a stabilisation or reinforcement function. In its stabilisation function the geotextile provides the soil with tensile strength and thus complements the soil's lack of tensile strength when subject to vertical loads.

There are three distinct mechanisms by which a geotextile can stabilise the aggregate base and improve its resistance to permanent deformation under repetitive loading (see Fig. 4.29).

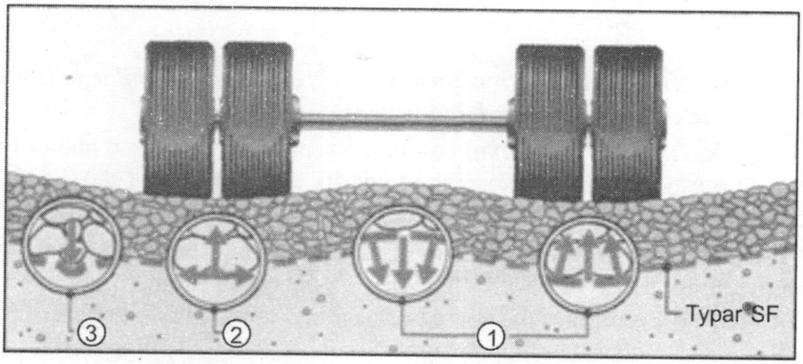

① ® Restraint + Confinement ② Membrane mechanism ③ Local reinforcement

Fig. 4.29: Three stabilisation mechanisms

The higher the initial modulus of the geotextile the more effective these mechanisms are geotextiles with a low initial modulus will have large deformations and provide little restraint, membrane mechanism or local reinforcement. A high initial modulus and high elongation are important to withstand large local deformations and resist puncturing.

4.15.3 Restraint and Confinement

As indicated in the Fig. 4.29, there are two types of restraint. One is related to the reverse curvature of the geotextile outside the wheel path where a downward pressure is created. This has the effect of a surcharge load, which levels out the deformation and enforces the compression of the subsoil. The other is the restraint the geotextile provides when aggregate particles attempt to move away from under the load. The geotextile provides tensile reinforcement to the aggregate layer. This confinement of the aggregate increases its strength and modulus, which in turn decreases the compressive stress on the sub-grade by spreading the load better underneath the wheel load.

4.15.4 Membrane mechanism

The membrane mechanism is effective when a geotextile is laid on deformable soil and vertical loads are applied. In-plane tensile stress develops in the geotextile, relieving the soil which is not capable of absorbing it. This in-plane force induces a component of stress perpendicular to the plane of the geotextile sheet, in the direction of the force. Therefore, this is of great significance in temporary road constructions, where it can reduce rutting tremendously. The higher the initial modulus of the geotextile, the higher the possible reduction of rutting.

4.15.5 Local reinforcement

Loads on individual stones can cause spot failures in the sub-grade. A geotextile with a high initial modulus allows to distribute the load, to reduce the stress and to provide resistance to displacement. A high elongation avoids the local puncturing of the geotextile because it allows the geotextile to stretch around a penetrating stone.

For more information and details on the use of geotextiles in reinforced earth structures please refer to the Typar HR product and design guide. Typical applications for earth reinforced structures are retaining walls, steep slopes, landslide repairs, soft-soil embankments, reinforcement under foundations, reinforcement or bridging over karstic areas or cavities etc.

4.16 FILTRATION

Filtration is defined as "The restraining of soil or other particles subjected to hydrodynamic forces while allowing the passage of fluids into or across a geotextile or a geotextile-related product". Typically the opening size and the permeability are used to describe the geotextile's filtration properties. The pore size of an efficient geotextile should be small enough to retain larger soil particles to prevent soil erosion. Small soil particles initially have to pass through the geotextile in order to support the build-up of a bridging network of larger particles which act as a natural soil filter adjacent to the geotextile (Fig. 4.30). If the pore size of a geotextile is too small, then the small particles cannot be drained away and a small diameter bridging network will be formed. This will produce a natural soil barrier with a lower permeability.

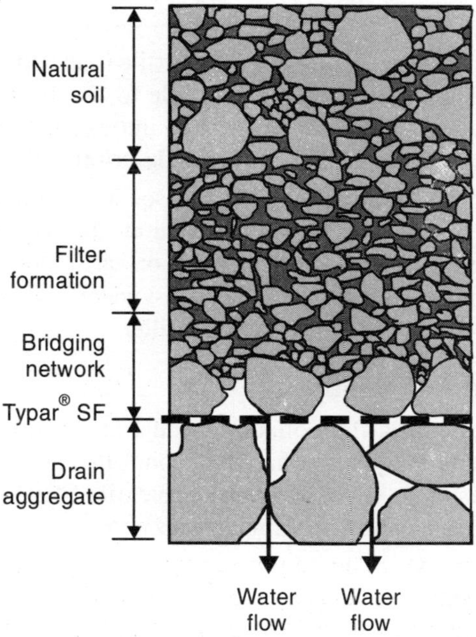

Fig 4.30: Natural soil filter adjacent to geotextile

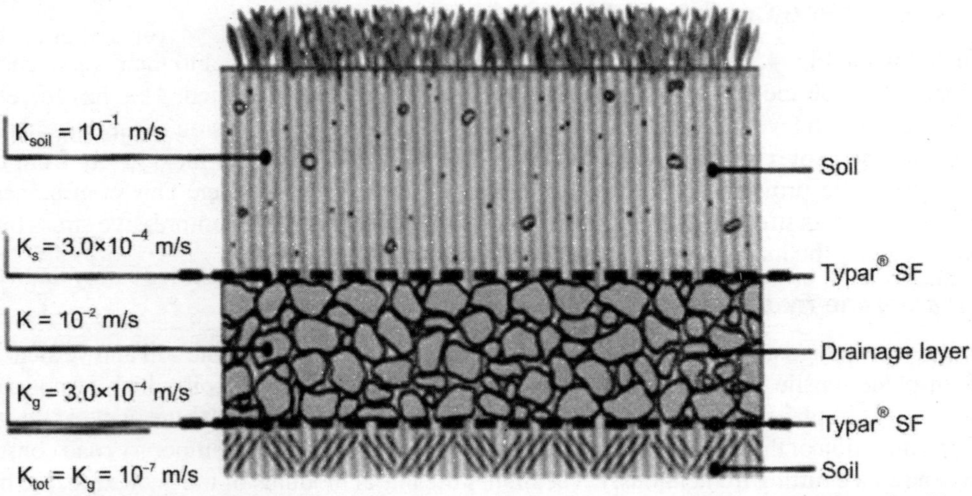

Fig. 4.31: Drainage system, soils and geotextile with different permeabilities, K_{tot} is determined by the least permeable soil layer

The per efficient geotextile filters must have pores of different shape, size and size distribution similar to the particle size distribution of soil. It is often ignored that in an aggregate sub-base system (Fig. 4.31) the permeability of the least permeable layer determines the system's permeability. Usually the soil has a permeability which is significantly lower than that of the geotextile.

Typical Soil Permeabilities	
Gravel 3×10^{-2} m/s	Sand 10^{-8} m/s
Silt $10^{-9} - 10^{-7}$ m/s	Clay $10^{-8} - 10^{-10}$ m/s

The permeability of a geotextile is also influenced by the compressibility of the geotextile. Generally thick geotextiles are susceptible to compression. When installed under pressure, the permeability is reduced. This needs to be taken into account when specifying the required permeability of a geotextile. The thickness itself is rather a descriptive than a design property.

The filtration function is associated with dam construction, erosion control, road drainage and sub-soil drainage. In these constructions the geotextile replaces a conventional granular filter. In an erosion control system of a riverbank or earth slope, coarse material (gabions/riprap) or concrete slabs are commonly used to protect against waterflow or wave action. The erosion of the fine particles is prevented by the use of a geotextile as a filter.

4.16.1 Drainage

Drainage is the collecting and transporting of precipitations, ground waters and/or other fluids. Traditionally, water has been controlled and evacuated using graded natural materials. Over the past 30 years or so, more and more geotextile filters have replaced the aggregate filters to increase the natural drainage capacity of impervious soils.

A geotextile should not be used as a (direct) drainage layer by itself because even though its drainage capacity can be measured in a laboratory using dean water, under realistic site conditions (soil trapped inside the structure) the drainage capacity is unpredictable. It is also important for drainage systems to be capable of maintaining adequate drainage capacity for long-term performance even when subjected to high earth pressures. To avoid clogging and contamination of the drainage layer a filter must always be incorporated in a drainage system.

Synthetic drains incorporating a geotextile filter have proved to be a more economic alternative to traditional sand drains, soak ways and other drainage systems. Typically geosynthetic drainage mats or drains are made of a core sandwiched between geotextile filters. (Fig. 4.32)

The filter material is required to show consistent quality and physical properties, outstanding strength and durability, good resistance to installation stresses and long-term filtration performance.

The malfunction or premature failure of a drainage system can create serious safety and functional problems with the earth structure concerned. At the very least, a drainage failure will necessitate costly remediation and attendant disruption. It is vital that a filter material is used that can function effectively over the long-term even with the most critical soils.

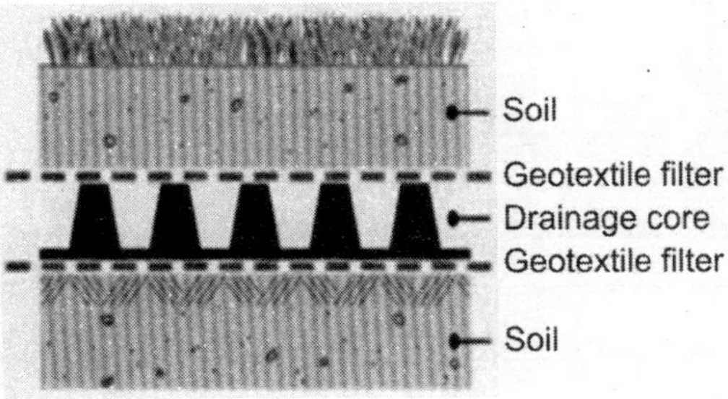

Fig. 4.32: Composite drainage element

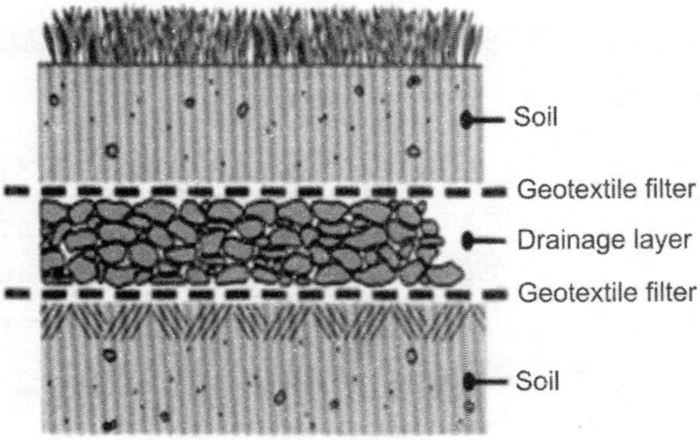

Fig. 4.33: Conventional aggregate drainage layer

4.16.2 Protection

Protection is defined as: *"The preventing or limiting of local damage to a given element or material by the use of a geotextile or a geotextile-related product"*.

Typically geotextiles are used for protection of geosynthetic barriers in landfill, roofing, tank and water projects.

The most important properties of a geotextile in a protection function are puncture resistance and product uniformity (i.e., no weak spots). Nail puncture resistance tests have shown that properties such as thickness and unit weight of the product alone do not provide good protection efficiency.

4.16.3 Resistance against damage during installation

A geotextile will not perform any function if it is destroyed during or immediately after installation. Analyses indicate that the critical period in the life-cycle of a geotextile is during the construction process rather than during the service life. Thus 95% of the damage usually occurs during installation, very often simply the result of impact damage during the off-loading and compaction of aggregates. Usually, if the geotextile survives these installation-related stresses, it will also withstand the in-service stresses. Considerable work has been undertaken to understand the relationship between the physical properties of a separation geotextile and its actual performance in the field. Test research has confirmed a close correlation between the ability of a geotextile to absorb impact energy and its susceptibility to damage during installation.

The following figures demonstrate the different forms of failure of a geotextile and the importance of a high energy absorption potential.

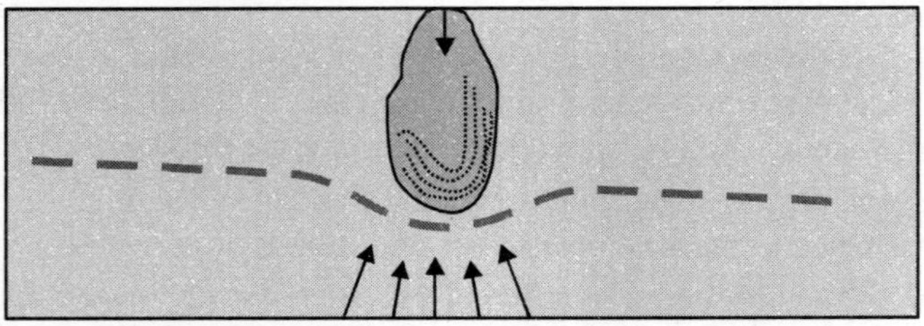

Fig. 4.34: High elongation allows the geotextile to stretch around penetrating stone

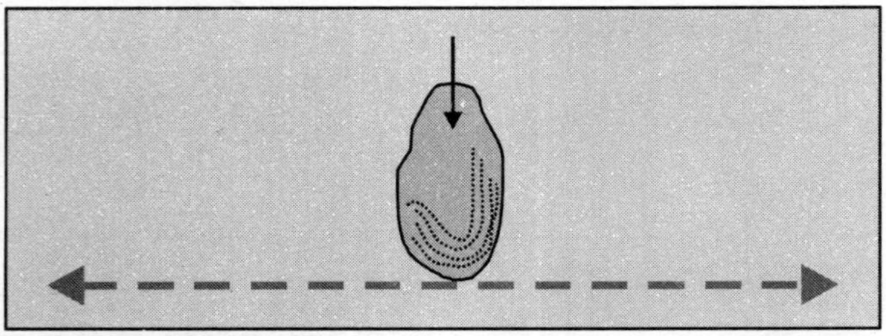

Fig. 4.35: High strength allows the geotextile to withstand force from falling stone

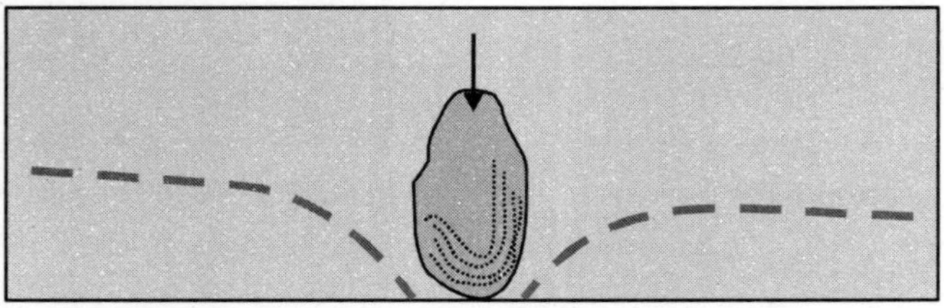

Fig. 4.36: Failure of the geotextile due to lack of strength or lack of elongation

4.16.4 Energy absorption

Definition: Energy is the capacity of a physical system to do work. The standard unit is the joule, symbolised by J. One joule (1 J) is the energy resulting from the equivalent of one Newton (1 N) of force acting over one metre (1 m) of displacement. There are two main forms of energy, called potential energy and kinetic energy. Potential energy, is energy stored in a material.

The energy absorption potential (W) of a geotextile can be described as the combination of its elongation and its applied strength. It is determined using the stress-strain curve of a geotextile obtained in a tensile strength test. The area under the curve ($\int$) is the energy absorption potential of the geotextile.

Several national specifications are in the process of adopting the energy absorption concept. Some specification however are based on theoretical values or energy absorption index rather than calculating the surface under the curve $W = \int (T \times \varepsilon)\, d\varepsilon$. The calculation is simplified to $W_{index} = \frac{1}{2} T \times \varepsilon$.

As a result the energy absorption index (W_{index}) of some products is significantly higher while for others the theoretical energy absorption is lower than the actual energy absorption potential measured during the tensile strength test (EN ISO 10319). The following graph (Fig. 4.37) illustrates this, it shows the different shapes of the actual energy absorption and the theoretical energy absorption potential.

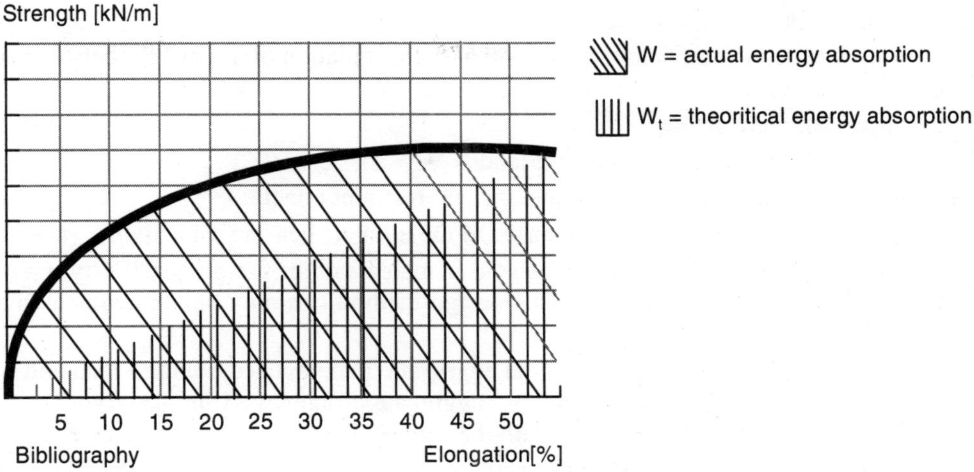

Fig. 4.37: Comparison of actual and theoretical energy absorption potential

4.17 DESIGNING AGGREGATE BASES WITH TYPAR® SF

The main causes of pavement degradation are:

1. The contamination of the aggregate base by a fine-grained sub-grade under dynamic loading 'pumping effect' that causes a substantial reduction of the shear resistance of the aggregate. The thickness of the 'clean' aggregate and, therefore, the bearing capacity of the structure is reduced down to unacceptable levels.
2. Contamination of the aggregate base as described above which will make the aggregate sensitive to frost, with subsequent reduction of bearing capacity during thaw periods.
3. Lack of subsurface drainage.
4. Unpredicted traffic increase.

The use of Typar® SF prevents aggregate contamination and, therefore, results in an increased service life.

This guide uses the CBR value as a measure of soil strength. The correlation factors between CBR, C_u (undrained shear strength), E_v (rigidity modulus) and M_E (compressibility modulus) are given in the following Table 4.7. The design properties presented here for unpaved and paved roads are based on a standard Typar® SF style with an energy level 2. Depending on the installation and traffic conditions a geotextile with a higher energy level may be chosen.

Table 4.7: Correlation chart for estimating the sub-grade CBR value

Scale	Values
Soil	Very poor, Poor, Fair, Medium, Good
CBR	0.5, 1, 1.5, 2, 2.5, 3, 4, 5, 6, 7, 8, 9
C_u [kN/m^2]*	20, 40, 60, 80, 100, 120, 140, 160, 180, 200, 220, 240, 260, 280
C_u [psi]	2, 6, 10, 14, 18, 22, 26, 30, 34, 38, 42
M_E [MN/m^2]**	2, 4, 6, 8, 10, 12, 14, 16, 18
E_v [MN/m^2]	2, 4, 6, 8, 10, 12, 14

4.17.1 Unpaved roads

An unpaved road that gives temporary or permanent access (i.e., construction or gravel road) normally consists of a simple, unbound aggregate base.

The proposed design method below assumes that the installation of Typar® SF between sub-grade and aggregate base allows for:

1. Better aggregate compaction
2. Sub-grade consolidation under dynamic loads
3. Reinforcement of the structure by membrane and restraint effect
4. Admissible pressure on sub-grade increased to the ultimate bearing capacity $p = (\pi + 2) * C_u$

The combination of these benefits is equivalent to an increase of the sub-grade CBR by approximately 3 per cent points based on empirical data. This design method can only be applied to designs using Typar® SF.

The procedure is first to determine the initial aggregate thickness according to load and sub-grade conditions, and then consider the service life and aggregate efficiency. After specifying the effective aggregate thickness T_{eff} a style of Typar® SF with the suitable energy level needs to be chosen.

Attention needs to be paid to axle loads > 130 kN. The appropriate curve for the determination of the initial aggregate thickness T_0 needs to be chosen and the actual number of passes N is used to determine the service life adjustment factor C.

4.17.2 Design method of unpaved road

Initial aggregate thickness, T_0

Soil bearing capacity = CBR, C_u

Axle load = P_i

Enter Fig. 4.38 using sub-grade CBR and axle load P_i to determine T_0, the compacted crushed stone thickness for 1000 axle loads. Alternatively, Table 4.8 lists the formula to calculate T_0.

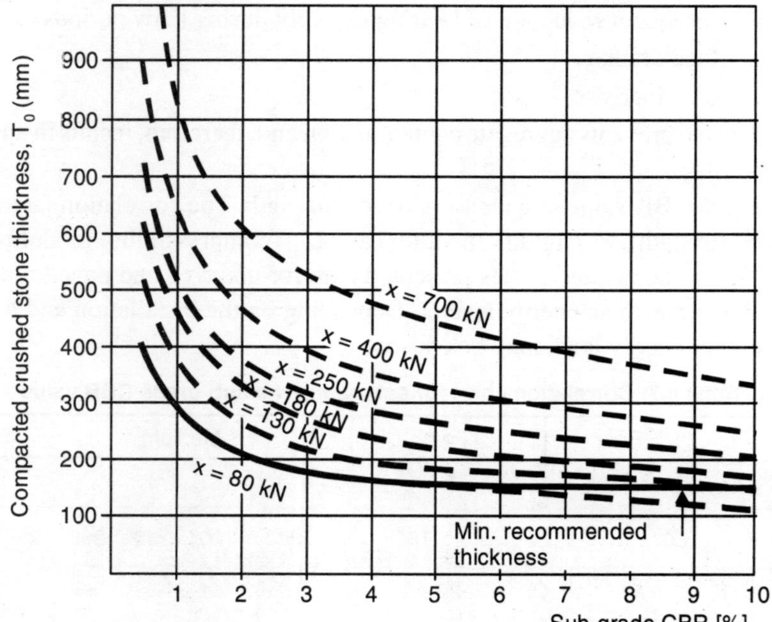

Fig. 4.38: Unpaved road: Compacted crushed stone thickness for 1000 axle loads

Table 4.8: Factors to determine curve P_i

CBR [%]	*P_1 [kN]*	*P_2 [kN]*
0.5	45.31	0.119
1	32.37	0.085
1.5	25.89	0.068
2	22.47	0.059
3	20.56	0.054
4	18.66	0.049
5	17.14	0.045
6	16.00	0.042
7	14.85	0.039
8	13.71	0.036
9	12.95	0.034
10	12.19	0.032

$$T_0\ (\text{mm}) = P_1 \sqrt{\text{Axle load (kN)}}$$

$$T_0\ (\text{in}) = P_2 \sqrt{\text{Axle load (lbs)}}$$

Adjustment of T_0 for service life

$$\text{Axle load} = P_i$$

$$\text{Actual number of passes} = N_i$$

$$\text{Compacted crushed stone thickness} = T_0$$

$$N_e = T = C \times T_0 = (0.27 \times \log(\Sigma N_i \times \text{ESAL}) + 0.19) \times T_0 \qquad ...(4.7)$$

- If the most frequent axle loads are heavier than 130 kN (e.g., quarry haul roads), the use of the $N_e = \Sigma N_i \times \text{ESAL}$ (Number of Equivalent Axle Loads) is not appropriate. The service life adjustment factor C must be determined using the actual number of passes N_i.
- The service life is expressed as the total number of 80 kN axle load application. The actual axle load is first converted to an equivalent standard axle load (P_0= 80 kN) using the equivalence factor (ESAL):

$$\text{ESAL} = \left(\frac{P_i}{P_0}\right)^{3.95} \qquad ...(4.8)$$

Lists of the equivalence factor ESAL for different axle loads are given in table 4.9.

Table 4.9: Equivalence factor (ESAL)

Axle load (kN)	*ESAL*	*Axle load (kN)*	*ESAL*
10	0.0003	140	9.12
20	0.004	150	11.98
30	0.021	160	15.45
40	0.065	170	19.64
50	0.16	180	24.61
60	0.32	190	30.47
70	0.55	200	37.31
80	1.0	250	90.08
90	1.59	300	185.10
100	2.41	400	576.70
110	3.52	500	1392.30
120	4.96	600	2860.80
130	6.80	700	5259.30

- By multiplying the actual number of axle passes (N_i) with ESAL, N_E the number of passes of equivalent standard axle loads is determined.

$$N_E = \Sigma N_i \times ESAL_i \qquad ...(4.9)$$

Since T_0 is indexed to a service life of 1000 axle load application, it must be adjusted by a factor C that depends on the actual number of standard loads N_E. The relationship between N_E and C is shown in Fig. 4.39.

Then the aggregate thickness T becomes:

$$T = C \times T_0 = (0.27 \times \log(\Sigma N_E) + 0.19) \times T_0 \qquad ...(4.10)$$

Adjustment of T for aggregate efficiency

$$T_{eff} = \frac{\Sigma T_i}{\alpha_i} \qquad ...(4.11)$$

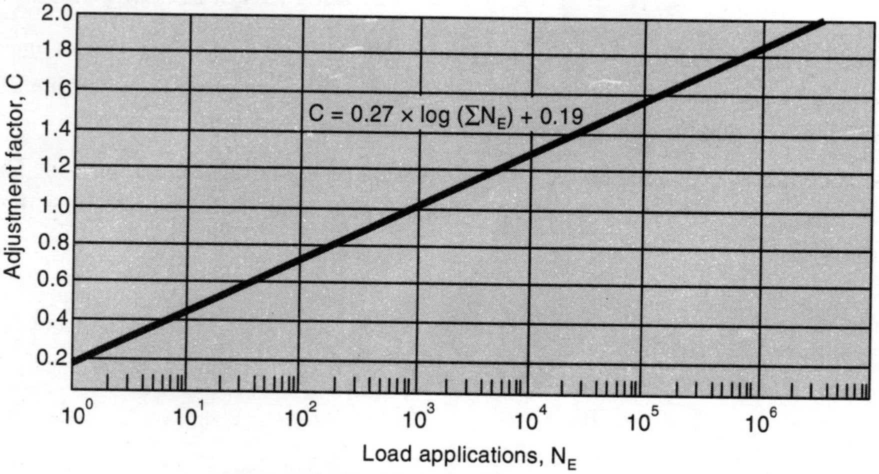

Fig. 4.39: Adjustment factor for service life

The different materials used for aggregate bases are more or less effective. This difference is accounted for by using the aggregate efficiency factor, α. The chosen aggregate should be compactable. The idea is to lock the whole mass together under load in order to take advantage of the reinforcement mechanisms of Typar® SF. Angular crushed aggregate is the best because it interlocks well and provides a high bearing capacity. Depending on availability, other materials or blends can be used. Table 4.10 indicates typical thickness efficiency factors of various surfacing and base materials.

Table 4.10: Adjustment for aggregate efficiency

Material	*Efficiency, α*
Paving stone	2
Hot mix (Dense-Macadam)	2
Dense surfacing course	2
Soil-cement (> 5 MPa compression)	1.5
Soil bitumen	1.5
Hard crushed stone aggregate—'standard'	**1.0**
Medium crushed stone aggregate (CBR > 80%)	0.8
Hard round stone aggregate (CBR > 80%)	0.8
Medium round stone aggregate	0.5
Sandy gravel (CBR = 20–30%)	0.5
Crushed limestone	0.5
Loose gravel, compactable sand	0.4

Example: 10 mm hot mix = 20 mm hard crushed stone 'standard'.

The original designed thickness T_i of crushed stone can therefore be replaced by the superposition of materials of thicknesses T_i and of efficiency α_i to obtain the final design value of aggregate thickness T_{eff} (efficient thickness).

$$T_{eff} = \frac{\Sigma T_i}{\alpha_i}$$

4.17.3 Paved roads

Permanent paved roads generally consist of an aggregate base, a bituminous road base and a concrete or bituminous surface coating.

The proposed design method assumes that the installation of Typar® SF between sub-grade and aggregate base of paved structures results in:

- Better aggregate compaction
- Sub-grade consolidation under dynamic loads
- Prevention of long-term aggregate contamination

These benefits mean a prolonged service life or, in other words, the ability to carry more traffic loads with a given aggregate base thickness. In addition, by using part of the aggregate base as an access road for construction traffic, one can take advantage of the stabilising effect of Typar® SF. Separation and filtration functions will favour sub-grade consolidation under static and dynamic loading. This tends to be as efficient as soil stabilisation itself.

The design procedure is similar to that of unpaved roads (see previous section). However, the compacted crushed stone thickness T'_0 for 1000 axle loads for paved roads is determined from Fig. 4.40. This thickness should be adjusted for service life and aggregate efficiency as for the unpaved structure.

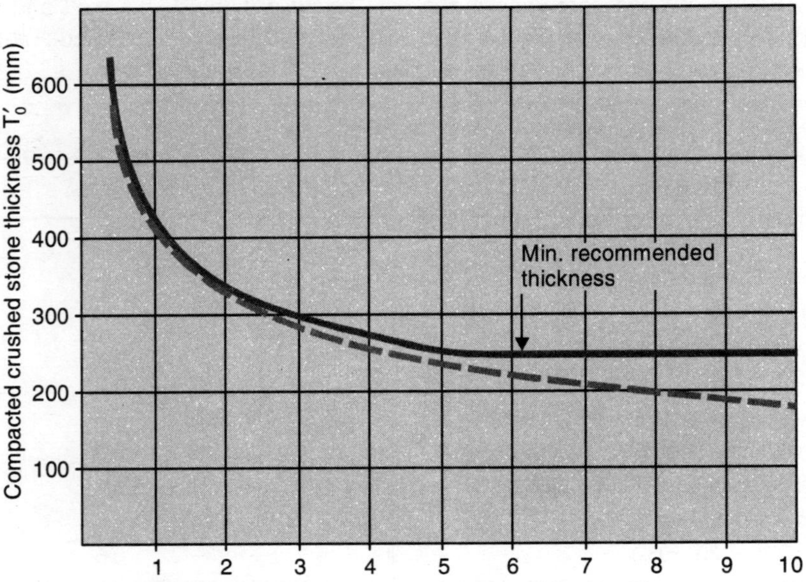

Fig. 4.40: Compacted crushed stone thickness T'_0

4.17.4 Paved roads with construction traffic sub-base

Take full advantage of the reinforcement mechanism of Typar® SF using Fig. 4.38 to determine the minimum aggregate thickness for temporary construction traffic roads. Then integrate this structure in the final paved road construction by adding the remaining aggregate to build up the necessary thickness as determined with Fig. 4.39. The design steps are summarised as follows.

Paved structure

A. *Initial Aggregate Thickness* T'_0 (Fig. 4.40)

Soil bearing capacity — CBR

Axle load — P_i

Unpaved construction road

A. *Initial Aggregate Thickness* T_0 (Fig. 4.38)

Soil bearing capacity — CBR

Axle load — P_i

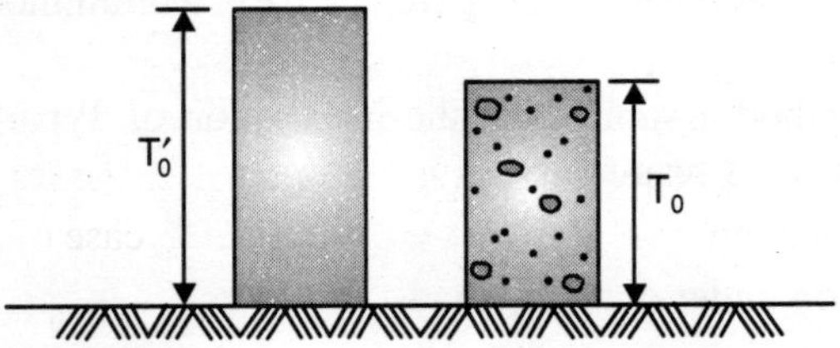

B. *Adjustment of* T'_0 *for service life*

Axle load — P_i

Actual number of passes — N'_I

Compacted crushed stone thickness — T'_0

B. *Adjustment of* T_0 *for service life*

Axle load — P_i

Actual number of passes — N_i

Compacted crushed stone thickness — T_0

$$\text{ESAL} = \left(\frac{P_i}{P_0}\right)^{3.95}$$

$$N'_E = \Sigma N'_I \times \text{ESAL} = C \text{ (Fig. 4.39)} \qquad N_E = \Sigma N_I \times \text{ESAL} = C \text{ (Fig. 4.39)}$$

$$T' = C \times T'_0 \qquad T = C \times T_0$$

$\alpha = 1$

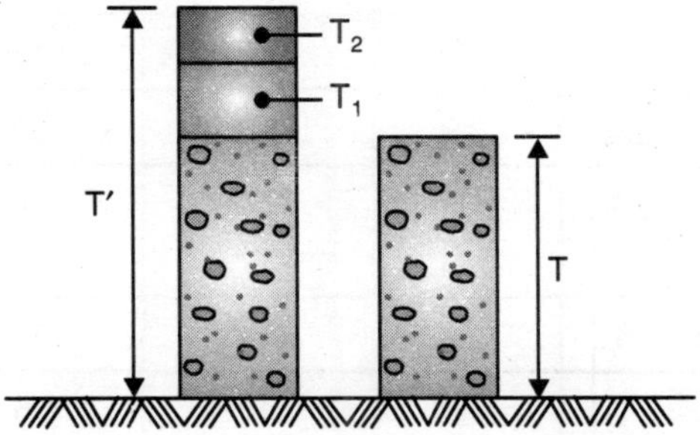

C. Adjustment of T′ for aggregate efficiency | *C. Adjustment of T for aggregate efficiency*

$$T'_{eff} = T_{eff} + \frac{\Sigma T_i}{\alpha_i} \qquad T_{eff} = \frac{\Sigma T_i}{\alpha_i}$$

With T_{eff} effective minimum thickness for construction traffic

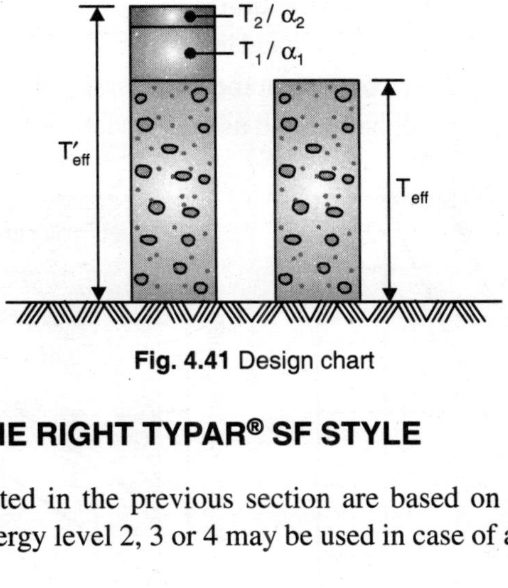

Fig. 4.41 Design chart

4.18 SELECTION OF THE RIGHT TYPAR® SF STYLE

The design guidelines presented in the previous section are based on a standard Typar® SF energy level 1. Higher performing energy level 2, 3 or 4 may be used in case of additional design requirements to withstand the:

- Effect of traffic
- Effect of installation conditions
- Effect of compaction

Determine the required level according to Figs. 4.42 to 4.44 and select the equivalent energy level of Typar® SF from Table 4.11 below.

Table 4.11: Minimum values for different Typar® SF energy levels						
Energy level			***Level 1***	***Level 2***	***Level 3***	***Level 4***
Test	Standard	Unit				
Energy absorption (actual)	EN ISO 1 031 9/ASTM D4595	kJ/m²	2	5	8	11
Tensile strength		kN/m	7	12	20	25
Elongation		%	50	50	50	50
Strength at 5% elongation		kN/m	2.5	5	7.5	10
Puncture CBR	EN ISO 12236	N	1000	1500	2500	3250
Cone penetration	EN 918	mm	40	35	30	20
Minimum recommended Typar® SF			SF32	SF49	SF77	SF94

4.18.1 Effect of traffic

Higher fabric properties are required to withstand

Fatigue caused by large number of equivalent standard axle loads (ESAL).

Additional stresses caused by heavy duty equipment (generally with axle loads larger than 130 kN). The correct Typar® SF energy level can be selected using Fig. 4.42 according to sub-grade CBR and the number of axle load applications.

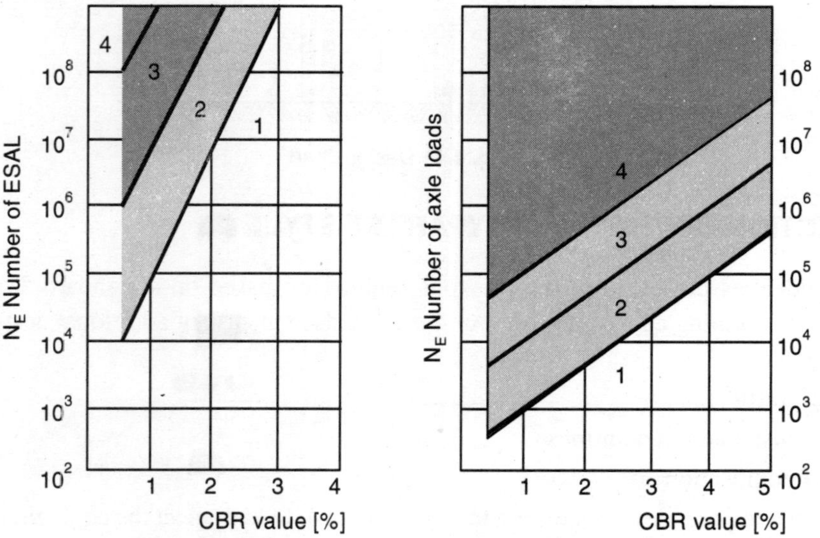

(a) Axle loads < 130 kN **(b)** Axle loads > 130 kN

Fig. 4.42: Recommended energy levels as a function of traffic

($N_E = \Sigma N_i \times ESAL_i$)

4.18.2 Effect of installation conditions

To fulfill its long-term functions, the geotextile should withstand installation stresses, particularly aggregate dumping and compaction. Fig. 4.43 indicates the recommended Typar® SF energy level as a function of aggregate size and drop height. It is evident that aggregate backdump and push-ahead over an existing layer instead of dumping directly on the geotextile allows the use of styles with a lower energy level.

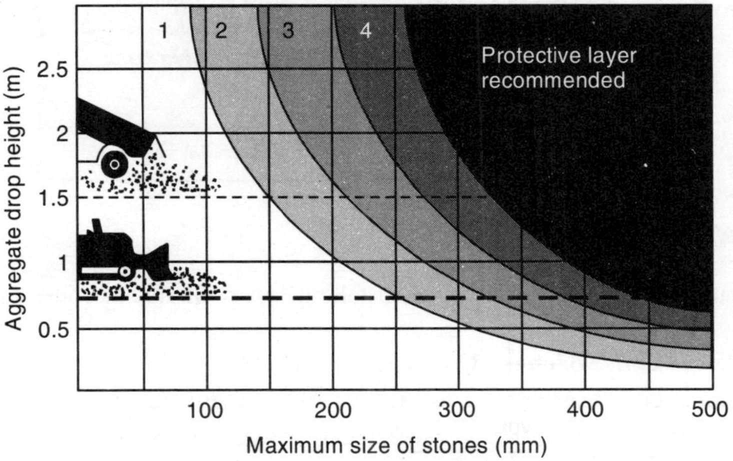

Fig. 4.43: Recommended energy level as a function of aggregate size and drop height

4.18.3 Effect of compaction

Puncture by sharp stones during aggregate compaction is detrimental to the long-term separation function. Fig. 4.44 indicates the recommended Typar® SF energy levels as a function of soil CBR and D_{90} (90% passing size) of the crushed aggregate in contact with Typar® SF.

> **Remark:** Styles with a lower energy than 2 kJ/m^2 may be used if there is light traffic (cars) only and maximum aggregate size does not exceed 50 mm.

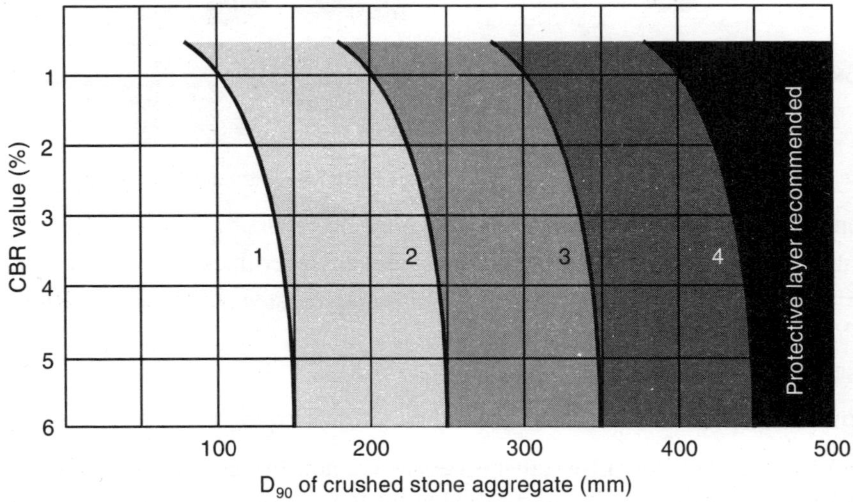

Fig. 4.44: Recommended energy level as a function of aggregate size and drop height

4.18.4 Filter requirements

Very fine, cohesive soils $D_{85} < 0.06$, $D_{10} < 0.002$ non cohesive soils $O_{95} \leq 0.200$ mm $O_{95} \leq 2 \times D_{85}$

To perform efficient long-term separation and filtration functions, the geotextile should meet simplified criteria of Table 4.12 in which the O_{90} is measured by wet sieving test method (EN 12956).

Table 4.12: General filter requirement

Very fine, cohesive soils *$D_{85} < 0.06$, $D_{10} < 0.002$*	*Non cohesive soils*
$O_{95} \leq 0.200$ mm	$O_{95} \leq 2 \times D_{85}$

4.19 INSTALLATION GUIDELINES

The following measures should be followed when installing Typar® SF in road constructions and aggregate bases:

1. Remove all large debris which might puncture Typar® SF.
2. Typar® SF should be at least as wide as the base of the aggregate layers.
3. When using two or more rolls, ensure sufficient overlap (usually min. 30 cm).
4. If it is windy, use shovelfuls of coarse aggregate at regular intervals to hold Typar® SF in place.
5. Back dump aggregate without driving directly on the geotextile (Fig. 4.45).

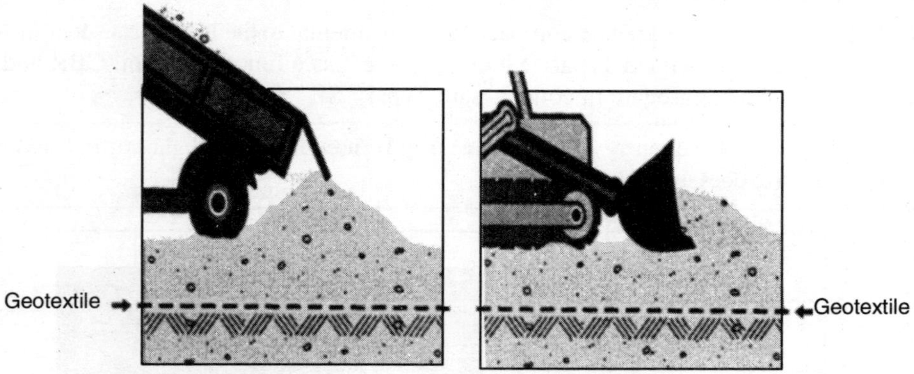

Fig. 4.45: Backdumping aggregate on Typar® SF without driving on it

6. Level and compact aggregate before any heavy traffic occurs.
7. Avoid aggregate size in excess of 1/3 of the aggregate layer thickness.
8. Fill up ruts, if any, as soon as they exceed 1/3 of the aggregate layer thickness. Rutting will then be stopped.
9. First aggregate layers must be at least 250 mm thick.

4.19.1 Design Examples

Example 4.4: A contractor desires all-weather access to a remote bridge construction site over an organic clay with a CBR of 2.5%. About 6 trucks (3 axles) will enter the site daily over a period of 5 months. A source of inexpensive gravel is close ($\alpha = 0.4$, $D_{max} = 100$ mm).

Initial aggregate thickness T_0

Soil bearing capacity

$$CBR = 2.5$$

Axle load $P_i = 80$ kN

From Fig. 4.38, $T_0 = 190$ mm

Adjustment of T_0 for service life

Axle load $P_i = 80$ kN

Actual number of passes $N_i = 6$ trucks/day

Compacted crushed stone thickness

$$T_0 = 190 \text{ mm}$$

$$ESAL = \left(\frac{P_i}{P_0}\right)^{3.95} = 1$$

$$N_E = \Sigma N_i \times ESAL_i$$

$$N_E = 5 \text{ months} \times 30 \text{ days/month} \times 6 \text{ trucks/day} \times 3 \text{ axles} \times 1 = 2700$$

Fig. 4.39, $C = 1.12$

$$T = C \times T_0 = 1.12 \times 190 = 212 \text{ mm}$$

Adjustment of T for aggregate efficiency

$$T_{eff} = \frac{\Sigma T_i}{\alpha_i} = \frac{212}{0.4} = 530 \text{ mm}$$

Selection of the suitable Typar® SF energy level

CBR = 2.5%; $N_E = 2700$, (From Fig. 4.42, level 1)

Drop height = 1 m; $D_{max} = 100$ mm, (From Fig. 4.43, level 1)

(Fig. 4.44, only applicable to crushed aggregate)

From Table 4.12, for cohesive soils,

$$O_{max} \leq 0.200 \text{ mm}$$

Use SF 37 Type of geotextile of DuPont make or equivalent brand of any other make.

Installation: Install two layers of gravel each 330 mm and compact to 265 mm.

Example 4.5: A transport company will build a terminal and parking area with an expected life time of fifteen years. Twenty (20) trucks per day will use the facility and 8 of these will be empty one way. The trucks have 4 loaded axles.

The site is in low area and on uneven ground. A CBR of 1% was obtained during a site investigation. The access road and parking area will be paved with a 70 mm (= $T_{hot\ mix}$) surfacing course of hot mix. A sandy gravel will be used for the base of the construction traffic road ($\alpha = 0.5$) and then topped off by a good quality round stone aggregate ($\alpha = 0.8$, $D_{max} = 100$ mm) for the final structure. Initially the contractor will create a stable working and assembly area to and around the terminal. This aggregate structure will be incorporated in the final paved structure which will save time and money.

Referring Fig. 4.41 design chart:

Paved Structure and Unpaved Construction Road

Paved Structure	Unpaved Construction Road
Initial aggregate thickness T'_0	*Initial aggregate thickness T_0*
Soil bearing capacity CBR = 1%	Soil bearing capacity CBR = 1%
Axle load P_i = 80 kN	Axle load P_i = 80 kN
From Fig. 4.40, T'_0 = 420 mm	From Fig. 4.39, T_0 = 280 mm
Adjustment of T'_0 for service life	*Adjustment of T_0 for service life*
Axle load P_{full} = 80 kN P_{empty} = 30 kN	Axle load P_i
Actual number of passes N'_I	Actual number of passes N_I
Compacted crushed stone thickness T'_0	Compacted crushed Stone thickness T_0
$ESAL_{full} = (P_i/P_0)^{3.95} = 1$ $ESAL_{full} = (30/80)^{3.95} = 0.021$	$ESAL_{construction\ estimate} = 3000$
$N'_{full} = 32 \times 6 \times 52 \times 15 \times 4$ axles $= 599040$ $N'_{empty} = 8 \times 6 \times 52 \times 15 \times 4$ axles $= 149760$	$N_{E\ construction\ estimate} = 3000$
$N'_E = 599040 \times 1 + 149760 \times 0.021 = 602185$	
From Fig. 4.39, C = 1.75 $T' = C \times T'_0 = 1.75 \times 420 \cong 740$ mm	From Fig. 4.39, C = 1.13 $T = C \times T_0 = 1.13 \times 280 \cong 320$ mm
($\alpha = 1$)	
Adjustment of T′ for aggregate efficiency $T'_{eff} = T_{eff} + \Sigma T_i/\alpha_i$	*Adjustment of T for aggregate efficiency* $T_{eff} = \Sigma T_i/\alpha_i$

(With T_{eff} effective minimum thickness for construction traffic)

Of the total thickness T′ of 740 mm, 320 mm ($\alpha = 1$) were used to support the construction traffic. 70 mm of wearing course are equivalent to 140 mm of a material with an efficiency of $\alpha = 1$. The remaining 280 mm (T_{rem}) can be provided by 350 mm (= 280/0.8) of round stone aggregate. This results in an effective minimum thickness of 790 mm.

$$T'_{eff} = T_{eff} + T_{hot\ mix}/T'_{rem}/\alpha'_{rem}$$

$$T_{eff} = \frac{320}{0.5} = 640 \text{ mm}$$

$$T'_{rem} = T' - T - T_{hot\ mix} (\alpha = 1) = 740 - 320 - 140 = 280 \text{ mm}$$

$$T'_{eff} = 640 + \frac{140}{2} + \frac{280}{0.8} = 1060 \text{ mm}$$

> **Note:** T'_{eff} is the thickness of the unpaved construction road.
> $T_{hot\ mix}$ is the thickness of the wearing course.
> T_{rem} is the additional thickness necessary to fulfill thickness requirements for permanent paved road.

Selection of the suitable Typar® SF energy level

CBR = 1.0%; N'_E = 602185 (Fig. 4.42 = > level 2)

Drop height = 1 m; D_{max} = 100 mm (Fig. 4.43 => level 1)

(Use Fig. 4.44 only applicable to crushed aggregate)

For DuPont brand of geotextile, use SF49. The equivalent material of other brands can also be used.

Installation

- Install 640 mm of round stone aggregate for construction traffic in two layers.
- Install 350 mm of round stone aggregate and 70 mm hot mix wearing course.

The typical cross-section is shown below. (Fig. 4.46)

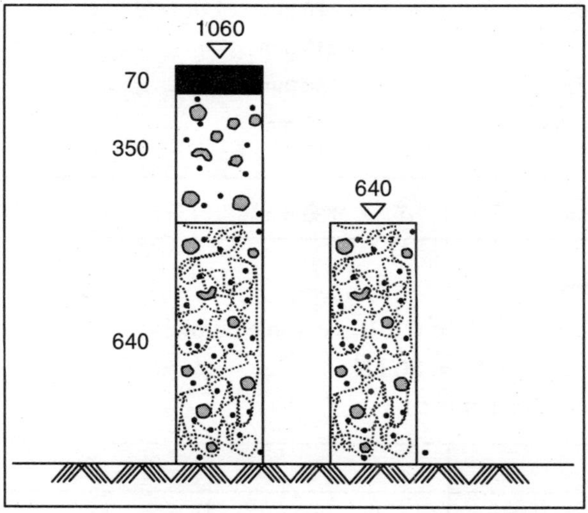

Fig. 4.46: Cross-section for proposed road

Example 4.6: A contractor wants to suggest a refined road design to the road authority in order to show possible savings by using a geotextile. The original design presented by the road authority to tender for it as follows:

Road structure:

1. Surface wearing corase	$\alpha = 2$	50 mm
2. Binder corase	$\alpha = 2$	60 mm
3. Bituminuous road-base	$\alpha = 2$	140 mm
4. Gravel crushed/stone CBR > 30%	$\alpha = 2$	150 mm
5. Sand/gravel mix CBR > 15%	$\alpha = 2$	300 mm
Total design thickness T_{design}	α_i	700 mm

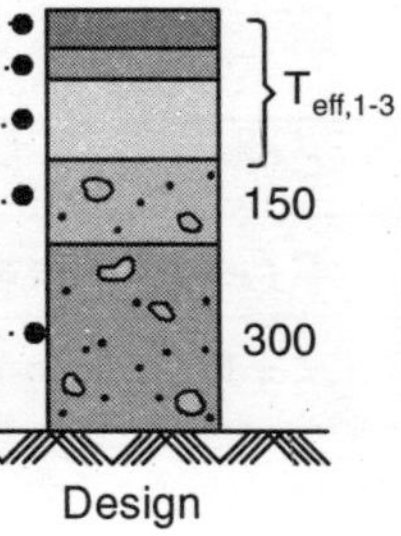

This design is based on the following estimates for traffic:

Traffic: Axle load is 8 tons or 80 kN

10 years design life

Total of 15×10^6 axle loads / lifetime of the road.

Bearing capacity: Existing roadbed CBR 1–5%.

As the CBR of the existing roadbed varies a new road structure is determined for the CBR = 1%, 3% and 5%. Furthermore layers 1, 2, 3 will remain unchanged with the current design thickness of $T'_{effl\ 1\text{-}3}$ = 250 mm and an equivalent thickness $T_{1\text{-}3}$, $\alpha = 1$ $(T_1 + T_2) \times \alpha_{1,2} + T_3 \times \alpha_3 = 430$ mm using the aggregate efficiency factors $\alpha_{1,2} = 2$ and $\alpha_3 = 1.5$. The equivalent thickness for layer 4 is $150/(\alpha = 1) = 150$ mm, for layer 5 the thickness is $300/(\alpha = 0.5) = 600$ mm. All of the following comparisons are based on an aggregate efficiency of $\alpha = 1$. The equivalent road structure is outlined below:

1. Surface wearing corase .... 100 mm
2. Binder corase .... 120 mm
3. Bituminuous road-base .... 210 mm
4. Gravel crushed/stone CBR > 30% .... 150 mm
5. Sand/gravel mix CBR > 15% .... 150 mm

Total equivalent thickness $\alpha = 1$ 730 mm

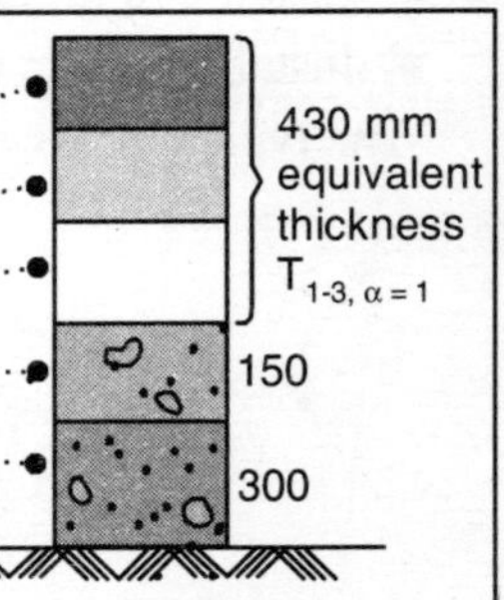

Initial aggregate thickness T_0

Soil bearing capacity = CBR, see Table below:

Axle load P = 80 kN

From Fig. 4.38, T'_0 as per Table below:

CBR	1%	3%	5%
T_0' (thickness) (Fig. 4.40) [mm]	420	300	250

Adjustment of T_0 for service life

Axle load P = 80kN

Number of passes (ESAL) N'_E = 15×10^6 axle loads

Compacted crushed stone thickness

T'_0 = see table above

CBR	1%	3%	5%
C (service life adjustment)	2.1	2.1	2.1
$T = T'_0 * C$ (min with $\alpha = 1$) [mm]	880	630	525

Adjustment of T for aggregate efficiency

CBR	1%	3%	5%
T_{remain} *(=* $T - T_{1-3,\,\alpha=1}$*) [mm]*	450	200	95

The remaining thickness T_{remain} can be divided between the two available materials in the following way:

T_4 *(standard aggregate) [mm]*	150	100	–
T_5 *(sand/gravel mix) [mm]*	300	100	95
Reduction (= T – 730 mm) [mm]	+150	–100	–205

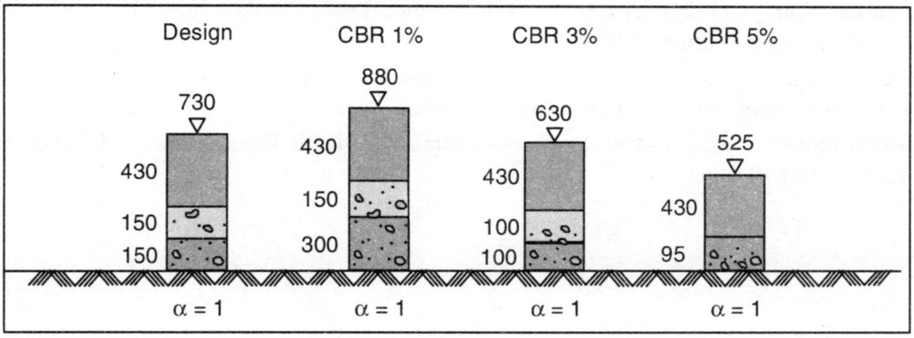

This leads to savings in the effective thickness for CBR = 3% and 5% and to an augmentation of thickness for CBR = 1%.

CBR	1%	3%	5%
$T_{4,\,eff}$ *(standard aggregate) [mm]*	150	100	-
$T_{5,\,eff}$ *(sand/gravel mix) [mm]*	600	200	190
Eff. reduction (=$T_{design} - T_{eff}$*) [mm]*	+300	–150	–260

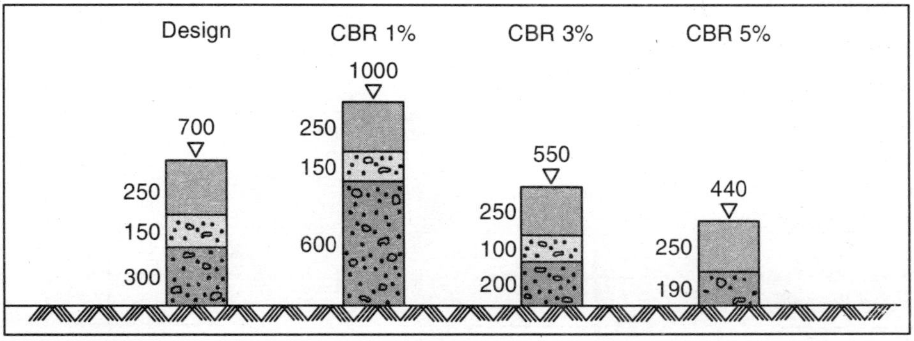

"Yes, making good roads is expensive, but not making good roads is still more expensive".

References

1. A Manual on use of Jute Geotextiles in Civil Engg., JMDC.
2. DuPont Typar Geotextile–Technical Handbook, Egypt.
3. Mittal, Satyendra, "Road Sections for Design of Hyde Park Estate at Mullanpur in Chandigarh", Report submitted to M/s DLF India Ltd., Gurgaon, 2011.
4. Mittal, Satyendra, Choudhary, P.K., Sanyal, T. and "Application of Jute Geotextile in Improving Permeability of Subsoil", National Jute Board, Ministry of Textiles, Govt. of India.
5. Mittal, Satyendra & Kannan, V. (2012). "Use of Chemical Stabilisers in Flexible Pavement Design", Unpublished paper.
6. "RBI Grade-81 Pavement Material Soil Stabiliser"– Alchemist Touchnology Ltd. brochure.
7. Giroud and Noirays, (1994). "An Overview of Pavement Design Methods with Geosythetics' *Geosynthetic World,* Wiley Eastern Limited, Mumbai.
8. Mandal, J.N. & Mhaiskar, S.Y. (1994), "An Overview of Pavement Design Methods with Geosynthetics", *Geosynthetic World,* Wiley Eastern Limited, Mumbai.
9. "Specifications for Roads and Bridge Works", Ministry of Road Transportation (MORTH), Indian Roads Congress, Govt. of India.

Grouting Technology

CHAPTER 5

5.1 INTRODUCTION

Grouting is the method of stabilising granular soil. The invention of pressure grouting is credited to a French Engineer Mr. Charlies Berigny. In 1802, while repairing a masonry wall in Port City of Dieppe by pumping a suspension of clay and lime in the wall, he named this process as "INJECTION PROCESS". The first cement grout had been used by one Mr. Marc Brunel in the year 1838. Between 1880 to 1905 in Belgium & France cement grouting was increasingly used to control water in—flow from fissures, water bearing rock strata, encountered during shaft sinking for coal mines. By middle of 20th century grouting which was known as an art had become a systematic science. Over the last two decades grouting had grown into one of the disciplines of Geotechnical Engineering and became one of the major tools for foundation treatment of dams and other hydraulic structures and to rehabilitate the dams structurally.

This process of grouting is used where rock strata is having cracks, crevices and joints. To improve the mechanical properties of poor rock and to make it impermeable, conventional drilling and grouting is restored, in this process, drilling is done in the foundation rock in stages of 3–5 metres and cement grout with admixture is pumped in the drill hole in descending or ascending order. As the cement grout is pumped under pressure, it spreads and permeates in the cracks and joints of rock formation and seals them to improve the mechanical properties and reduces the permeability. This conventional drilling and grouting is done invariably in the rock formations only as drilling is possible in rocks only. However, for granular soils, other types of grouting are possible as discussed below.

Grouting process is the injection of special liquid or slurry materials called grout into the ground for the purpose of improving the soil or rock. There are two primary kinds of grout e.g., *cementitious grouts* and *chemical grouts*. Cementitious grouts are made of Portland cement that hydrates after injection, forming a solid mass. Chemical grouts include a wide range of chemicals that solidify once they are injected into ground. These include silicates, resins etc. Chemical grouts can be used where cement grouts are ineffective. Chemical grouts are expensive, toxic and corrosive.

5.2 METHODS OF GROUTING

There are four principal grouting methods as explained below:

***(i)* Intrusion grouting (or slurry grouting):** It consists of filling joints or fractures in rocks or soil by injecting grout through pipes. The pipes may be inserted from ground surface or from tunnels. (Fig. 5.1)

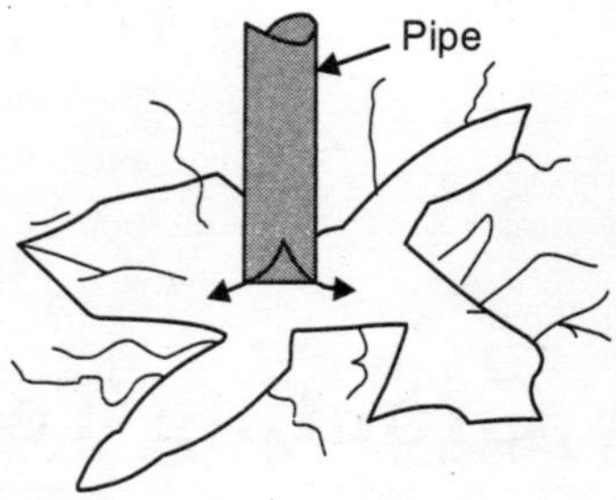

Fig. 5.1: Intrusion grouting

The primary benefit from this work is a decrease in hydraulic conductivity. The method is often used to prepare the foundations and abutments for dams. It usually is done using cementitious grouts.

***(ii)* Permeation grouting (also known as chemical grouting):** It is the injection of thin grouts into soil such that they permeate into voids. Once the grout cures, the porous soil is transformed into a near solid mass. Although this can sometimes be done using cement grouts, the void space in most soils is much too small to permit passage of the Portland cement particles.

Hence most permeation grouting is performed using chemical grouts. Thus it is sometimes called chemical grouting also. The treated soil had a much lower hydraulic conductivity, and is stronger and less compressible than before. It is often used to form groundwater barriers and to stabilise soils in advance of making excavations or tunnels (Fig. 5.2).

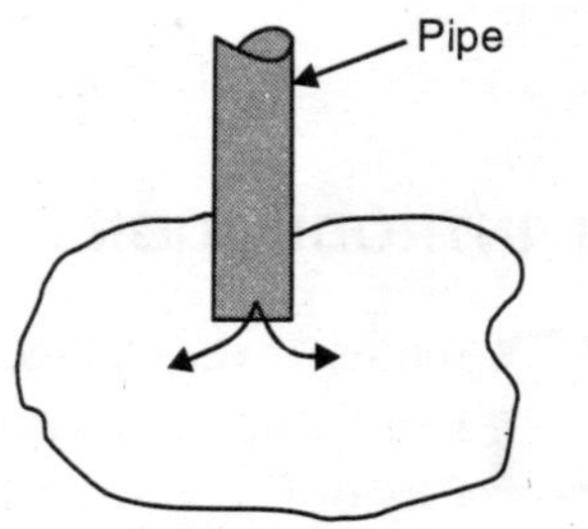

Fig. 5.2: Permeation grouting

***(iii)* Compaction grouting (also known as displacement grouting)**: It uses a stiff grout (about 25 mm slump) that is injected into the ground under high pressure through a pipe to form a series of inclusions. This grout is too thick to penetrate significantly into the soil, but the grout inclusions compact the adjacent soil. Compaction grouting is often used to repair structures that had experienced excessive settlement, since it both improves the underlying soils and raises the structure back into position.

***(iv)* Jet grouting:** It is the newest method. It was developed in Japan. The method uses a special pipe equipped with horizontal jets that inject grout into the soil at high pressure. The pipes are first inserted to the desired depth, then they are raised and rotated while the injection is in progress, thus forming a column of treated soil. Because of the high pressures, this method is usable on a wide range of soil types. This method had been used for ground water control, underpinning, stabilisation in advance of tunnelling and many other applications.

5.3 CASE STUDY OF GROUTING TECHNOLOGY

Curtain grouting is an important component of foundation treatment of a dam and other hydraulic structures. The purpose of curtain grouting is to form a zone of low permeability (*k*) up to a designed depth on the upstream of dam. This grout curtain along with downstream drainage system controls the uplift pressure and piping which are potential hazards. The case study discussed here deals with providing grout curtain to Almatti Masonry dam on Krishna River at Almatti village in Bijapur distt. of Karnataka, India, by M/s Ferro Concrete Construction Pvt. Ltd., Indore (India). This case study also deals with treatment of weak zone (unconformity zone) of thickness up to 6 m which was existing in the foundation below the joint of base granite rock and overlying quartzite foundation. The single line grout curtain of permeability less than 3 Lugeon had been effectively formed below the foundation, despite very poor strata having maximum pre-grout permeability of 90 Lugeon (1 Lugeon = water loss in litres/minute/m of a hole under a pressure of 10 kg/cm^2. If water pressure is say 2 kg/cm^2, then multiply the Lugeon value by 5 to get its standard value).

5.3.1 Curtain grouting and treatment of unconformity zone

Curtain grouting is an important component of foundation treatment for all the high dams and water retaining structures. The purpose of curtain grouting is to form a zone of low permeability up to a designed depth below a specified portion of the upstream of the dam. The main purpose of this work was to treat the weak zone (unconformity zone) which was existing in the foundation between joint of base granite rock and overlying quartztic formations.

5.3.2 Design of grout curtain and drainage holes

Design of grout curtain is based on factors like geology of dam foundation, hydraulic head, the potential of leakage and piping and of course it is based on the observation and performance of grout curtain of other dams. Parameters like spacing of holes, depth of grout curtain, limit of reduction of permeability ('*k*') are selected beforehand. Some of the parameters adopted were:

- Permeability limit of grout curtain = < 3 lugeon
- Depth of grout curtain = 0.5 H (H = hydraulic head from foundation)
- Spacing of single line grout curtain = 3 m c/c (for primary holes). The secondary and tertiary holes were also planned to achieve desired permeability.

In few blocks heavy artisan conditions were observed, hence even tertiary holes were required to be done to bring down the permeability below 3 lugeon.

5.3.3 Geology of dam foundation

In all, 43 holes were drilled (NX-size) at 15 m c/c. The bores had fresh and hard migmatite granite gneiss associated with later intrusions of pneumatic bands, micro granite and dolerite dykes. In bores, core recovery was very good ranging from 8% to 100% rock was very hard and abrasive in nature. The pre-grout permeability was normal and ranging from 3 Lu (lugeon) to 24 Lu in foundation area. The general sequence of drilling and grouting operations are: mobilisation of drilling equipments, drilling of primary grout holes, cleaning of grout holes, packer setting, water loss measurement, mobilisation of grouting packer setting again, water loss measurement, demobilisation.

5.3.4 Drilling and curtain grouting

The holes are washed thoroughly after drilling. The washing is done by alternate jets of air and water. In all holes, pre-grout '*k*' was computed as 12 Lu, and in some locations as 29 Lu. It had been observed that grout intake on an average, came to 35 kg/m of drilling. However, maximum grout intake was 1065 kg. in first stage of one hole in at one location.

After completion of primary curtain holes, secondary holes were drilled at 1.5 m c/c. In secondary holes, pre-grout '*k*' was on an average 7 Lu and also at some locations as 21 Lu. In secondary holes, grout intake came as 20 kg/m of drilling with a maximum intake of 210 kg/m at one location.

The consistency of grout had been taken from 1 : 10 to 1 : 1 cement to water by weight. The result revealed that permeability values had come within permissible limit (less than 3 to 5 Lu), in all the locations. Post-grout permeability had reduced to less than 3–5 Lu.

The drilling is normally done in sizes as:

NX–17/8″ (54–56 mm) with bore size 1.5 times the sample size, BX–13/8″ (41–44 mm), AX–9/8″ (28 mm).

5.4 TREATMENT OF UNCONFORMITY (WEAK) ZONES

A weak zone lying between quartzite and the basement granite on the right bank had been established. A typical type of yellow soil weathered as well as fused formations had been observed.
At such locations, the primary holes are taken up 3.0 m c/c initially. At these locations, the '*k*' was of the order of 90 Lu and maximum grout consumed at these locations was 3200 kg in first stage.

5.4.1 Secondary hole curtain grouting

Secondary holes drilling and grouting were taken up at 1.5 m c/c i.e. between the already done primary holes. The pre-grout permeability (*k*) of 78 Lugeon was observed with average '*k*' as 22 Lu.

5.4.2 Tertiary hole curtain grouting

It was observed that maximum pre-grout '*k*' was 19 Lu, with average of 10 Lu. Here, average grout intake as 54 kg/mt.

5.4.3 Post grout test holes

Lugeon value had come down well within limit i.e., maximum 3.0 Lugeon and minimum 1.62 Lu, by above treatment of grouting.

Thus, it is seen in above case study that the grout curtain of permeability less than 3 Lugeon had been effectively formed below the foundations, despite of very poor strata (unconformity zone) having maximum pre-grout permeability of 90 Lu.

5.5 GROUT MATERIALS

There are many grout materials which can be used for grouting. Selection of material depends on site conditions and strength of solution grout. Karol (1982) has presented guidelines for selection of solution grout, which are explained in Table 5.1.

Table 5.1: Relative ranking of solution grouts according to their toxicity, viscosity and strength (After Karol, 1982)

Grouts	*Corrosivity or toxicity*	*Viscosity*	*Strength*
Silicates			
Joosten process	Low	High	High
Siroc	Medium	Medium	Medium High
Silicate-bicarbonate	Low	Medium	Low
Lignosulphates			
Terra Firma	High	Medium	Low
Blox-all	High	Medium	Low
Phenoplasts			
Terranier	Medium	Medium	Low
Geoscal	Medium	Medium	Low
Aminoplasts			
Herculox	Medium	Medium	High
Cyanaloc	Medium	Medium	High

Contd...

Grouts	*Corrosivity or toxicity*	*Viscosity*	*Strength*
Acrylamides			
AV-100 Rocagel BT Nitto–SS	High High High	Low Low Low	Low Low Low
Polyacrylamide			
Injectite 80	Low	High	Low
Acrylate			
AC-400	Low	Low	Low
Polyurethane			
CR-250	High	High	High

Particle size ranges over which each of these grout types are effective is presented in Fig. 5.3 (after Mitchell, 1976).

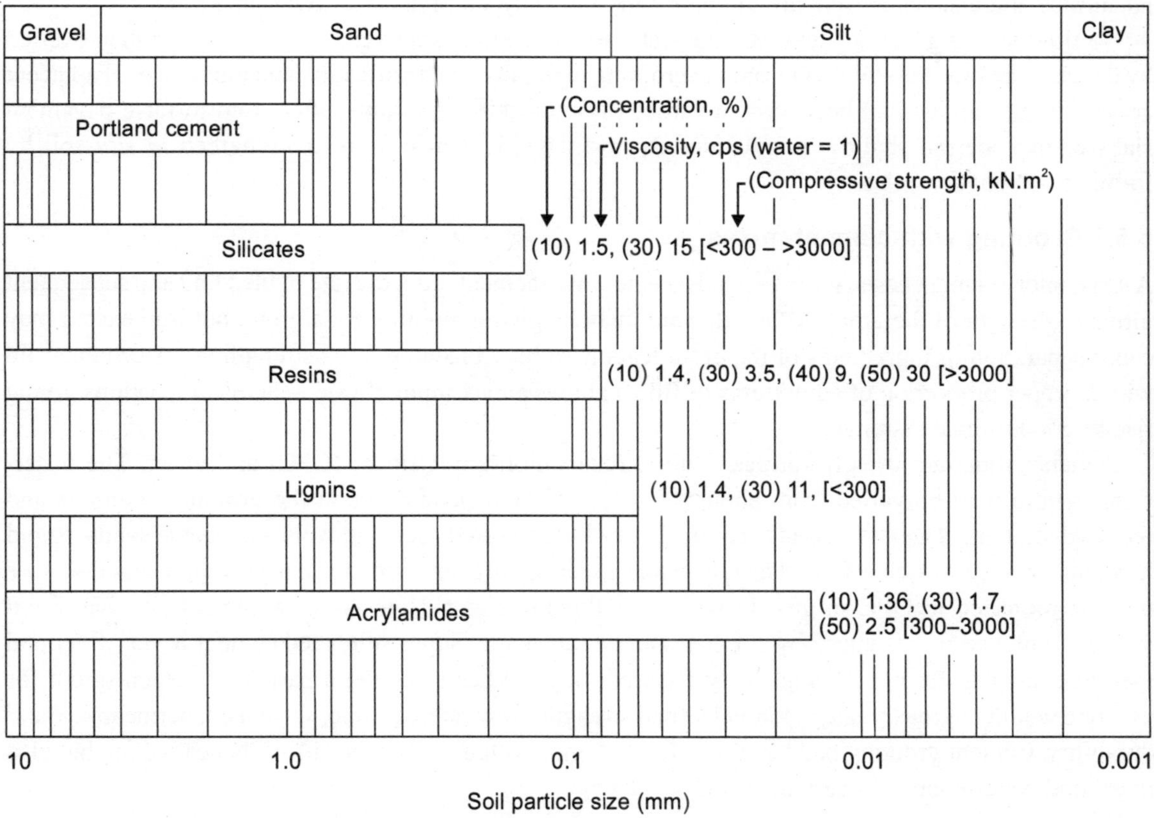

Fig. 5.3: Soil particle sizes suitable for different grout types (After Mitchell, 1976)

5.6 SUSPENSION GROUTS

When grouts (particularly suspensions) are injected into the soil formation the relationship between the grout particle size and the soil void size should be considered. A groutability ratio (GR) had been defined as a rough guide (Kravetz, 1958).

$$GR = \frac{D_{15}\text{ (formation)}}{D_{85}\text{ (grout)}} > 20$$

D_{15} = particle size at which 15% of the soil is finer (of the formation being grouted)

D_{85} = particle size at which 85% of the soil is finer (of the grout being injected)

This criterion decides the limit for suspension.

Soil itself can also be used to fill up some of the voids in coarse-grained soils. Even fine sands and silts may be used for this purpose which would settle out quite quickly after injection. The soil to be used as grout should be a very fine-grained soil. Clay is a complex compound with particles < 0.002 mm and is thus suitable for injection into medium coarse sands and other soils with permeability of the order of 1 to 10^{-1} mm/s. This type of grout will exhibit low shear strength and hence can be used to reduce permeability. Kaolinite based clays produce low viscosities and are preferred as filler grouts. Bentonite clay is a commonly used material whose structure is such that water is readily adsorbed on its surface and enables reasonably to control the viscosity, strength and flow properties.

No flow of soil-grout occurs when the water-to-soil ratio is kept very low. Pressure is then exerted by the grout against the soil mass from the grouting pipe causing densification and movement of adjacent areas. This technique had been known earlier as mudjacking and quite often used to raise pavement slabs or to underpin shallow foundations. Now-a-days, it is also used to strengthen *in situ* soil by forming compaction piles.

5.6.1 Grouting with cement mixes

A satisfactory cement based grout depends on the water-cement ratio, the rate of bleeding and subsequent ultimate strength of the grout. Although water may be giving mobility for a grout, but if in excess may cause separation of ingredients of the grout leading to high bleeding. Low strength water-cement ratio with a proper proportion of admixtures or fillers allows a maximum physico-chemical reactions giving rise to good ultimate strength.

Cement grouts are usually formed from Ordinary Portland Cement (OPC) and water. The water-cement ratio may be varied from about 0.5 : 1 to 5 : 1 depending upon the ground conditions and required strength. The OPC grouts are suitable only in fissured rocks, gravels and coarse sands. Rapid hardening cement which is finer than OPC had a quick setting time and high early strength and therefore may be preferred to OPC in ground with high flowing water. High alumina cement also had rapid strength gain and resists attack by sulphates and dilute acids. Super-sulphated cement is very finer and therefore suitable for penetrating finely fissured rocks. When selecting a particular cement grout for use, one would obviously like to know its final strength, flow rate, set time, shrinkage, permeability and durability. Cement grouting had been widely used, more often in seepage cut-off beneath dam, but also in ground water control in certain cases.

Other ingredients sometimes used in cement mixes are clay, fine sand, fly ash, fluidisers, accelerators, and retarders or expansion additives. Clay is considered as a filler in cement grouts when used in amounts less than 3% by weight. It can also be used to control bleeding by holding the cement in suspension while it is setting. Sand is a true filler in cement grouts and is used to reduce the overall cost of the grout material. Pozzolans such as fly ash or ground slag are also sometimes used as fillers. Additives or admixtures are meant to impart controlled features to the final composition of the grout.

Soil in combination with a stabilising material, e.g., cement would do better than soil alone. Grouts may have different properties depending on the amount and type of soil (say clay, silt or sand), cement, and water they contain. Further the viscosity of grout depends on solid-to-water ratios and different cement-to-soil ratios. In soil-cement systems, volumes of soil between four and six times the loose volume of cement are common. The volume of mixing water varies from about three-fourth to twice the volume of clay per bag of cement bag of cement-sand grouts. Water-cement ratios in the range of 0.5 : 1 to 5 : 1 had been in use. The lower this ratio, the less likely will be cement segregation and filtering, but the more difficult will be injection and the greater will be friction losses in the pumping system. Bentonite clay-cement mixes has an interesting synergistic properties and had been widely used as permeation grout. Low-water-content soil-cement mixtures, mixes over soil alone is permanence of the grout and their disadvantages is increased cost. The other factors, such as equipment, pressure and pumping rates, however, are roughly the same for that of soil alone.

Water-insoluble and chemically active materials, natural possolans comprises volcanic glass, opal, pumicite, clay minerals, zeolites and hydrated oxides of aluminium.

The most useful from the grouting point of view is pulverised fuel ash or fly ash which is often employed in conjunction with cement. PFA can retard alkali aggregate reactions and reduce the heat generation. The chemical interaction produced by PFA with cement is much more effective than that between sand and cement. Addition of fly ash markedly reduces the consistency at a constant water-cement-fly ash ratio. Fly ash with OPC provides lower viscosity than special cements. It had been reported (Somanathan, 1968) that about 20 to 30% of fly ash reduce gelation time, afflux time, bleeding and strength. Certain additives are used with the cement in order to vary its characteristics depending on the individual job. Some are water-soluble, a few are colloidal and some produce gas. For example, calcium chloride is used as an accelerator, carbohydrates can be employed to reduce the water-cement ratio.

5.7 APPLICATIONS

The major application of grouting is seepage control and strengthening. However, grouting had been in use for other allied engineering works such as underpinning, vibration control of machine foundations, slab jacking, anchor systems, etc. Some of the major application of grouting are briefly explained below.

5.7.1 Seepage control

Seepage control in rock under dams is the oldest area of grouting application which has had considerable exposure. Based on the degree of fracturing in the rock near its surface a shallow blanket grout is placed, through which a deep curtain grout injected (Fig. 5.4).

In order to prevent hazardous materials and toxic waste landfill leaches from environmental pollution, grouting technique had been adopted. The isolation of pit with such materials from its surrounding environment is done by drilling a row of closely spaced grouts and continuous grouting is performed

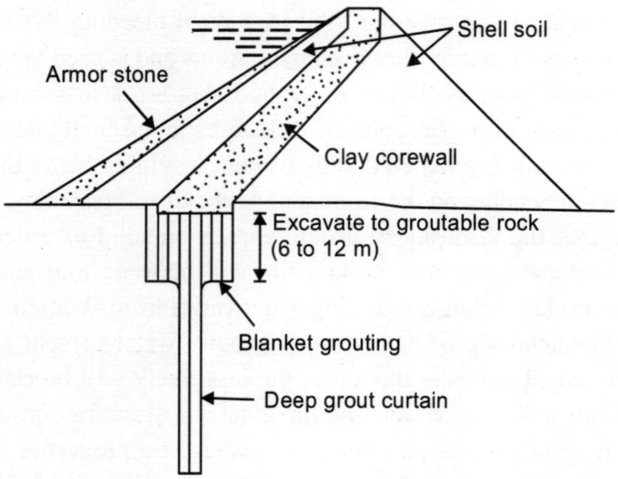

Fig. 5.4: Grout pattern typically used in fractured rock where upper surface is highly fractured

from their bottom upward. Generally chemical grouts of low viscosities are adopted such that the grout curtain is tight. However, jet grouting is well suited to address the problem of toxic waste containment.

References

1. IS: 6066–1994 (reaffirmed in 2004). "Recommendations for Pressure Grouting of Rock Foundation in River Valley Projects", *Bureau of Indian Standards*, New Delhi.
2. Karol, R.H. (1982). Chemical Grouts and their Properties, *Proc. Conf. on Grouting in Geotech. Engg.*, New Orleans, Wallace Hayward Baker, (ed.) ASCE, New York, pp. 359–378.
3. Kravetz, G.A. (1958). Cement and Clay Grouting of Foundations–The Use of Clay in Pressure Grouting, *JSMFD*, ASCE, Vol. 84, No. SM1, pp. 1546–1 to 1546–30.
4. Mittal, S., Shukla, J.P. (2012). "Soil Testing for Engineers", Khanna Publishers, New Delhi, India.
5. Mitchell, J.K. (1976). Stabilisation of Soils for Foundations of Structures, Report prepared for U.S. Army Waterways Experiment Station, pp. 1–34.
6. Personal discussions with Dr. Mahavir Bidasaria of Ferro Concrete Construction Ltd., Indore.
7. Raj, Purushothama, P. (2009). "Ground Improvement Technique", University Science Press, New Delhi.
8. Somanathan, C. (1968). Neyveli Fly ash as an Admixture in Cement Grout, *Ind. JSMFE*, Vol. 7, No. 3.

Soil Nailing

CHAPTER 6

6.1 INTRODUCTION

Ground improvement and soil reinforcement have become necessary globally in view of shortage of space available for construction. For soil reinforcement and ground improvement techniques, many investigators have done commendable work in the past (Mandal and Divshikar, 1988, Tatsuoka et al., 1995, Rao, 2001, Pinto, 2002, Ambily and Gandhi, 2004, Chu Jian, 2004, Puppala et al., 2004, Saran et al., 2006, etc.). Soil nailing is an emerging technology for *in situ* stabilisation and reinforcing the soil with steel bars or other materials. It can be used for the construction of underground basements, prevention of landslides and stabilisation of foundations. This is also a relatively simple technique for ground improvement which has made it popular over a period of time.

6.2 SOIL NAILING

Soil nailing consists of reinforcing the soil mass by the introduction of a series of thin elements called nails to resist tension, bending and shear forces. The reinforcing elements are made of steel round cross-section bars called as nails. Nails are installed sub-horizontally or horizontally into the soil mass in pre-bored holes, which are grouted along their total length to form 'Grouted nails', or simply driven into the ground, called as 'Driven nails' . Soil nailing technique is used to stabilise both in natural and steep cut slopes. In urban areas, this technique is also used to construct temporary or permanent retaining structures to support the ground close to neighbouring structures that are sensitive to deformation with appropriate measures to reduce ground movement. The idea of reinforcing new cut slopes is inspired by New Austrian Tunneling (NAT) method, which was primarily a rock tunneling system, using a combination of shotcrete and fully bonded steel inclusions to provide efficient excavation stability. In 1975 the idea of *in situ* reinforced soil was taken up in Germany, when the German research and development project 'Boden-Vernagelung' started to study and develop the new technique for stabilising slopes and excavation cuts. The technical term soil nailing derived from two German words 'Boden' meaning *soil* and 'Vernagelung' meaning *nailing*. In USA a simultaneous, but independent development of *in situ* reinforced earth took place in the year 1976 to 1981, which was called 'lateral earth support system'. In France the national project 'Clouterre', 'Clou' means *nail* and 'terre' means *soil* was started in 1986. Since then, some more research work had been done on soil nailing technique by various researchers from various parts of the world and still it is continued (Gassler, 1997).

6.3 FUNCTIONS AND APPLICATIONS OF SOIL NAILING

Soil nailing has been used for both temporary and permanent work. Various applications of the soil nailing technique for new constructions are shown vide Fig. 6.1.

Soil nailing has also been used for remedial works including:

1. Repair of unstable ancient masonry gravity retaining walls.
2. Stabilisation of failed soil slopes.
3. Repair of reinforced soil wall, construction abutments.

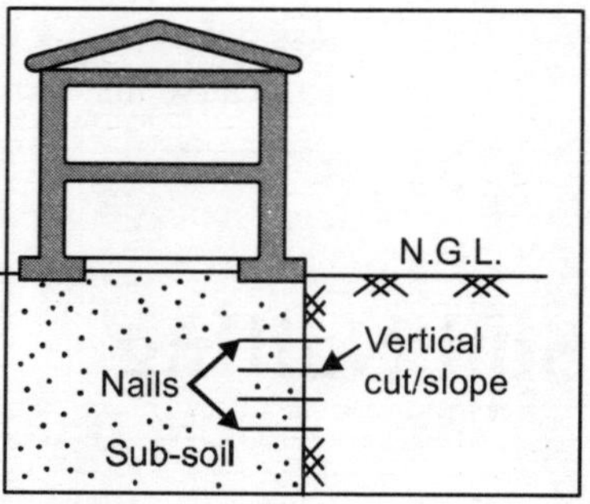

Fig. 6.1: (a) Housing construction on cut/slopes

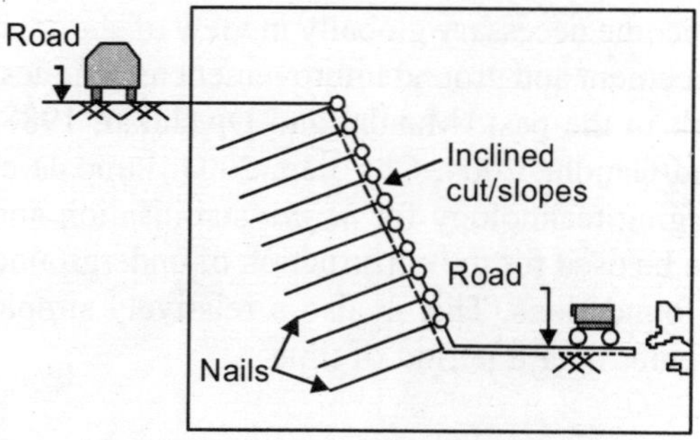

Fig. 6.1: (b) Laying of roads on cuts/slopes

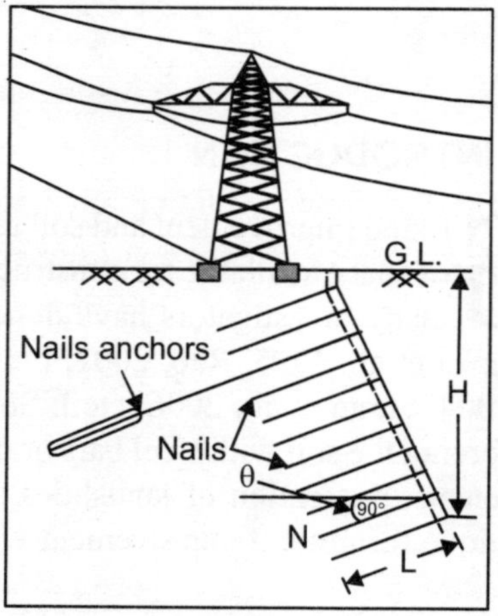

Fig. 6.1: (c) Transmission tower foundations

Fig. 6.1: Application of soil nailing (after Patra and Basudhar, 2001)

6.3.1 Advantages of soil nailing

The soil nailing technique offers forthcoming advantages (Juran and Elias, 1992) when compound to other methods of retaining excavations.

(*i*) Soil nailing results in saving in cost when compared to other methods/cost effective technique.

(*ii*) The equipments required for execution of soil nailing are relatively of small scale, easily movable, and produce little noise/simple and equipments.

(*iii*) Soil nailing operation is fast and systematic/time bound.

(*iv*) The maximum lateral displacement of the soil nailed cut at the end of excavation was generally not more than 0.3% of excavation depth/performance.

(*v*) Soil nailing provides an obstruction free working environment. This technique also requires lesser working space for the construction of soil nail wall.

(*vi*) Soil nailing performs well even in seismically active regions/suitability during earthquakes (Gosavi, 2006).

6.3.2 Limitations of soil nailing

(*i*) The requirement of soil nailing is to have the cuts which can stand unsupported for the depth of about 1 to 2 m at least for a few hours prior to shotcreting and nailing which though requires some cohesion or apparent cohesion in the natural soil.

(*ii*) The soil mass deforms due to nailing, thus leading to displacements in the surroundings of the wall. This can cause unacceptable deformation to a sensitive structure, if any in the vicinity of the wall.

(*iii*) In corrosive ground, durability considerations rule out the use of soil nails as permanent support further in soft cohesive soils, pull out capacity of inclusions cannot be economically mobilised (Juran and Elias, 1992)

6.4 METHOD OF CONSTRUCTION OF SOIL NAILED CUT

For a typical soil nailing method, construction steps are broadly divided into four steps and these steps are repeated in stages till the complete wall (Fig. 6.2) is constructed.

Step (*a*) Excavation of slope [Fig. 6.2 (*a*)]

Step (*b*) Positioning of facing material [Fig. 6.2 (*b*)]

Step (*c*) Drilling, nailing and grouting [Fig. 6.2 (*c*)]

Step (*d*) Reinforcement tightening [Fig. 6.2 (*d*)]

Sometimes steps (*b*) and (*c*) may be interchanged.

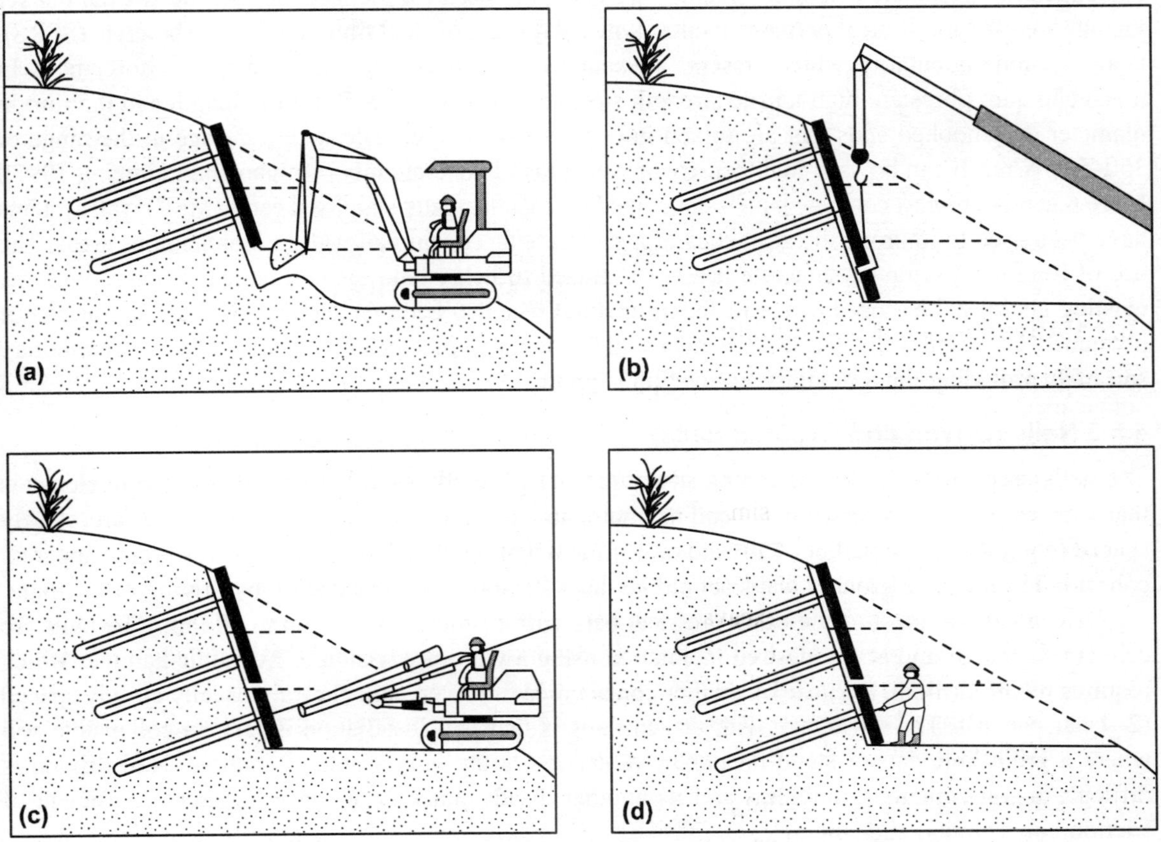

Fig. 6.2: Steps involved in soil nailing **(a)** Excavation **(b)** Positioning of nailing **(c)** Nail driving **(d)** Reinforcement tightening (after Pokharel, 2001)

6.5 COMPONENTS OF SOIL NAIL SYSTEM

6.5.1 Excavation

Excavation is carried out usually using conventional earth work equipments, from top to bottom. The maximum excavation depth in each stage depends on the soil type and the inclination of the slope, which must be kept stable during the stage of excavation until the reinforcement is placed and it starts to mobilise its effects. The excavation depth is generally kept as 1–1.5 m. Seepage of ground water, if any, towards the excavation face must be avoided by providing suitable drainage.

6.5.2 Facing

The function of facing in soil nailing technique is to ensure local stability of the soil between the nails and to limit decompression of the soil. The facing protects soil from surface erosion and weathering effects. It also gives an aesthetic appearance and prevents moisture loss from the face of the excavated soil. The most commonly used facing types are the shotcrete facing. For permanent structures, steel or concrete pre-fabricated facing panels are preferred. Shotcrete is very common and standard type of facing in soil-nailed structures.

The thickness of the shotcrete facing varies between 50–150 mm with one or two steel meshes embedded in the shotcrete. Thinner facings are generally used for inclined slopes whereas the thicker facings are used for vertical permanent cuts. Now a days, use of steel fibre reinforced shotcrete (SFRS), is also gaining popularity, which presents advantages in relation to the use of mesh reinforcement. In this technique, fibres are high tensile strength steel elements with 30–50 mm in length and 0.5 mm in diameter duly hooked ends that are mixed in concrete as an aggregate with a dosage in the range of 350–600 N/m^3. It can be used in either dry or wet sprayed concrete mix (Ortigao and Palmeira, 1997). More recently pre-cast concrete panels developed for mechanically stabilised earth (MSE) wall systems have been used to improve the aesthetics of a shotcrete faced soil nail wall in mountainous areas. The use of reinforced sculpted shotcrete has been utilised to help blend the soil nailed structure with the existing geology. The top-down wall facing method has also been developed for the construction of permanent basement walls. These permanent wall facings are typically 250–300 mm thick with temporary soil nails providing the temporary excavation support.

6.5.3 Nails (Driven and Grouted nails)

The nails used in soil nailing retaining structures, are generally steel bars or other metallic elements that can resist tensile stresses and bending moments. The nails are not pre-stressed but are closely spaced (e.g., one driven nail per 1 m^2 and one grouted nail per 4 m^2) to provide an anisotropic apparent cohesion to the native ground. Nails should be installed soon after excavation by driving or drilling.

Driven nails in most of the cases are steel bars with diameter 20 mm up to 50 mm. The nails are driven into the ground at any desired inclination using a vibro percussion or hydraulic hammer which requires no pre-drilling. Because of smaller contact area of driven nails they are usually closely spaced (2–4 bar per m^2). The advantage with driven nails is that nail installation is fast. In collapsible soil where a grout hole cannot stand unsupported, driving is the only viable alternative. Nail driving is difficult in soil containing boulders and its another disadvantage is of lower apparent coefficient of friction between nail and soil.

Soil nails are usually grouted to increase the bond and to enhance corrosion protection. A grouted nail consists of a central reinforcing bar (15–46 mm diameter) with a few centimeters of grout. To ensure a uniform grout cover, centralisers are placed [Fig. 6.3 (*a*)] at intervals of 1.5–2.0 m along the nail. Alternatively, the centraliser could be just 3 pieces of tor steel bar duly welded with nail [Fig. 6.3 (*b*) with suitable overlap length. These bars can be provided at 2–3 locations along length of nail, depending upon nail's length. The provision of these pieces facilitates movement of concrete from tip of nail to its full length (Mittal, 2010). The author has provided such centralisers in Lucknow Project (Table 6.2). The borehole diameters generally range from 75–150 mm. in soils possessing high cohesion. The rate of installation can be increased by drilling boreholes without casing. The common method used for drilling uncased boreholes are rotary boring, percussive rotary boring with the use of air flush, and dry auger boring. The life of the driven or the fired nails can be increased by using established corrosion protection techniques such as galvanising and cathodic protection. Alternatively a corrugated outer sheath of PVC enclosing an inner cement annulus can be used as a corrosion protecting [Fig. 6.3 (*c*)] permanent nails (Gassler, 1995, Warner and Barley, 1997). The latest and the most significant technological innovations have been the development and use of the jet-grouted nails and launched soil nails (Hanna et al., 1998). Fibre reinforced plastic (FRP) nails (Ortigao and Palmeira, 1997) are a composite material obtained by an industrial process called pultrusion that combines the fibre reinforced elements, usually glass fibres, with different types of resins. The FRP nails are with high tensile strength, usually three times steel strength, low unit weight, very high corrosion resistance but its cost is still high as compared to steel.

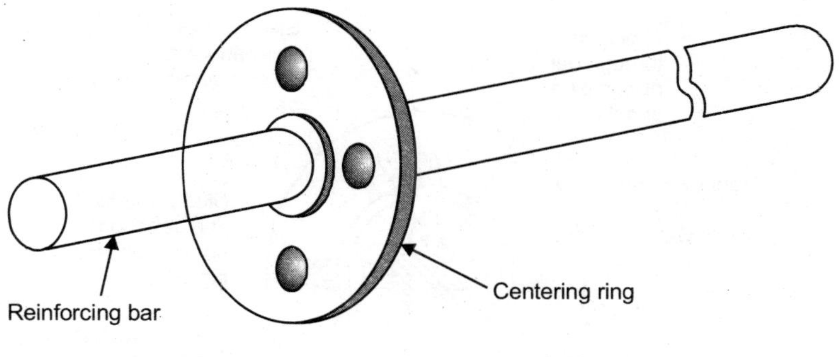

Fig. 6.3: (a) A typical centraliser used for grouted soil nails (after Juran and Elias, 1992)

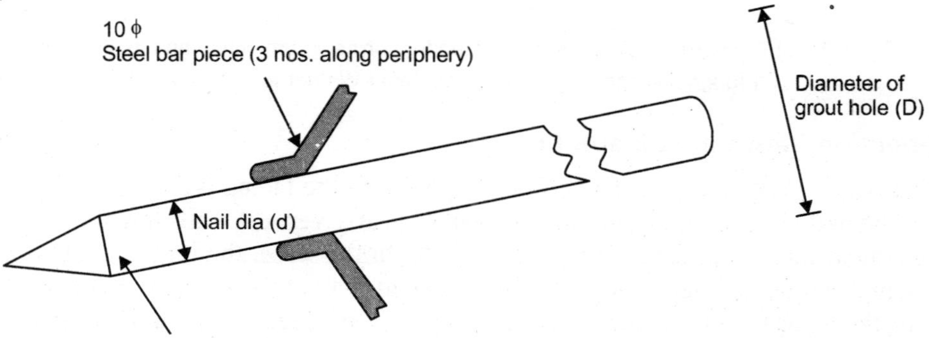

Fig. 6.3: (b) Another type of centraliser (after Mittai, 2012)

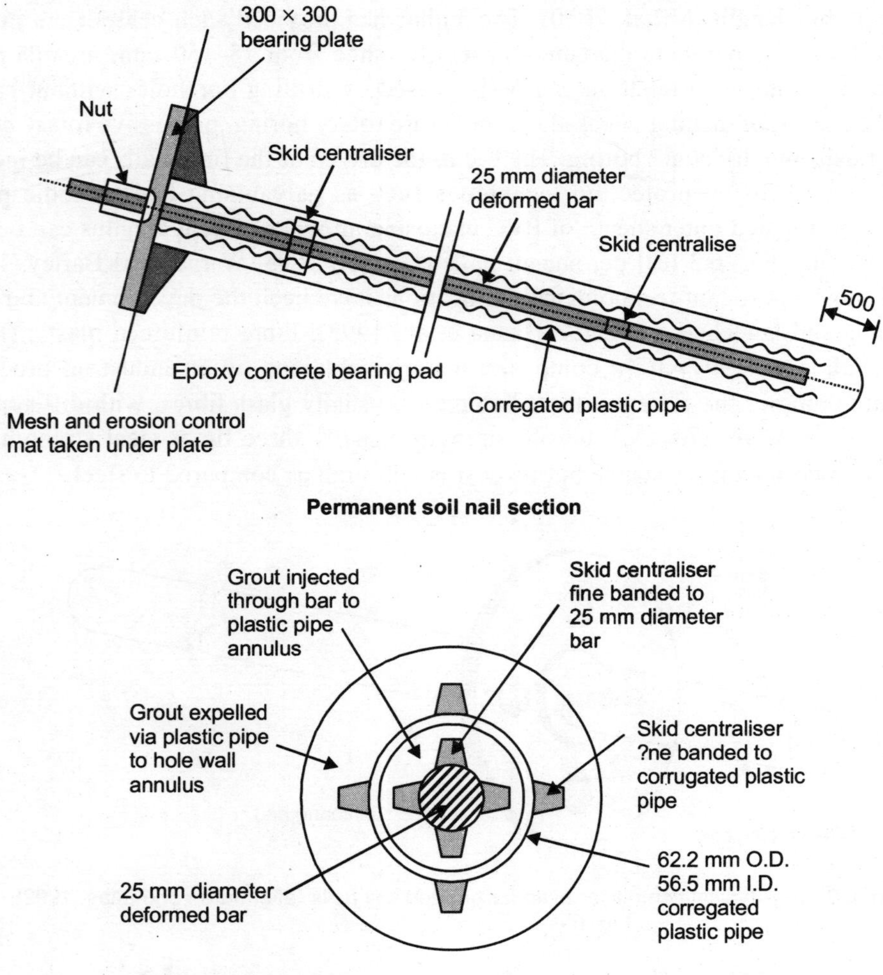

Fig. 6.3: (c) Permanent soil nail section used for corrosion protection and its section through nail showing components (after Warner and Barley, 1997)

6.5.4 Connection between nail and facing

In soil nailed structures the connection between the nail and the facing is important in transmitting the forces at the nail head to the facing. For vertical or nearly vertical walls nut bolt system is used as a reinforcement tightening technique [Fig. 6.4 (*a*)]. In this case, the torque applied to the bolt and nut system, applies a small loading of the order of 5 kN to each nail in order to maintain contact of the facing against the soil. In the stabilisation of inclined slopes in which small diameter nails (less than 20 mm in diameter) are used, the steel bar is laterally bent at right angles to anchor the mesh and shotcrete [Fig. 6.4 (*b*)]. The bearing plate, wiremesh and shotcrete used in soil nailing technology are shown in [Fig. 6.4 (*c*)].

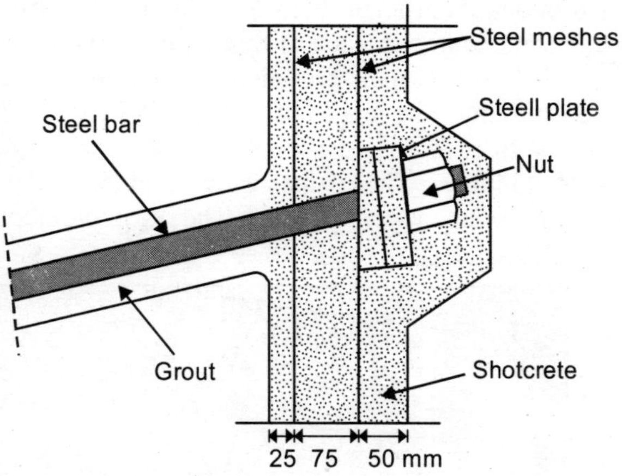

Fig. 6.4: (a) Corrosion protection for permanent nails

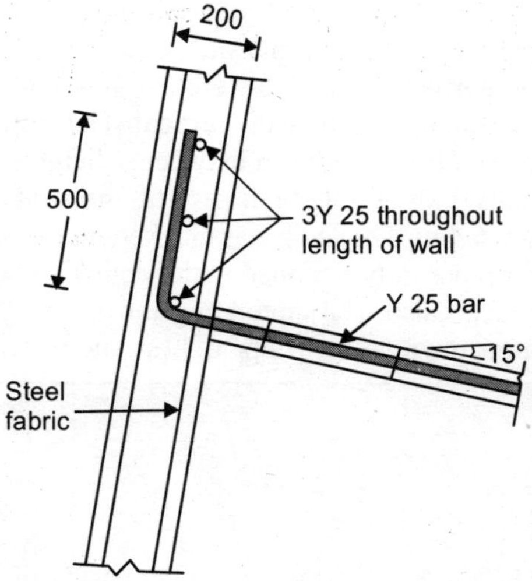

Fig. 6.4: (b) Bending the nails on to wale bars

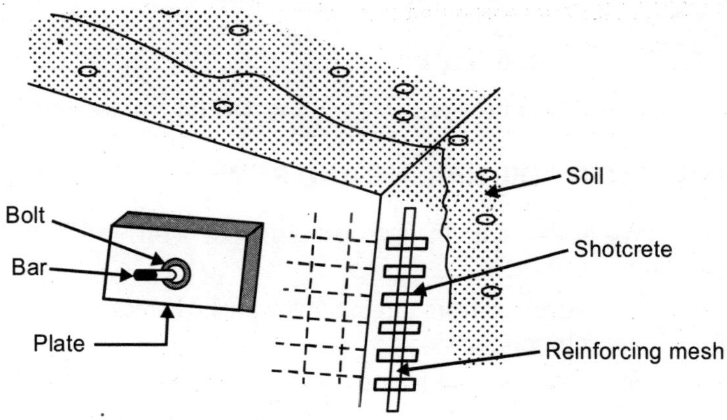

Fig. 6.4: (c) Using steel plate and nut

6.5.5 Drainage

An appropriate drainage system is important in nailed soil structures for preventing the development of excessive hydrostatic pressures on the facing, to protect the facing from deterioration due to contact with water and to avoid frost or ice action behind the facing in cold regions as presence of water weakens the bond between the nail and soil by reducing the friction and presence of pore water pressures can affect the structure performance and may cause instability during and after excavation stages. The drainage of soil nailed walls can be achieved (Bruce and Jewell, 1987) by using the following methods:

(*i*) Shallow drains, to prevent accumulation of water behind the facing.

(*ii*) Deep drains, to prevent saturation of the wall.

(*iii*) Drains, to discharge water into a collector drain at the foot and at the top of the wall or slope (Raju, 1996).

The author used soil nailing technique in landslide control in Nainital (India) in the year 2010 (Mittal, 2010). At this site, a 16 m high landslide prone area has been treated by soil nailing. The soil properties at this site were $c = 0$, f = 42°, apparent frictional coefficient between soil and nail = 0.6, slope angle = 10° with vertical. The slope was modified into 2 slopes, each of 8 m with a provision of berm of 3 m width in between 2 heights of 8 m each. The types of nails provided at site were grouted nails with bore hole dia as 100 mm and steel bar of 25 mm dia. The length of nail was 0.8 H¢ = 0.8 ′ 8 = 6.4 m. The facia was chicken mesh followed by shotcreting. To ensure a proper drainage, perforated GI pipes duly wrapped with geotextiles were placed into soil mass. These pipes were filled up with pee gravels also. The shotcreting and grouting was done with 1 : 2 cement-concrete mix. The site photographs are shown below [Figs. 6.5 (*a*) and 6.5 (*b*)].

Fig. 6.5: (a) Perforated pipes used for drainage

Fig. 6.5: (b) A view of shotcreting on soil nails

6.6 COMPARISON OF BOTH SOIL NAILING AND REINFORCEMENT OF EARTH WALL

Soil nailed slopes also behave in the same manner as reinforced earth wall. Some major differences between two techniques are:

1. Soil nailed slopes are constructed from top to downwards whereas reinforcement soil walls are constructed from bottom to top.
2. Some soil nails such as grouted steel bars have a much higher bending stiffness than strip or sheet soil reinforcement.
3. Soil nailing is applied to existing soil slopes and may therefore involve more cohesive soils than the select fills used for reinforced soil walls.
4. Soil reinforcement sheets or strips are usually laid horizontally whereas soil nails are normally driven at an inclined angle.

6.7 MECHANISM OF SOIL REINFORCEMENT

Schlosser (1982) found that active failure zone for nailed slopes was similar to but larger than that of reinforcement soil wall. In both cases, the active failure zone is similar than standard column active wedge assumed with other retaining structure. It has been suggested by Schlosser (1982) that this difference is due to inclination of soil mass. The reduction in lateral earth pressure as a result of inclined face should enable the length of nails to be slightly less than that for the reinforcement to a vertically faced reinforcement soil wall. It is, therefore, often possible to use a soil nail length of about 60–70% of height of embankment. The FOS in such structures is taken as 2.5. In case of nailed wall, the potential failure plane can be considered to be inclined at 90 – (α + ϕ/2) to the vertical where α is the inclination of face to the horizontal. At first sight, it might seem to be logical to take the friction developed on a soil nail as simply the product of vertical stress by active length by the soil nail perimeter by tan δ.

Refer Fig. 6.6 below, the nail friction F_N is given by:

$$F_N = L_a \tan \delta \gamma h'd [2 + (\pi - 2) K_o] \quad ...(6.1)$$

where,

L_a = active length of nail beyond failure plane

d = dia of soil nail

h' = depth of soil above active adherence length of nail

K_0 = coefficient of earth pressure at rest

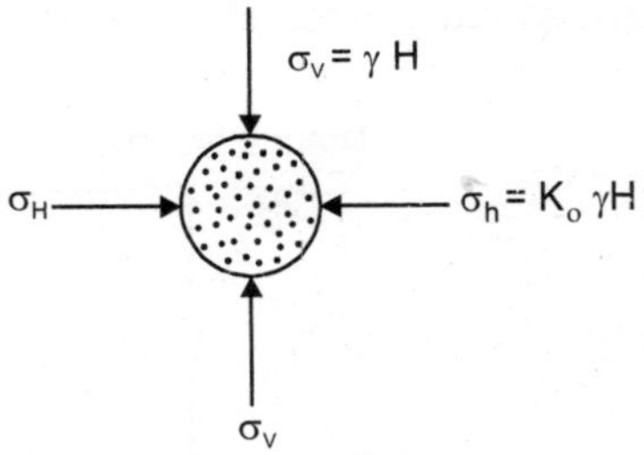

Fig. 6.6: Mechanism of earth reinforcement

The tension developed in each soil nail can be assumed to be lateral earth pressure acting over the area of the slope face that is supported by the soil nail. Schlosser observed that in the case of both nailed slopes and reinforce soil walls, the reinforcement restrain the lateral movements, particularly at the top of structure. In this region, the lateral earth pressure coefficient is, therefore, close to K_o. However, these structures tend to have a fairly uniform reinforcement distribution, making the lower most reinforcement the most critical level in the tension design calculations. For this location, the earth pressure coefficient can safely be assumed to be K_a. Therefore tension (T) in the nail can be taken as:

$$T = K_a \cdot \gamma h_m \cdot S \cdot V \quad ...(6.2)$$

where,

γ = unit weight of soil

h_m = depth of lowest soil nail

S = horizontal spacing between soil nails

V = vertical spacing between soil nails

For cohesive soils, lab tests should ideally be carried out to determine surface adhesion, using a specially modified shear box. In the case of geotextiles ties with a ribbed surface, the adhesion coefficient can often be assumed to be equal to lesser of 50 kN/m^2 or the soil cohesion value. During construction this assumption should be checked by pull out tests. If it indicates that the adhesion between nail and soil has been over estimated, then the length of the nail should be increased to compensate for this deficiency.

The spacing of nails obtained by Gassler and Gudehus (1983) design charts is generally fairly close to that obtained with the adapted reinforced soil wall procedure.

Third alternative is developed by Schlosser. This considers the equilibrium of a circular slip surface through the nailed soil slope, using method of slices rather than similar to the Fellenius and Bishop methods used with unreinforced slopes.

A very simple design of soil nailed slope is given as follows.

EXAMPLE 6.1: It is proposed to use soil nails to form an 8 m high soil nailed slope excavated in a cohesionless soil with $\phi = 36°$ and $\gamma = 18$ kN/m³. The face of the slope is to be inclined at 10° to the vertical and the soil nails to be inclined 10° below the horizontal. The ties (nails) have a short-term ultimate strength of 50 kN and a safety factor of 3 on this strength may ensure a design life of 120 years. Their diameter is 15 mm and their surface friction coefficient is 80% of that of soil. If the layers of soil nails are to be spaced at vertical interval starting at a depth of 0.5 m below upper soil surface, determine the required horizontal spacing and length of soil nails. Refer Fig. 6.7 below.

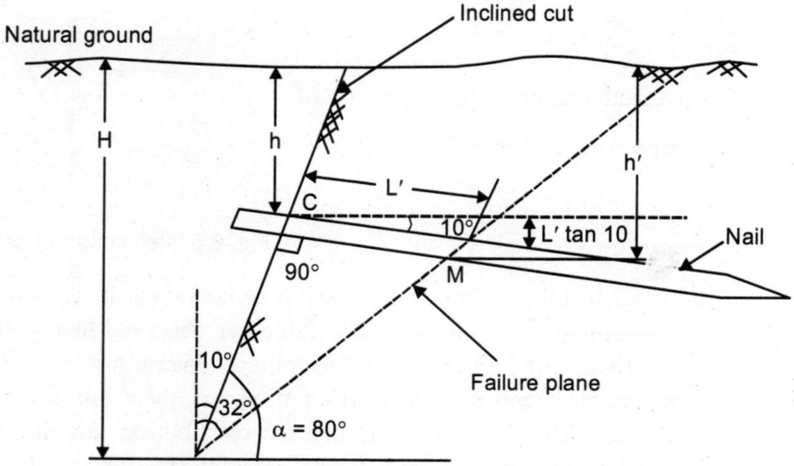

Fig. 6.7: Section showing a single nail

SOLUTION:

$$K_a = \left[\frac{\sin(\alpha - \phi)}{(\sin\alpha)^{1.5} + \sin\phi\,(\sin\alpha)^{0.5}}\right]^2$$

$$= \left[\frac{\sin(80 - 36)}{(\sin 80)^{1.5} + \sin 36\,(\sin 80)^{0.5}}\right]^2$$

$$= 0.1981$$

$$K_0 = 1 - \sin\phi = 0.4122$$

Neglecting anchoring force developed on small driving heads at end of soil nail and taking the bond as equal to the shaft friction, gives

$$F_N = L_a \tan\mu\,\gamma\,h'd\,[2 + (\pi - 2)\,K_o] \quad \ldots(6.3)$$

Take $\tan\mu = 0.8 \tan 36°$ (as it is given that friction coefficient is 80% of that of soil)

$$F_N = L_a \times 0.8 \tan 36° \times 18\,h' \times 0.015\,[2 + (\pi - 2)]\,0.4122$$

$$= 0.388\,(L_a)\,h'$$

Inclination of the potential failure plane to the vertical = $90 - \frac{1}{2}(\alpha - \phi) = 32°$. From the above figure, length of soil nail within the potential failure wedge L′ is given by

$$L' = (H - h)\,\frac{\tan 22}{\cos 10°}$$

Nail depth at M, $h' = h + L' \sin 10°$.

Trying a soil nail length of 70% H, i.e., 5.6 m, cautiously ignoring the increase in nail depth beyond point M, and tabulating the overall adherence calculations as follows: (Table 6.1)

Table 6.1: Computation of friction in nails

Level	*Depth (h)*	*L′ = (8 – h) 0.41*	*h′ = h + L′ sin 10°*	*L_a = 5.6 – L′*	*Friction = 0.388 L_a × h′ (kN)*
1	0.5	3.08	1.04	2.52	1.02
2	1.5	2.67	1.97	2.93	2.24
3	2.5	2.26	2.90	3.34	3.76
4	3.5	1.85	3.83	3.75	5.57
5	4.5	1.44	4.75	4.16	7.67
6	5.5	1.03	5.68	4.57	10.07
7	6.5	0.62	6.61	4.98	12.77
8	7.5	0.21	7.54	5.39	15.77
					Σ = 58.87 kN

Permissible tension force = 50/3 = 16.67 kN (FOS is taken as 3). Therefore, as none of the friction forces (e.g., given in the last column of the above table) developed exceed this value, the total frictional bond for a single vertical row of soil nails = 58.87 kN.

$$\text{Lateral earth pressure per m} = \frac{1}{2} K_a \gamma H^2$$

$$= \frac{1}{2}(0.1981) \times 18\,(8)^2$$

$$= 114.1 \text{ kN}$$

Trying a horizontal soil nail spacing of 0.2 m gives the lateral earth pressure per vertical row of soil nails

$$= 114.1 \times 0.2 = 22.82 \text{ kN}$$

Hence the overall adherence safety factor

$$= \frac{58.87}{22.82} = 2.6$$

The tension in soil nails is given as

$$T = K_a \gamma h_m . S.V.$$

$$= 0.1981 \times 18 \times 7.5 \times 0.2 \times 1$$

$$= 5.3 \text{ kN}$$

Therefore the tensile safety factor $= \frac{50}{5.3}$

$$= 9.4 \text{ (adequate).}$$

6.8 FIELD APPLICATIONS OF THE TECHNIQUE

Soil nailing is gaining popularity day by day. In India alone, the author of this book has successfully carried out soil nailing work in Hero Cycles Factory Ludhiana, landslide control in Nanital, hotel construction Meerut, construction of underground car parking Lucknow, construction of via-duct in New Delhi etc. The salient features of these projects are given in the following table (Table 6.2).

Table 6.2: Salient features of some of noted soil nailing works, carried out by Mittal (2005- 2010) in India

Sl. No.	*Name of project*	*Depth cut treated by soil nailing*	*Type of nail*	*Remarks*
1	Foundation for annealing plant in Hero cycles factory, Ludhiana, Panjab (Year 2005)	6.25 m	Driven	This project was tested successfully in earthquake of magnitude 7.4, occurred on October 8, 2005 (Figs. 6.8 to 6.12)
2	Soil stabilisation after occurrence of landslide in Nainital (Year 2006)	13.00 m	Grouted	Nailing done after occurrence of major iandslide. About 450 m length of hill slope successfully stabilised by soil nailing (Figs. 6.13 to 6.14).
3	Construction of via-duct below railway track in existing embankment in New Delhi (Year 2010)	10.00 m	Driven	Due to common wealth games, via-ducts (3 nos.) had to be constructed by puncturing existing rail embankment by using 'box push' technology. Soil nailing only helped construction work (Fig. 6.15).
4	Underground car parking near historical protected monument in Lucknow (Year 2010)	6.50 m	Driven	Due to existence of historical protected monument in very close proximity, no other solution was possible (Figs. 6.16 to 6.17).

Author has guided Ph.D. and M. Tech. thesis on soil nailing work. Research works have proved that even in saturated conditions, soil nailing can perform extremely well in wet as well as in dry conditions. (Mittal et al., 2008) Softwares have been developed in these research works by which nailing can be designed in almost no time. Gosavi (2006) developed design charts for ready design of soil nailed walls. One typical chart is shown vide Table 6.3. More such charts are available elsewhere (Gosavi, 2006).

Fig. 6.8: View of gantry crane resting on isolated RCC footing

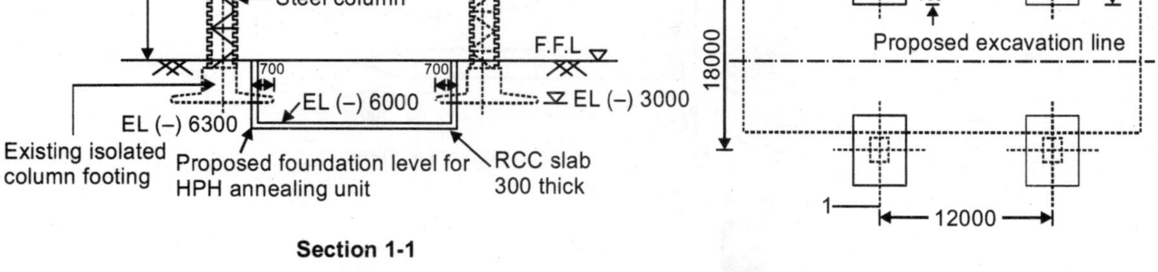

Fig. 6.9: Site plan and sectional view of the proposed HPH annealing plant excavation site

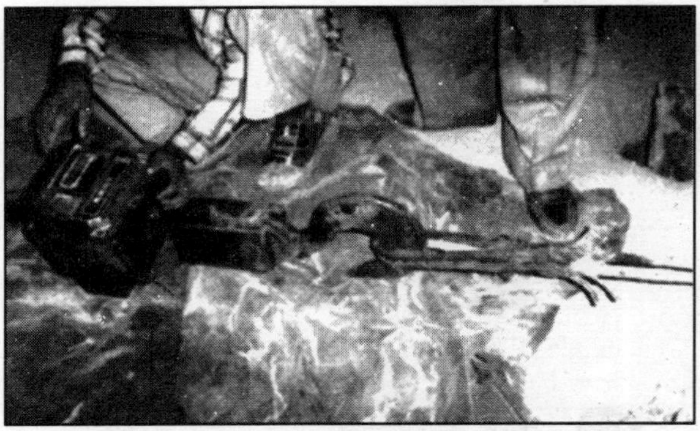

Fig. 6.10: **(a)** Load cell used for measurement of pull-out load

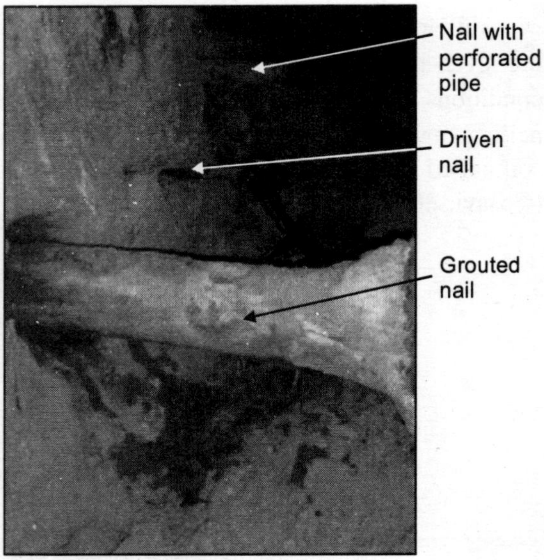

Fig. 6.10: **(b)** View of pulled-out nails

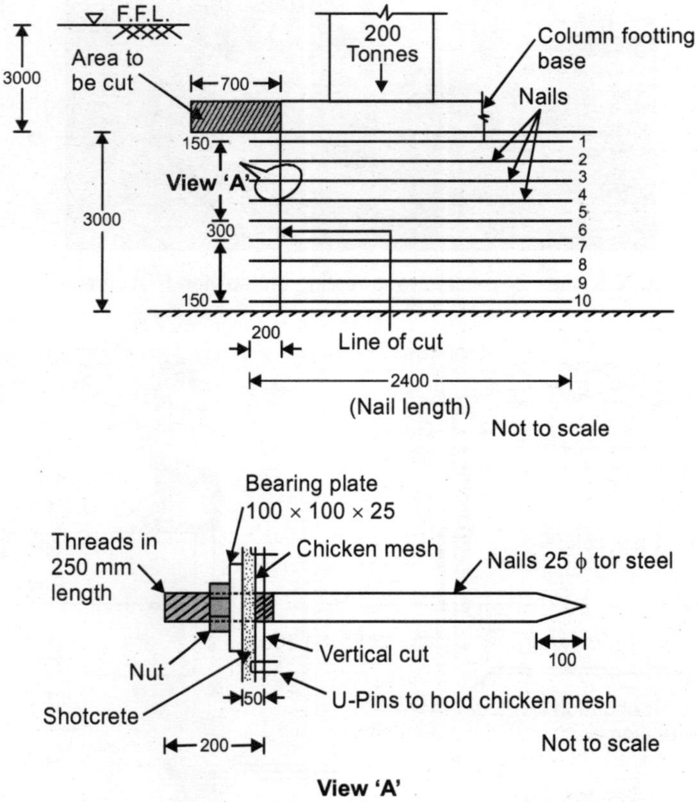

Fig. 6.11: Details of proposed nailed cut in hero cycle factory

Fig. 6.12: View of completed nailed cut

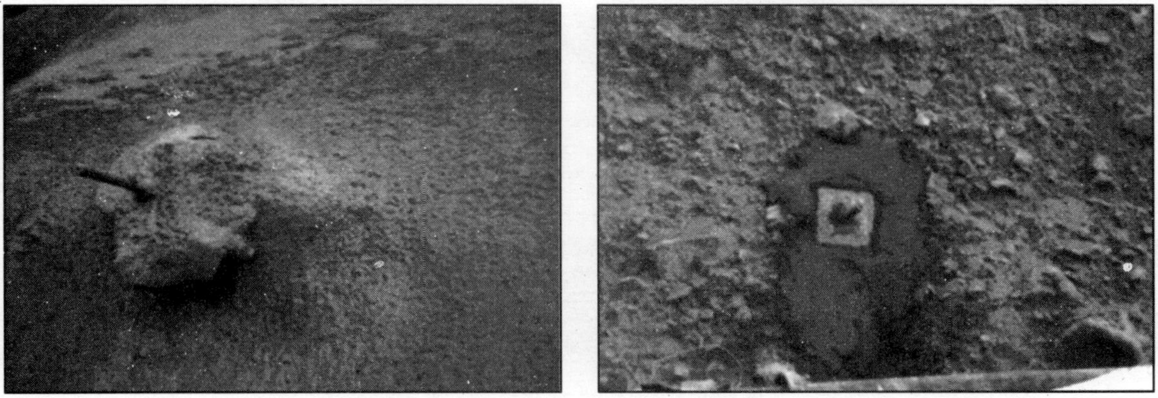

Fig. 6.13: A close view of grouted nail in landslide area of Nainital

Fig. 6.14: View of nail and drainage pipes in Nainital landslide project

Fig. 6.15: Overall view of driven nails for construction of via-duct below rail track in Delhi

Fig. 6.16: Pull-out tests conducted on soil nail at car parking site at Lucknow

Fig. 6.17: Historical monument near proposed underground car parking where soil nailing has been used

Table 6.3: Parametric study for H = 6 m, γ = 16.5 kN/m³, ϕ = 25°, N = H/S_v, L/H = 0.8, d = 25 mm, ε = 0°, $\alpha_v = \alpha_h/2$, $f_y = 4.15 \times 10^5$ kPa

		FOS																	
α_h	α	N	c=0, q=0	N	c=10 kPa, q=0	N	c=20 kPa, q=0	N	c=0 q=80 kPa	N	c=10 kPa, q=80 kPa	N	c=20 kPa, q=80 kPa	N	c=0 q=120 kPa	N	c=10 kPa, q=120 kPa	N	c=20 kPa, q=120 kPa
0	0	10	1.34	8	1.51	5	1.61	12	1.47	11	1.56	10	1.64	13	1.60	12	1.65	11	1.67
0	0	11	1.62	9	1.74	6	1.76	13	1.72	12	1.80	11	1.86	14	1.86	13	1.89	12	1.90
0	0	12	1.93	10	2.00	7	1.94	14	2.00	13	2.06	12	2.11	15	2.13	14	2.14	13	2.14
0	0	13	2.27	11	2.28	8	2.15	15	2.29	14	2.33	13	2.36	16	2.42	15	2.42	14	2.40
0	5	10	1.59	7	1.51	3	1.56	11	1.46	10	1.58	9	1.68	12	1.61	11	1.68	10	1.71
0	5	11	1.93	8	1.75	4	1.67	12	1.74	11	1.84	10	1.92	13	1.89	12	1.94	11	1.96
0	5	12	2.29	9	2.02	5	1.82	13	2.04	12	2.12	11	2.19	14	2.19	13	2.22	12	2.23
0	5	13	2.69	10	2.33	6	2.00	14	2.37	13	2.43	12	2.47	15	2.52	14	2.52	13	2.51
0	10	9	1.52	7	1.73	2	1.65	10	1.41	9	1.56	8	1.69	11	1.57	10	1.66	8	1.48
0	10	10	1.87	8	2.02	3	1.75	11	1.70	10	1.84	9	1.95	12	1.86	11	1.94	9	1.72
0	10	11	2.26	9	2.34	4	1.89	12	2.02	11	2.14	10	2.23	13	2.19	12	2.24	10	1.98
0	10	12	2.69	10	2.70	5	2.06	13	2.37	12	2.46	11	2.53	14	2.53	13	2.57	11	2.26
0	15	8	1.40	6	1.70	2	1.85	9	1.31	8	1.51	6	1.46	10	1.47	9	1.60	7	1.45
0	15	9	1.77	7	1.99	3	1.97	10	1.61	9	1.79	7	1.68	11	1.78	10	1.89	8	1.69
0	15	10	2.18	8	2.32	4	2.13	11	1.95	10	2.11	8	1.94	12	2.11	11	2.20	9	1.96
0	15	11	2.64	9	2.70	5	2.34	12	2.32	11	2.45	9	2.23	13	2.48	12	2.55	10	2.26
0.05	0	11	1.46	9	1.59	6	1.62	13	1.55	12	1.65	10	1.52	14	1.66	12	1.51	11	1.54
0.05	0	12	1.74	10	1.82	7	1.78	14	1.79	13	1.88	11	1.72	15	1.91	13	1.72	12	1.75

Contd...

FOS																			
α_h	α	*N*	*c*=0, *q*=0	*N*	*c*=10 *kPa*, *q*=0	*N*	*c*=20 *kPa*, *q*=0	*N*	*c*=0 *q*=80 *kPa*	*N*	*c*=10 *kPa*, *q*=80 *kPa*	*N*	*c*=20 *kPa*, *q*=80 *kPa*	*N*	*c*=0 *q*=120 *kPa*	*N*	*c*=10 *kPa*, *q*=120 *kPa*	*N*	*c*=20 *kPa*, *q*=120 *kPa*
0.05	0	13	2.04	11	2.08	8	1.97	15	2.06	14	2.13	12	1.94	16	2.17	14	1.95	13	1.97
0.05	0	14	2.37	12	2.36	9	2.18	16	2.34	15	2.40	13	2.18	17	2.44	15	2.20	14	2.21
0.05	5	10	1.42	8	1.59	4	1.53	12	1.55	11	1.68	9	1.55	13	1.67	11	1.52	10	1.58
0.05	5	11	1.72	9	1.83	5	1.66	13	1.82	12	1.93	10	1.77	14	1.94	12	1.76	11	1.80
0.05	5	12	2.05	10	2.11	6	1.83	14	2.11	13	2.20	11	2.01	15	2.23	13	2.01	12	2.04
0.05	5	13	2.40	11	2.41	7	2.02	15	2.42	14	2.49	12	2.26	16	2.53	14	2.28	13	2.30
0.05	10	9	1.34	7	1.56	3	1.59	11	1.50	10	1.66	8	1.55	12	1.63	11	1.75	9	1.57
0.05	10	10	1.65	8	1.81	4	1.71	12	1.78	11	1.93	9	1.78	13	1.91	12	2.01	10	1.81
0.05	10	11	2.00	9	2.10	5	1.87	13	2.09	12	2.22	10	2.03	14	2.21	13	2.30	11	2.06
0.05	10	12	2.38	10	2.42	6	2.06	14	2.42	13	2.53	11	2.30	15	2.54	14	2.61	12	2.33
0.05	15	9	1.55	6	1.52	2	1.68	10	1.40	9	1.61	7	1.53	11	1.53	10	1.69	8	1.54
0.05	15	10	1.91	7	1.78	3	1.78	11	1.69	10	1.89	8	1.76	12	1.82	11	1.97	9	1.78
0.05	15	11	2.31	8	2.07	4	1.92	12	2.02	11	2.19	9	2.02	13	2.14	12	2.26	10	2.04
0.05	15	12	2.75	9	2.40	5	2.11	13	2.37	12	2.52	10	2.31	14	2.48	13	2.59	11	2.33
0.1	0	8	1.40	10	1.67	7	1.64	13	1.38	12	1.51	11	1.59	14	1.49	13	1.57	12	1.61
0.1	0	9	1.78	11	1.90	8	1.81	14	1.61	13	1.72	12	1.79	15	1.71	14	1.77	13	1.82
0.1	0	10	2.20	12	2.15	9	2.00	15	1.84	14	1.95	13	2.01	16	1.95	15	2.00	14	2.03
0.1	0	11	2.66	13	2.43	10	2.22	16	2.09	15	2.19	14	2.24	17	2.19	16	2.23	15	2.26
0.1	5	12	1.57	9	1.67	5	1.53	13	1.61	11	1.53	10	1.63	13	1.49	12	1.59	11	1.65
0.1	5	13	1.84	10	1.91	6	1.68	14	1.87	12	1.75	11	1.84	14	1.72	13	1.82	12	1.87
0.1	5	14	2.14	11	2.18	7	1.85	15	2.14	13	2.00	12	2.08	15	1.98	14	2.06	13	2.10
0.1	5	15	2.45	12	2.48	8	2.05	16	2.43	14	2.26	13	2.33	16	2.25	15	2.31	14	2.35
0.1	10	11	1.54	8	1.64	4	1.57	12	1.56	10	1.51	9	1.63	13	1.67	11	1.57	10	1.65
0.1	10	12	1.83	9	1.90	5	1.71	13	1.83	11	1.74	10	1.86	14	1.94	12	1.81	11	1.88
0.1	10	13	2.15	10	2.18	6	1.88	14	2.12	12	2.00	11	2.10	15	2.22	13	2.06	12	2.12
0.1	10	14	2.49	11	2.49	7	2.08	15	2.43	13	2.28	12	2.37	16	2.53	14	2.33	13	2.39
0.1	15	10	1.47	7	1.60	2	1.53	11	1.46	10	1.69	8	1.61	11	1.32	10	1.51	9	1.62
0.1	15	11	1.77	8	1.86	3	1.62	12	1.74	11	1.96	9	1.84	12	1.57	11	1.75	10	1.86

Contd...

FOS																			
α_h	α	*N*	*c=0, q=0*	*N*	*c=10 kPa, q=0*	*N*	*c=20 kPa, q=0*	*N*	*c=0 q=80 kPa*	*N*	*c=10 kPa, q=80 kPa*	*N*	*c=20 kPa, q=80 kPa*	*N*	*c=0 q=120 kPa*	*N*	*c=10 kPa, q=120 kPa*	*N*	*c=20 kPa, q=120 kPa*
0.1	15	12	2.11	9	2.15	4	1.75	13	2.04	12	2.25	10	2.10	13	1.84	12	2.01	11	2.11
0.1	15	13	2.47	10	2.47	5	1.91	14	2.37	13	2.56	11	2.37	14	2.13	13	2.29	12	2.38
0.15	0	9	1.36	10	1.53	7	1.52	15	1.66	13	1.57	11	1.48	15	1.54	14	1.62	12	1.49
0.15	0	10	1.67	11	1.74	8	1.68	16	1.88	14	1.77	12	1.66	16	1.75	15	1.82	13	1.68
0.15	0	11	2.03	12	1.97	9	1.85	17	2.12	15	1.99	13	1.86	17	1.97	16	2.03	14	1.87
0.15	0	12	2.41	13	2.22	10	2.05	18	2.38	16	2.22	14	2.07	18	2.21	17	2.26	15	2.07
0.15	5	9	1.54	9	1.52	6	1.55	13	1.43	12	1.59	10	1.51	14	1.53	13	1.64	11	1.52
0.15	5	10	1.90	10	1.75	7	1.71	14	1.66	13	1.81	11	1.70	15	1.76	14	1.86	12	1.72
0.15	5	11	2.30	11	1.99	8	1.89	15	1.90	14	2.04	12	1.91	16	2.00	15	2.09	13	1.92
0.15	5	12	2.74	12	2.56	9	2.09	16	2.16	15	2.29	13	2.14	17	2.25	16	2.33	14	2.14
0.15	10	12	1.42	8	1.49	5	1.57	13	1.60	11	1.58	9	1.50	14	1.70	12	1.62	10	1.52
0.15	10	13	1.66	9	1.72	6	1.72	14	1.85	12	1.80	10	1.71	15	1.95	13	1.84	11	1.72
0.15	10	14	1.93	10	1.98	7	1.91	15	2.13	13	2.04	11	1.93	16	2.22	14	2.08	12	1.94
0.15	10	15	2.21	11	2.25	8	2.11	16	2.42	14	2.30	12	2.16	17	2.50	15	2.33	13	2.17
0.15	15	12	1.64	8	1.68	4	1.60	12	1.50	10	1.52	8	1.48	13	1.58	12	1.78	10	1.69
0.15	15	13	1.92	9	1.93	5	1.74	13	1.76	11	1.75	9	1.69	14	1.84	13	2.02	11	1.91
0.15	15	14	2.22	10	2.22	6	1.92	14	2.04	12	2.00	10	1.91	15	2.11	14	2.28	12	2.15
0.15	15	15	2.55	11	2.54	7	2.13	15	2.34	13	2.27	11	2.15	16	2.39	15	2.56	13	2.40

References

1. Ambily, A.P. and Gandhi, S.R. (2004). "Analysis of Hydro Test Results for Steel Tank on Stone Column Improved Ground", *Ground Engineering: Emerging Techniques, Proc. of Indian Geotechnical Conference,* 2004, Vol.1, pp. 420–423.
2. Bruce, D.A. and Jewell, R.A. (1987). "Soil Nailing: Application and Practice", Part 1 & 2, *Ground Engineering,* Nov. 1986, Jan. 1987.

3. Chu, J. and Jian, S.W. (2004). "Case Studies of Two Soil Improvement Methods for Road Construction", *5th International Conference on Ground Improvement Technique,* Kuala Lumpur, Malaysia, pp. 1–13.
4. Gassler, G. (1995). "Stabilisation of a Cutting Slope along a High-Speed Railway Line using Extremely Long Nails", *The Practice of Soil Reinforcing in Europe,* Thomas Telford, London, pp. 214–226.
5. Gassler, G. and Gudehus, G. (1983). "Soil Nailing Statistical Design", *Proc. of 8th ECSMFE,* Helsinki, Vol. 2, pp. 491–494.
6. Gosavi, M. (2006). "Behaviour of Nailed Open Cuts", Ph.D. Thesis, Indian Institute of Technology, Roorkee.
7. Gupta, R.P. (2003). "A Study on Soil Nailing with Respect to Open Excavations and Slopes", M.E. Thesis, Indian Institute of Technology, Roorkee.
8. Juran, I. and Elias, V. (1992). "Ground Anchors and Soil Nails in Retaining Structures", *Foundation Engineering Handbook,* Van Nostrand Company, Second Edition, pp. 868–905.
9. Mandal, J.N. and Divshikar, D.G. (1988). "Large Scale Shear Box Tests with Inclined Reinforcement", *Proc. 1st Indian Geotextiles Conference on Reinforced Soil and Geotextiles,* Bombay, pp. C.73–C.77.
10. Mittal, Satyendra, Singh, K. and Mathur, S. (2008). "Behaviour of Vertical Cuts using Soil Nailing Techniques in Saturated Conditions", *Journal of the South East Asian Geotechnical Society,* pp. 113–120.
11. Mittal, S. (2010). "Landslide Control through Soil Nailing in Nainital", IIT Roorkee India, Report.
12. Mittal, S. (2012). "Recent Experiences of Ground Improvement Techniques" Keynote lecture delivered in Indian Institute of Technology, Guwahati.
13. Ortigao, J.A.R. and Palmeira, E.M. (1997). "Optimised Design for Soil Nailed Walls", *Ground Improvement Geosystems,* Thomas Telford, London, pp. 369–373.
14. Patra, C.R. and Basudhar, P.K. (2001). "Nailed Soil Structures: An Overview", *Indian Geotechnical Journal,* Vol. 31, No. 4, pp. 322–362.
15. Pinto, M.I.M. (2002). "Applications of Geosynthetics for Soil Reinforcement", *4th International Conference on Ground Improvement Technique*, Kuala Lumpur, Malaysia, pp. 147–162.
16. Pokharel G. (2001). "Stabilisation of Slopes by Soil Nailing Method: Theory Experiment and Practice", *Proc. of Int. Symposium on Geotechnical & Geoenvironmental Challenges in Mountainous Terrain,* Kathmandu, Nepal, pp. 237–245.
17. Puppala, A.J., Bhadriraju, V. and Porbaha, A. (2004). "Deep Mixing Technology to Stabilise Problematic Soils", *Ground Engineering: Emerging Techniques, Proc. of Indian Geotechnical Conference,* 2004, Vol. 2, pp. 25–30.
18. Raju, G.V.R. (1996). "Behaviour of Nailed Soil Retaining Structures", Ph.D. Thesis, Nanyang Technological University, Singapore.
19. Rao, G.V. (2001). "Geosynthetics–An Overview of Types, Functions and Applications", *Geosynthetic Applications in Civil Engineering: A Short Course*, Publication No. 283, pp. 1–14.
20. Saran, S., Mittal, S. and Gosavi, M. (2004). "Pseudo-static Analysis of Nailed Vertical Excavations", *Indian Geotechnical Journal.*
21. Schlosser, F. (1982). "Behaviour and Design of Soil Nailing", *Symposium on Recent Developments in Ground Improvement Techniques*, Bangkok, pp. 399–413.
22. Warner, M.F. and Barley, A.D. (1997). "Cliff Stabilisation by Soil Nailing, Bouley Bay, Jersey, C.I.", Ground *Improvement Geosystems,* Thomas Telford, London, pp. 468–476.

Wall with Reinforced Backfill

CHAPTER 7

7.1 INTRODUCTION

Since long, engineers have been attempting to improve the mechanical properties of soil. Henry Vidal (1966), a French engineer, presented the concept of reinforced earth (Fig. 7.1). He investigated the frictional effects of reinforcement in soil with the aim of improving the mechanical properties of the soil in the direction in which soil is subjected to tensile strain. Reinforced earth is a composite material where in soil is reinforced by the elements which can take tension. These reinforcing elements may be in different forms, e.g., metal sheets, strips, nets, mats, synthetic fabrics or fibre reinforced plastic, etc.

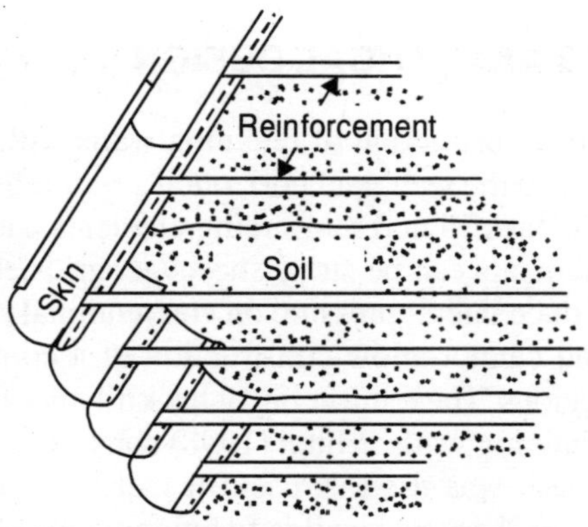

Fig. 7.1: (a) Schematic of reinforced earth wall

Most popular use of this technology has been made in retaining wall construction. Thousands of retaining walls have been constructed all over the world. The other advantage in this technology is that there is no restriction of height of wall. Many special structures, e.g., coal storage slots, rock crushers, ore transfer infrastructure, etc. have been constructed by using this technique (Vidal, 1978).

7.2 WALL WITH REINFORCED BACKFILL

Since the invention of reinforced earth (Vidal, 1966), intensive worldwide research is being made with particular emphasis on the performance of model and full scale retaining walls. In wall with reinforced backfill, the lateral pressure on wall is reduced by reinforcing the backfill with unattached horizontal strips or sheets. Broms (1977) was the first to report this concept. Broms (1977/1987) discussed internal and external stability analysis of wall retaining backfill reinforced with unattached continuous fabric reinforcement. According to his findings, (*a*) sufficient anchor zone which is capable of transferring a force more than allowable tension in the fabric, is needed just behind the wall elements for enabling the reinforcement thus provided to behave as attached reinforcement and (*b*) lateral earth pressure at any distance away from the wall face could also be computed.

Hausmann and Lee (1978) conducted small scale models tests to investigate the behaviour of rigid walls with reinforced backfill to establish effectiveness of unattached reinforcement in the fill.

Talwar (1981) developed the analysis for computation of lateral earth pressure in rigid retaining walls having cohesionless backfill reinforced with unattached reinforcing strips. Expressions of resultant earth pressures and its point of application had been derived in terms of soil properties and characteristics and distribution of reinforcement. Results were presented in the form of non-dimensional charts which indicated significant reduction in earth pressure with the increase in length of the reinforcing strips and decrease in their horizontal and vertical spacing. Theoretical results have been substantiated with carefully conducted model test data. Garg (1988) extended this work considering the uniformly distributed surcharge load on the backfill. He also developed a concept of economical placement of the reinforcement for vertical walls. Khan (1991) considered inclined retaining wall in the study and presented non-dimensional charts for design of wall with uniformly distributed surcharge. Analytical work has supported by model tests also. Line load surcharge was also considered in the study and the empirical relations were developed for determination of lateral earth pressure and moments on the retaining wall. The soil used by Garg (1988) and Khan (1991) was frictional soil.

7.3 ANALYTICAL DESIGN (After Mittal, 1998)

The construction of retaining walls with reinforced backfill employs reinforcing elements, that are not tied to the wall, laid horizontally to reinforce the backfill behind retaining walls. The interaction between the backfill and reinforcing elements causes appreciable reduction in the lateral thrust. The reinforcing elements can be strips/sheets/geogrids etc. This type of reinforcement can be beneficially employed to reduce lateral pressure on conventional retaining walls when the space behind the wall is very restricted and cannot allow construction of a conventional gravity wall with a wider base, particularly in hilly regions. The reinforcing of backfill may also be adopted as a corrective measure for inadequately designed retaining walls. Mittal (1998) constructed one such wall on Roorkee-Dehradun road [Fig. 7.1 (*b*)] where flyash was used as backfill material and netlon geogrid (CE-121) was used as reinforcing material. Since then, this wall is functioning absolutely well. Fig. 7.1 (*b*) (*i*) shows when reinforcement is placed in the form of strips Fig. 7.1 (*b*) (*ii*) shows when reinforcement is placed in the form of continuous sheet. Fig. 7.1 (*b*) (*iii*) shows completed view of retaining wall with movement of vehicle over it.

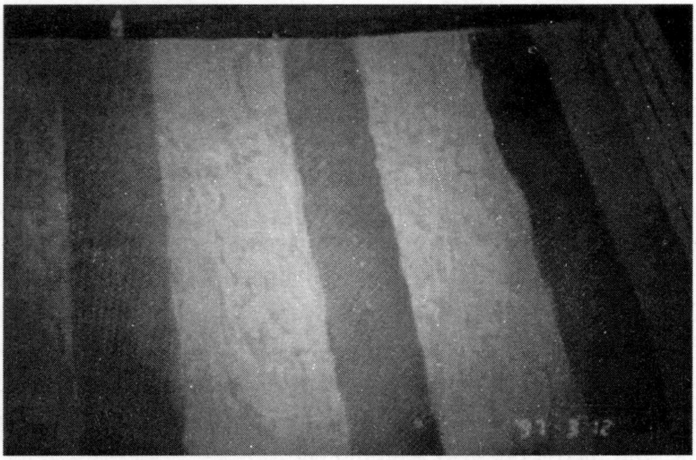

Fig. 7.1: (b) (i) Horizontal placement of reinforcement when placed in the form of strips

Fig. 7.1: (b) (ii) Horizontal placement of reinforcement when placed in the form of complete sheet

Fig. 7.1: (b) (iii) A completed view of wall with reinforced backfill

7.3.1 Analysis for retaining wall with inclined back face having reinforced cohesive frictional backfill with uniformly distributed surcharge load

The lateral earth pressure acting on a conventional retaining wall supporting reinforced earth backfill has been determined by considering the stability of a failure wedge. The analyses generally involve certain assumptions to arrive at simple solutions. Since soil is highly variable in nature, it needs to have some assumptions about its behaviour and composition to develop analytical solutions.

7.3.2 Assumptions

Following assumptions had been made in the analysis:

1. The backfill is homogeneous and isotropic.
2. The failure surface is a plane passing through the heel of retaining wall.
3. The coefficient of friction between soil and reinforcement is independent of the overburden pressure and the length of reinforcement and is constant for different beds of reinforcement.

4. The failure plane divides the length of reinforcing strip in two zones, one that lies within failure wedge and another outside. Only that part of strip which experiences movement of soil relative to itself is assumed to be contributing frictional resistance.
5. The frictional resistance to the lateral movement of wedge of backfill behind the retaining wall contributed by a reinforcing strip is assumed to be uniformly distributed over a fill height equal to vertical spacing of reinforcement encompassing that reinforcing strip.
6. The retaining wall undergoes an outward movement or rotation about the base which is sufficient to cause mobilisation of frictional resistance in the soil as well as reinforcing strips.

7.3.3 Analysis

Consider a retaining wall of height H with inclined back making an angle β with vertical, retaining a horizontal cohesive-frictional fill of dry density γ and an angle of internal friction φ supporting uniformly distributed surcharge load of intensity *q* [Fig. 7.2 (*a*)]. It is reinforced with unattached horizontally laid strips of length L and width *w* at vertical spacing S_z and horizontal spacing S_x. A failure plane BC making an angle θ with the vertical passes through the heel of the retaining wall. The frictional resistance to the lateral movement of the wedge ABC contributed by a reinforcing strip is computed from its effective length (l_e). Effective length is the portion of the strip that experiences movement of soil relative to itself. Reinforcing strip located completely within the moving wedge will not contribute any frictional resistance to the movement of the wedge.

Considering the equilibium of an element PQSR of thickness *dy* a failure wedge ABC, located at a depth *y* from the top of the wedge [Fig. 7.2 (*a*)]. Following forces per unit length of the wall act on the element of the wedge ABC.

σ_γ : pressure intensity acting on PQ in the vertical direction.

$(\sigma_\gamma + d\,\sigma_\gamma)$: pressure intensity acting on RS in the upward vertical direction.

R′ : reaction intensity on SQ acting at an angle φ to the normal to SQ

p : pressure intensity on PR acting at an angle δ with the normal to PR

w : weight of slice PQSR acting downwards

T : Tensile force in the strip assumed transmitted uniformly to soil layer of thickness S_z encompassing the strip

t : $\dfrac{T}{S} = \dfrac{2.w.f * l_e \sigma_v}{S_x \cdot S_z}$

c_s : cohesion of the soil

c_w : adhesion between the soil and wall acting per unit length

β : inclination of wall back with vertical

θ : angle of the rupture surface with vertical

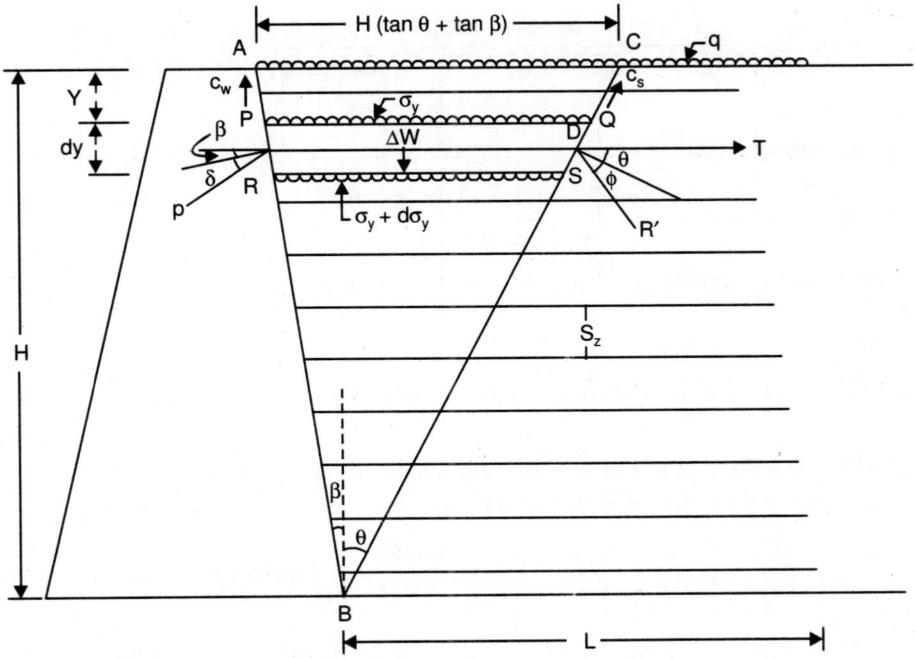

Fig. 7.2: (a) Wall details with reinforcement

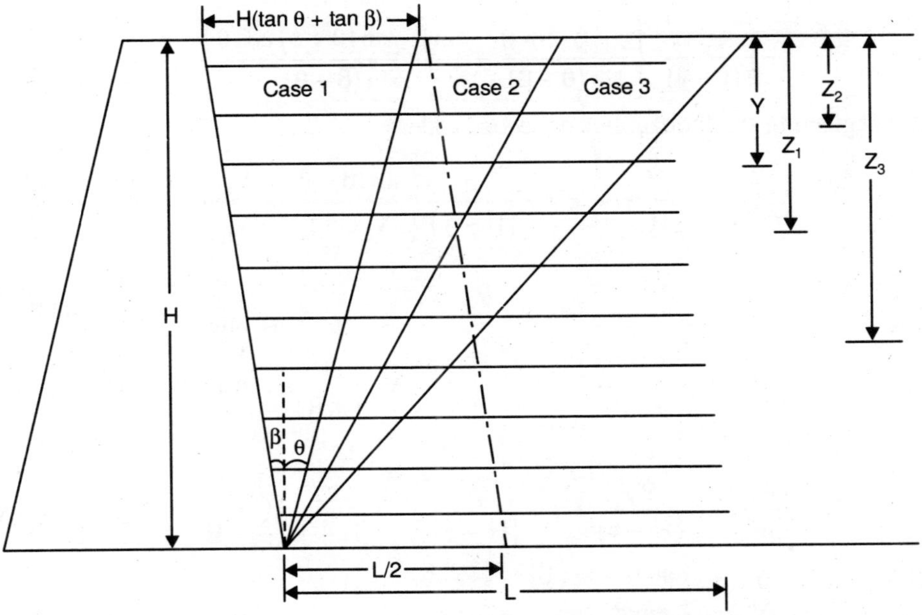

Fig. 7.2: (b) Schematic representation of three cases of analysis

(*i*) Considering static equilibrium of element under the action of all forces in the horizontal direction, i.e.,

$$\Sigma H = 0$$

$$p \cdot \cos(\delta + \beta) \cdot PR - R' \cos(\theta + \phi) \cdot QS + c_s \sin\theta \cdot QS + t \cdot dy - c_w \sin\beta \cdot PR = 0$$

Thus
$$R' = \frac{p \cdot \cos(\delta+\beta)\cos\theta \cdot \sec\beta}{\cos(\theta+\phi)} + \frac{c_s \sin\theta}{\cos(\theta+\phi)} - \frac{c_w \tan\beta \cdot \cos\theta}{\cos(\theta+\phi)} + \frac{t \cdot \cos\theta}{\cos(\theta+\phi)} \quad ...(7.1)$$

(*ii*) Weight of element PQSR is

$$\Delta W = \frac{\gamma \cdot \sec\beta \cdot dy\,(H-y)\sin(\beta+\theta)}{\cos\theta}$$

Considering static equilibrium of element under action of all the forces in the vertical direction i.e.,

$$\Sigma V = 0$$

$$p \cdot \sin(\delta+\beta) \cdot PR + R' \sin(\theta+\phi)\, QS + (\sigma_y + d\sigma_y) \cdot RS - \Delta W - \sigma_y \cdot PQ + c_w \cos\beta \cdot PR + c_s \cos\theta \cdot QS = 0$$

Substituting the value of PR, QS, RS, PQ and ΔW in above equation and on further simplification (after substituting the value of R′ from eqn 7.1).

$$\frac{d\sigma_y}{dy} = \gamma + \frac{\sigma_y}{(H-y)} - \frac{p\cos\theta}{\sin(\beta+\theta)(H-y)}\{\sin(\beta+\delta) + \cos(\delta+\beta)\cdot\tan(\theta+\phi)\}$$
$$- \frac{c_s}{(H-y)}\left\{\frac{\cos\beta\cdot\cos\theta}{\sin(\beta+\theta)} + \frac{\sin\theta\cdot\tan(\theta+\phi)\cos\beta}{\sin(\beta+\theta)}\right\}$$
$$- \frac{c_w}{(H-y)}\cdot\left\{\frac{\cos\beta\cdot\cos\theta}{\sin(\theta+\beta)} + \frac{\sin\beta\cdot\tan(\theta+\phi)\cos\theta}{\sin(\beta+\theta)}\right\} - \frac{t\cdot\tan(\theta+\phi)\cos\beta\cdot\cos\theta}{\sin(\beta+\theta)(H-y)}$$

(After neglecting small quantities of second order)

or,
$$\frac{d\sigma_y}{dy} = \frac{\sigma_y}{(H-y)} + \gamma - \frac{p}{(H-y)}\left\{\frac{\sin(\beta+\delta)}{V_1\cos\beta} + \frac{V_3}{V_1}\frac{\sin(\theta+\phi)\cos(\beta+\delta)}{\cos\theta\cos\beta}\right\}$$
$$- \frac{c_s}{H-y}\left\{\frac{1}{V_1} + \frac{V_3}{V_1}\tan\theta\frac{\sin(\theta+\phi)}{\cos\theta}\right\}$$
$$- \frac{c_w}{H-y}\left\{\frac{1}{V_1} - \frac{V_3}{V_1}\tan\beta\frac{\sin(\theta+\phi)}{\cos\theta}\right\} \frac{t}{H-y}\left\{\frac{V_3}{V_1}\frac{\sin(\theta+\phi)}{\cos\theta}\right\}$$

or,
$$\frac{d\sigma_y}{dy} = \frac{\sigma_y}{(H-y)} + \gamma - \frac{p}{(H-y)}V_4 - \frac{c_s}{H-y}V_5 - \frac{c_w}{H-y}V_6 - \frac{t}{H-y}V_7 \quad ...(7.2)$$

where,
$$V_1 = [\tan\theta + \tan\beta]$$
$$V_2 = -\tan\beta$$
$$V_3 = \frac{\cos\theta}{\cos(\theta+\phi)}$$
$$V_4 = \left\{\frac{\sin(\delta+\beta)}{V_1\cos\beta} + \frac{V_3}{V_1}\frac{\sin(\theta+\phi)\cos(\delta+\beta)}{\cos\theta\cos\beta}\right\}$$

$$V_5 = \left\{\frac{1}{V_1} + \frac{V_3}{V_1} \tan\theta \frac{\sin(\theta+\phi)}{\cos\theta}\right\}$$

$$V_6 = \left\{\frac{1}{V_1} - \frac{V_3}{V_1} \tan\beta \frac{\sin(\theta+\phi)}{\cos\theta}\right\}$$

$$V_7 = \frac{V_3}{V_1} \frac{\sin(\theta+\phi)}{\cos\theta}$$

(*iii*) Taking moments of all the forces about the midpoint (D) of element PQRS between Q and S, i.e.,

$$c_w dy \left\{c_w[H-y]\cos\beta\cdot A\right\} - \frac{\gamma A^2}{2\cos\beta}[H-y]^2 dy\cos\beta - \sigma_y[H-y]A\left\{\frac{(H-y)A}{2} - \frac{\sin\theta\, dy}{2}B\right\}$$

$$+ p\sin(\delta+\beta)(H-y)A\,dy + \left[\sigma_y + d\sigma_y\right]A(H-y-dy)\cdot\left\{\frac{(H-y-dy)A}{2} + \frac{\sin\theta}{2}dy\,B\right\} = 0$$

where, $A = \dfrac{\sin(\beta+\theta)}{\cos\theta}$ and $B = \dfrac{\cos\beta}{\cos\theta}$

On simplification and neglecting small quantities of higher order, the above equation reduces to

$$\frac{d\sigma_y}{dy} = \frac{\sigma_y}{(H-y)}\left\{1 - \frac{\sin(\theta-\beta)}{\sin(\theta+\beta)}\right\} - \frac{2p\sin(\delta+\beta)}{A(H-y)} - \frac{2c_w\cos\beta}{A(H-y)} + \gamma \quad ...(7.3)$$

Equating eqns. (7.2 and 7.3), we get

$$p = \frac{\sigma_y}{V_8}\left\{1 + \frac{2V_2}{V_1}\right\} - c_s\left\{\frac{V_5}{V_8}\right\} - \frac{c_w}{V_8}\left\{V_6 - \frac{2}{V_1}\right\} - \frac{t}{V_8}\{V_7\} \quad ...(7.4)$$

where, $V_8 = \left\{V_4 - \dfrac{2\sin(\delta+\beta)}{V_1\cos\beta}\right\}$

or, $p = \sigma_y(S_1) - c_s(S_2) - c_w(S_3) - t(S_4)$

where, $S_1 = \left\{1 + \dfrac{2V_2}{V_1}\right\}\dfrac{1}{V_8}$

$$S_2 = \frac{V_5}{V_8},\quad S_3 = V_6 - \frac{2}{V_1}\cdot\frac{1}{V_8},\quad S_4 = \frac{V_7}{V_8}$$

On differentiating eqn. (7.4) w.r.t. y,

$$\frac{dp}{dy} = \frac{d\sigma_y}{dy}\left\{1 + \frac{2V_2}{V_1}\right\}\frac{1}{V_8} - \frac{dt}{dy}\left\{\frac{V_7}{V_8}\right\} \quad ...(7.5)$$

$$\frac{dp}{dy} = \frac{d\sigma_y}{dy}(S_1) - \frac{dt}{dy}(S_4) \quad ...[7.5\,(a)]$$

Substituting the value of $\dfrac{d\sigma_y}{dy}$ from eqn. (7.3) into eqn. 7.5 (*a*) and simplifying further,

$$\frac{dp}{dy} = \frac{p}{(H-y)}\cdot(G_1)+(G_2)(\gamma)-(G_3)\cdot\frac{\sigma_\gamma}{H-y}+\frac{c_w}{H-y}(G_4)-\frac{dt}{dy}\cdot(G_5) \quad ...(7.6)$$

where, $G_1 = -\frac{2S_1\cdot\sin(\delta+\beta)}{V_1\cos\beta}$, $G_2 = S_1$, $G_3 = \frac{2V_2 S_1}{V_1}$, $G_4 = -\frac{2S_1}{V_1}$ $G_5 = S_4$

In eqn. (7.6), σ_y may be eliminated by substituting its value from eqn. (7.4). Thus

$$\frac{dp}{dy} = -\frac{p}{(H-y)}(C_1)+\gamma(C_2)-\frac{c_s}{H-y}(C_3)+\frac{c_w}{H-y}(C_4)-\frac{t}{H-y}(C_5)-\frac{dt}{dy}(C_6) \quad ...(7.7)$$

where, $C_1 = \left\{\frac{G_3}{S_1}-G_1\right\}$, $C_2 = G_2$, $C_3 = \frac{G_3\cdot S_2}{S_1}$,

$$C_4 = \left\{G_4-\frac{G_3\cdot S_3}{S_1}\right\}, C_5 = \frac{G_3\cdot S_4}{S_1},\ C_6 = G_5 \quad ...(7.8)$$

Neglecting C_5, being very small for small values of β, we get

$$\frac{dp}{dy} = -\frac{p}{(H-y)}(C_1)+\gamma(C_2)-\frac{c_s}{H-y}(C_3)+\frac{c_w}{H-y}(C_4)-\frac{dt}{dy}(C_6) \quad ...[7.8(a)]$$

Tension, T at limiting equilibrium can be taken as given below:

$$T = 2\{\gamma(y+dy/2)+q\}\cdot w\cdot l_e f^* + 2c_r\cdot w\cdot l_e \quad ...(7.9)$$

where l_e = effective length of reinforcing strip

w = width of reinforcing strip

c_r = coefficient of adhesion with reinforcement

f^* = apparent coefficient of friction between backfill soil and reinforcement

From eqn. (7.9),

$$t = \frac{2wf^*\cdot l_e\left[\gamma(y+dy/2)+q\right]+2\cdot c_r\cdot w\cdot l_e}{S_x S_z} \quad ...(7.10)$$

where, S_x and S_z are horizontal and vertical spacings of reinforcing strips respectively.

The values of l_e will vary from strip to strip and will depend on wedge angle θ and length L of the strip. There may be three cases of analysis [as shown in Fig. 7.2 (*b*)].

Case 1: $H(\tan\theta+\tan\beta) \le L/2$ ...[7.11(*a*)]

$l_e = (H-y-dy/2)(\tan\theta+\tan\beta)$ for all reinforcing elements ...(7.12)

Case 2: $L/2 \le H(\tan\theta+\tan\beta) \le L$ ...[7.11(*b*)]

$l_e = L-(H-y-dy/2)(\tan\theta+\tan\beta)$, for $y \le Z_1$ ...(7.13)

$l_e = (H-y-dy/2)(\tan\theta+\tan\beta)$, for $y > Z_1$ ...(7.14)

Case 3: $H(\tan\theta+\tan\beta) \ge L$ ...[(7.11(*c*)]

$l_e = 0$ for $y \le Z_2$ ...(7.15)

$l_e = L-(H-y-dy/2)(\tan\theta+\tan\beta)$, for $Z_2 \le y \le Z_3$ ...(7.16)

$l_e = (H-y-dy/2)(\tan\theta+\tan\beta)$, for $y > Z_3$ ...(7.17)

Z_1, Z_2, and Z_3 are shown in Fig. 7.2 (*b*).

7.4 PRESSURE INTENSITY ON WALL

Case 1: H (tan θ + tan β) ≥ L/2 ...[7.11(*a*)]

$$l_e = (\text{H} - y - dy/2)(\tan\theta + \tan\beta)$$

$$= (\text{H} - y)(\tan\theta + \tan\beta), \text{ neglecting } dy/2 \text{ being small}$$

$$t = \frac{2wf*\left[\gamma(y + dy/2) + q\right]\cdot(\text{H}-y)(\tan\theta+\tan\beta) + 2\cdot c_r\cdot w(\text{H}-y)(\tan\theta+\tan\beta)}{\text{S}_x\text{S}_z} \quad ...(7.18)$$

$$\frac{dt}{dy} = 2\,\text{D}_{p\cdot\gamma}(\tan\theta+\tan\beta)\left[[1-2y/\text{H}] - \frac{q}{r\text{H}}\right] - 2\text{D}_c(\tan\theta+\tan\beta)\cdot\gamma \quad ...(7.19)$$

where, $\text{D}_p = \dfrac{wf*\text{H}}{\text{S}_x\text{S}_z}$ and $\text{D}_c = \dfrac{c_r\cdot w\text{H}}{\gamma\text{H}\cdot\text{S}_x\text{S}}$

Here, D_p = non dimensional spacing coefficient, w = width of reinforcement, $f*$ = apparent coefficient of friction between soil and reinforcement, determined from pulloput test, H = height of wall. By this parameter, user can decide the mutual horizontal spacing (s_x) of reinforcement strips and vertical spacing (s_z) of reinforcement strips. If the reinforcement is placed like Fig. 7.1 (*b*) (*ii*), then $w = \text{S}_x$, in that case $\text{D}_p = f*\text{H}/(\text{S}_z)$.

D_c = non dimensional coefficient or adhesion with reinforcement

Substituting the values of *dt*/*dy* from Eqn. (7.19) to (7.8*a*).

$$\frac{dp}{dy} = -\frac{p}{(\text{H}-y)}(\text{C}_1) + \gamma(\text{C}_2) - \frac{c_s}{\text{H}-y}(\text{C}_3) + \frac{c_w}{\text{H}-y}(\text{C}_4) - (\text{C}_6)$$

$$\left[2\text{D}_p\cdot\gamma\,(\tan\theta+\tan\beta)\cdot\left\{\left(1 - \frac{2\gamma}{\text{H}} - \frac{q}{\gamma\text{H}}\right)\right\} - 2\text{D}_c\,(\tan\theta+\tan\beta)\gamma\right] \quad ...(7.8b)$$

The above equation can be solved like a common differential equation. The solution of this equation is given elsewhere (Mittal, 1998) for three different cases. The Case 2 and Case 3 are defined as below (also refer Fig. 7.2 (*b*):

Case 2(*a*): L/2 ≤ H (tan θ + tan β) ≤ L

$$l_e = \text{L} - (\text{H} - y - dy/2)(\tan\theta+\tan\beta) \text{ for } y < \text{Z}_1$$

$$t = \frac{2wf*\left\{\gamma\left[y + \frac{dy}{2}\right] + q\right\}\left\{\text{L} - \left(\text{H} - y - \frac{dy}{2}\right)(\tan\theta+\tan\beta)\right\}}{\text{S}_x\text{S}_z} + \frac{2\text{C}_r\cdot w\left\{\text{L} - (\text{H}-y-dy/2)(\tan\theta+\tan\beta)\right\}}{\text{S}_x\text{S}_z} \quad ...(7.20)$$

$$\frac{dt}{dy} = 2\gamma\cdot\text{D}_p(\tan\theta+\tan\beta)\left\{\frac{\text{L}}{\text{H}(\tan\theta+\tan\beta)} - 1 + \frac{2y}{\text{H}} + \frac{q}{\gamma H}\right\} + 2\text{D}_c\cdot\gamma\cdot(\tan\theta+\tan\beta)$$

$$\frac{dt}{dy} = \gamma\,\text{T}_1\left[\frac{\text{L}}{\text{H}\cdot\text{T}_1} - 1 + \frac{2y}{\text{H}} + \frac{q}{\gamma H}\right] + \lambda\cdot\text{T}_2 \quad ...(7.21)$$

where $T_1 = 2(\tan\theta + \tan\beta)\cdot D_p$

$T_2 = 2(\tan\theta + \tan\beta)\cdot D_c$

and, $T_3 = (\tan\theta + \tan\beta)$ ...(7.22)

Eqn. 7.8 (*a*) may be written as below:

$$\frac{dp_2}{dy} = -\frac{p}{(H-y)}(C_1) + \gamma(C_2) - \frac{c_s}{H-y}(C_3) + \frac{c_w}{H-y}(C_4) - C_6\frac{dt}{dy}$$

Substituting the value $\frac{dt}{dy}$ from eqn. (7.21) in above equation.

$$\frac{dp_2}{dy} + C_1\frac{p}{(H-y)}$$

$$= \gamma C_2 - \frac{c_s}{H-y}(C_3) + \frac{c_w}{H-y}(C_4) - (C_6)\cdot\lambda\cdot T_1\left[\frac{L}{HT_1} - 1 + \frac{2y}{H} + \frac{q}{\gamma H}\right] - C_6\cdot\gamma T_2 \quad ...(7.23)$$

Similarly equations for Case 2 (*b*), Case 3 (*a*) to (*c*) are evaluated (Mittal, 1998).

These are common differential equations, the solutions to which are done by using boundary conditions that at $y = 0$, $\sigma_y = q$, for $y > z$, the effective length = $(H - y)\tan\theta$. These equations were solved by computer and their results are presented in the form of charts presented at the end of this chapter.

7.5 METHOD OF COMPUTATION

The pressure distribution along the height of wall for few typical cases, obtained by using above equations are shown in Fig. 7.3 (*a*) and (*b*). It is evident that the pressure intensities become negative in some portion of the wall.

The value of the total pressure is obtained by numerical integration neglecting the negative pressure. The designed section of wall needs to be checked for its stability against sliding and overturning. The former needs the value of resultant lateral earth pressure which is obtained by maximising the lateral earth pressure with respect to wedge angle θ. Similarly the moments of the positive pressure intensities were taken about the heel of the wall and the same is maximised with respect to wedge angle θ. The maximised values of resultant earth pressure and moment are denoted by P and M respectively and are presented in the charts in non-dimensional form as $\frac{P}{\frac{1}{2}\gamma H^2}$ and $\frac{M}{\frac{1}{6}\gamma H^3}$ respectively.

7.5.1 Parametric study

A range of design parameters, e.g., D_p, angle of internal friction φ, L/H ratio, $c/\gamma H$, $q/\gamma H$ etc. likely to be used in practice has been considered and is given in Table 7.1.

Values of non-dimensional resultant pressure $\frac{P}{\frac{1}{2}\gamma H^2}$ for some typical cases are presented in Figs. 7.3 to 7.6 and values of non-dimensional moments $\frac{M}{\frac{1}{2}\gamma H^2}$ are given in Figs. 7.7 to 7.10. For various other cases the values of pressures and moments are given. The maximum value of D_p has been taken as 1.0 only as it was shown by the results of parametric study [Tables 7.2 (*a*) and 7.2 (*b*)] that there was no appreciable reduction in the resultant pressure with the increase in the values of $D_p > 1.0$. Similarly, it was observed from Tables 7.2 (*a*) and 7.2 (*b*) that for the values of L/H more than 0.6 pressure becomes almost constant.

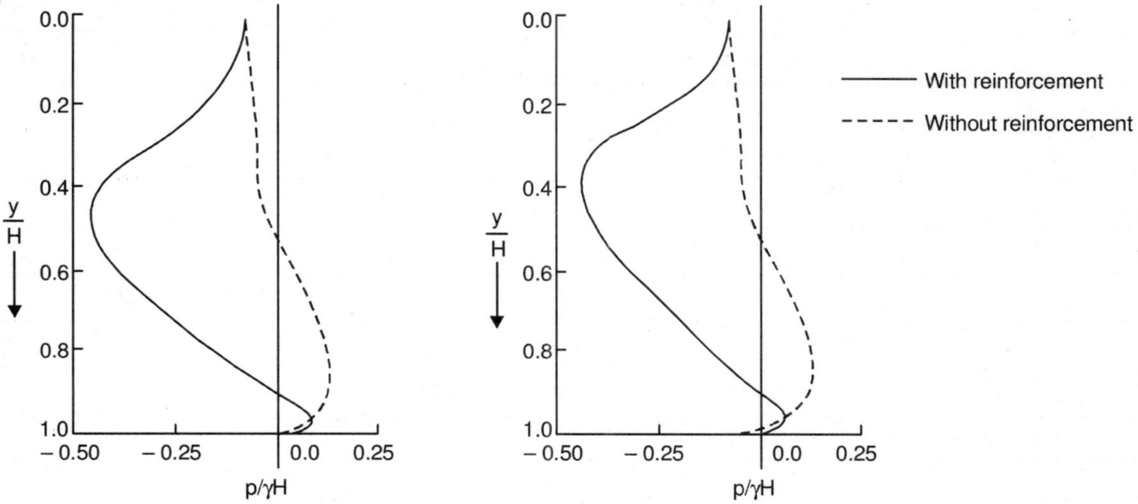

$\phi = 30°, \beta = 0, \frac{C}{\gamma H} = 0.05, D_p = 1.0, \frac{L}{H} = 0.8, \frac{q}{\gamma H} = 0.0$

(a) Pressure optimisation **(b)** Moment optimisation

Fig. 7.3: Pressure distribution along height of wall

Table 7.1: Parameters considered for parametric study		
Parameter	***Range***	***Interval***
ϕ	20°–35°	5°
δ	2 ϕ/3 for all cases	–
D_p	0.25–1.0	0.25
L/H	0.0–1.0	0.2
$c/\lambda H$	0.0–0.10	0.05
β	0°–15°	5°
$q/\gamma H$	0–0.50	0.25

Note: In the parametric study, (*i*) c_s is taken equal to c_w (*ii*) both c_s and c_w are taken numerically equal to $c/\gamma H$.

Table 7.2: (a) Values of D_p v/s Total pressure (Without cohesion)

$\phi = 30°$, $\delta = 20°$, $\beta = 10°$, $c/\gamma H = 0.0$, $q/\gamma h = 0.0$

D_p	Pressure $\left[\dfrac{P}{\frac{1}{2}\gamma H^2}\right]$		
	L/H = 0.6	*L/H = 0.8*	*L/H = 1.0*
(1)	(2)	(3)	(4)
0.5	0.244	0.235	0.235
0.75	0.215	0.215	0.215
1.00	0.202	0.202	0.202
1.25	0.192	0.192	0.192
1.50	0.185	0.185	0.185
1.75	0.178	0.178	0.178
2.00	0.173	0.173	0.173

Table 7.2: (b) Values of D_p v/s Total pressure (Without cohesion)

$\phi = 30°$, $\delta = 20°$, $\beta = 10°$, $c/\gamma H = 0.5$, $q/\gamma h = 0.0$

D_p	Pressure $\left[\dfrac{P}{\frac{1}{2}\gamma H^2}\right]$		
	L/H = 0.6	*L/H = 0.8*	*L/H = 1.0*
0.5	0.113	0.113	0.113
0.75	0.103	0.103	0.103
1.00	0.095	0.096	0.096
1.25	0.090	0.090	0.090
1.50	0.085	0.085	0.085
1.75	0.082	0.082	0.082
2.00	0.079	0.079	0.079

Following figures give the values of pressure (Figs. 7.4 to 7.7) and moments (Figs. 7.8 to 7.11) for all possible values of cohesion (c), surcharge (q), angle of inclination (β) and D_p factor. The users may use these charts for design of wall with reinforced backfill.

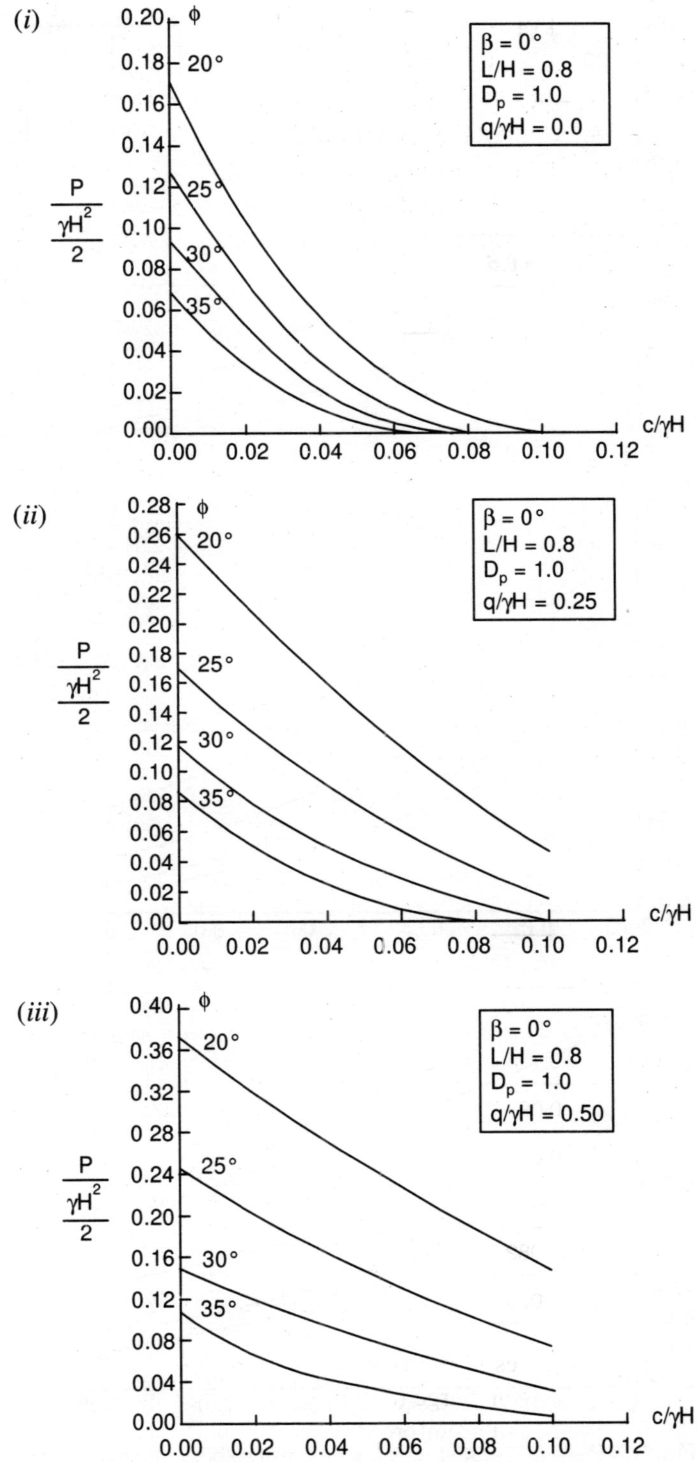

Fig. 7.4: Resultant earth pressure *vs.* $c/\gamma H$ (Stability against sliding)

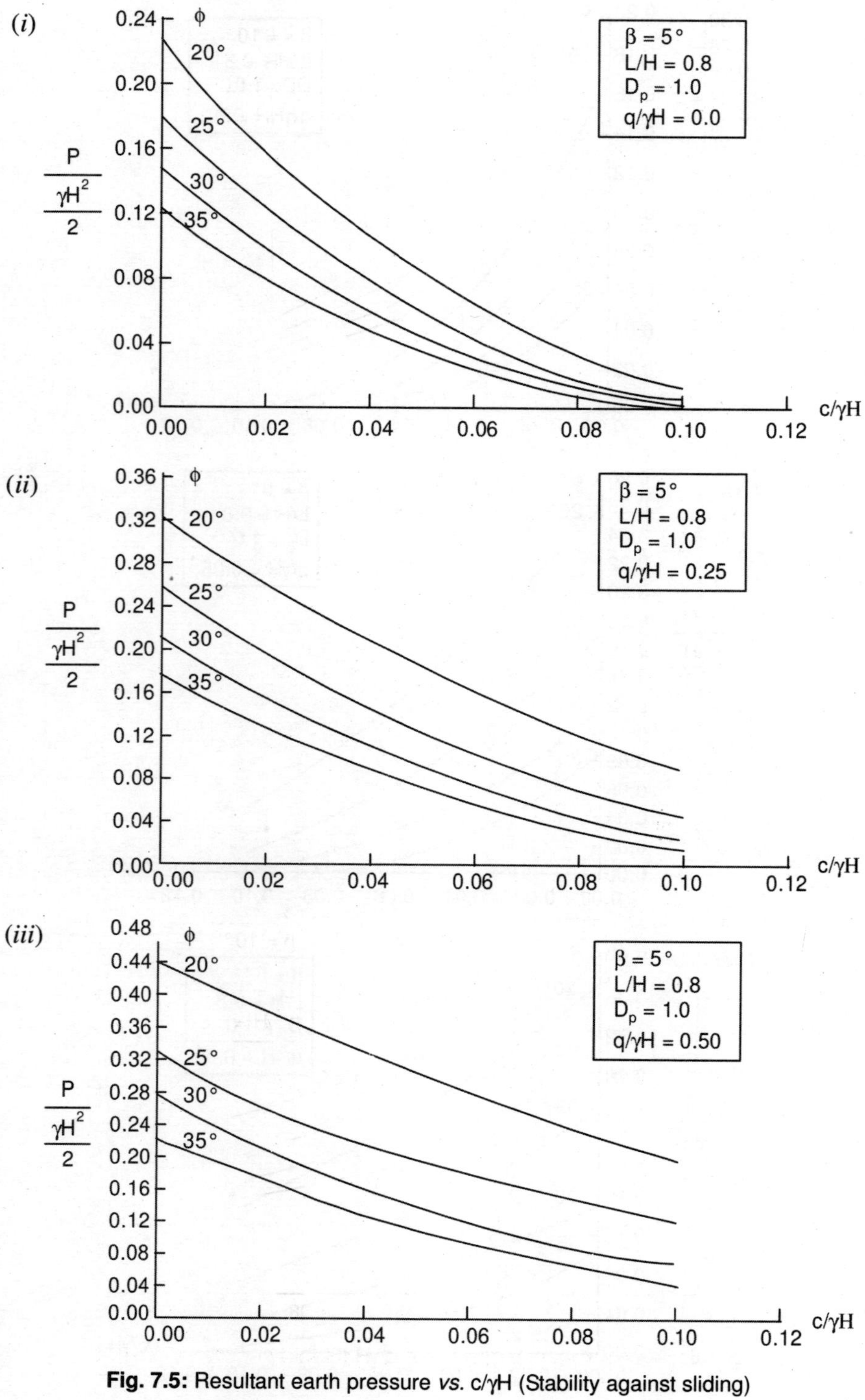

Fig. 7.5: Resultant earth pressure *vs.* c/γH (Stability against sliding)

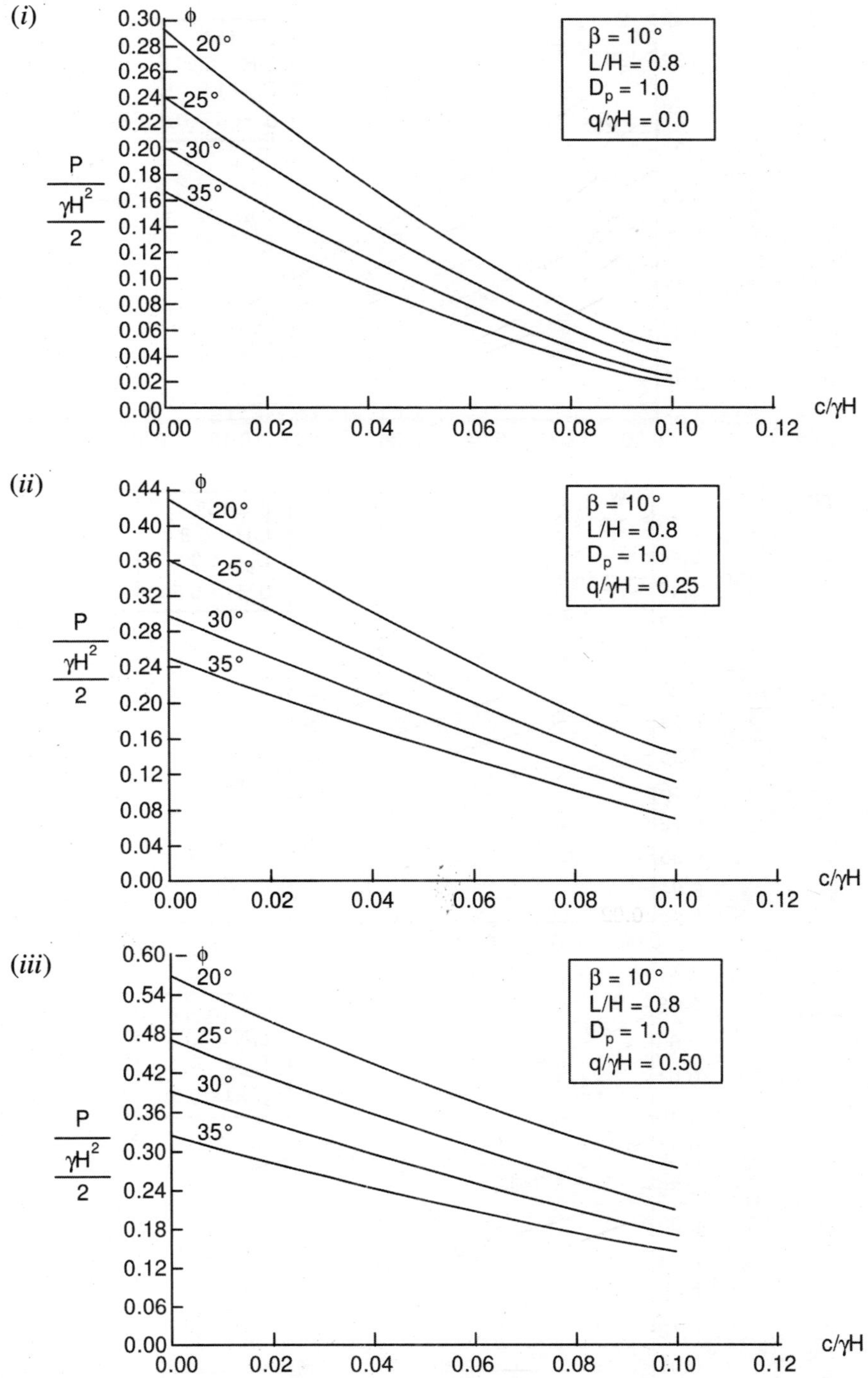

Fig. 7.6: Resultant earth pressure *vs.* c/γH (Stability against sliding)

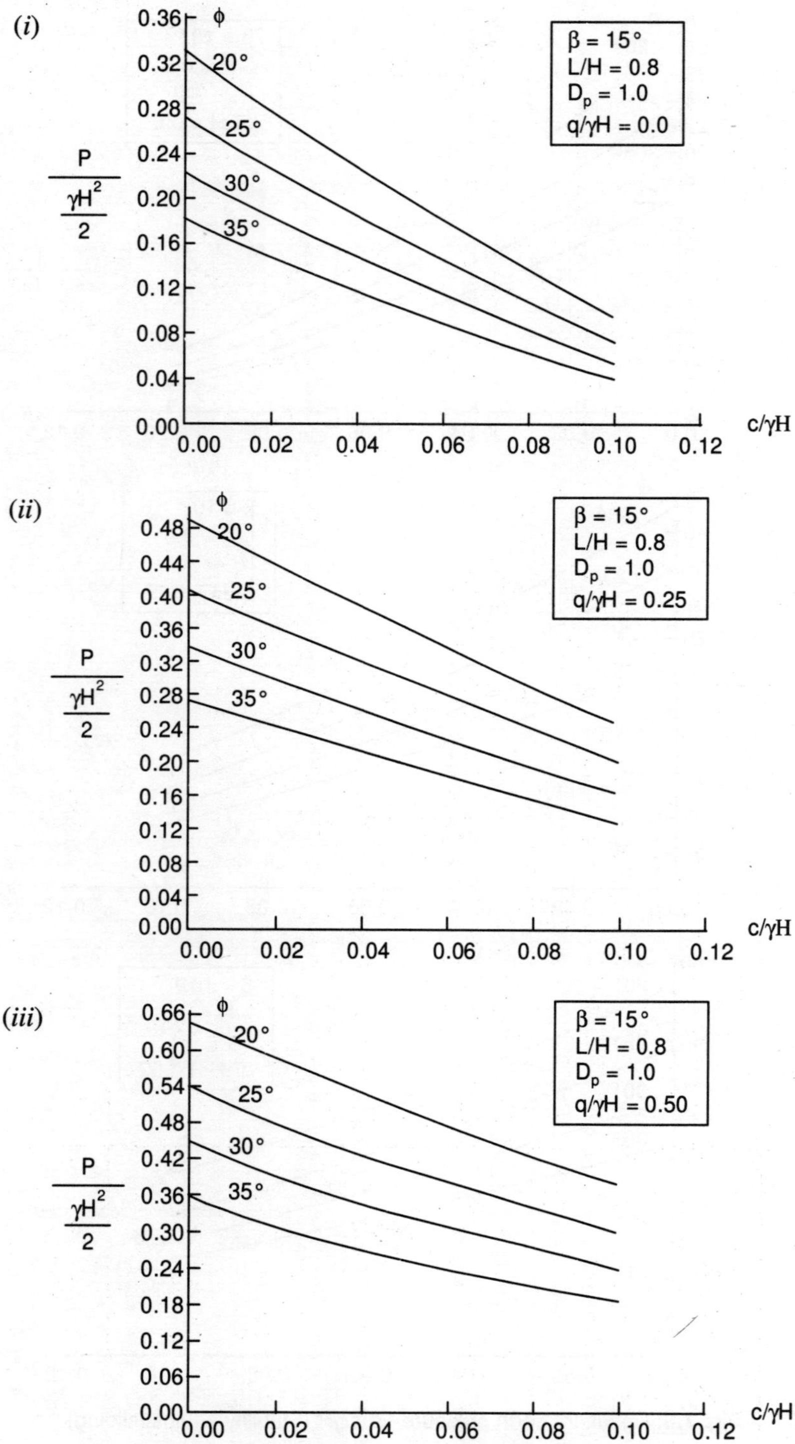

Fig. 7.7: Resultant earth pressure *vs.* c/γH (Stability against sliding)

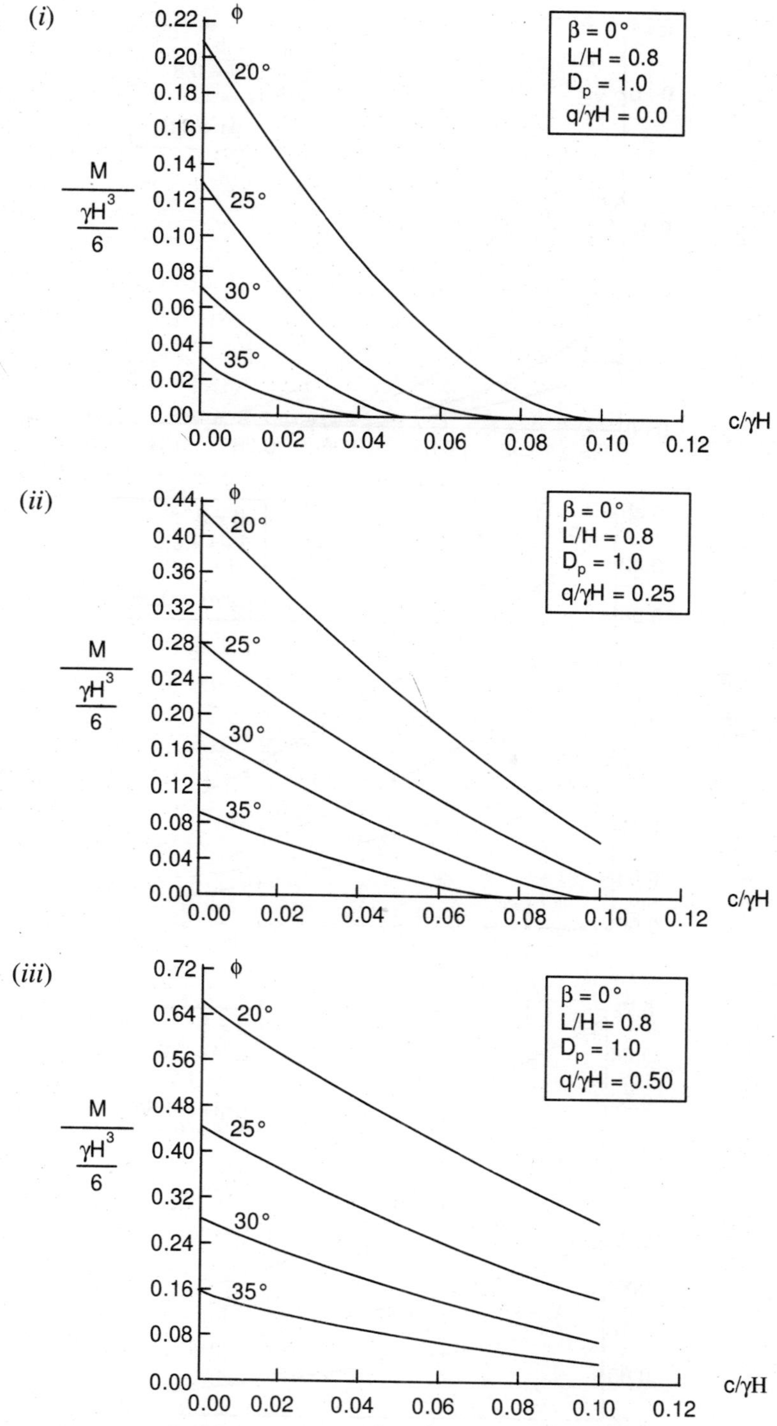

Fig. 7.8: Resultant moment *vs.* c/γH (Stability against turning)

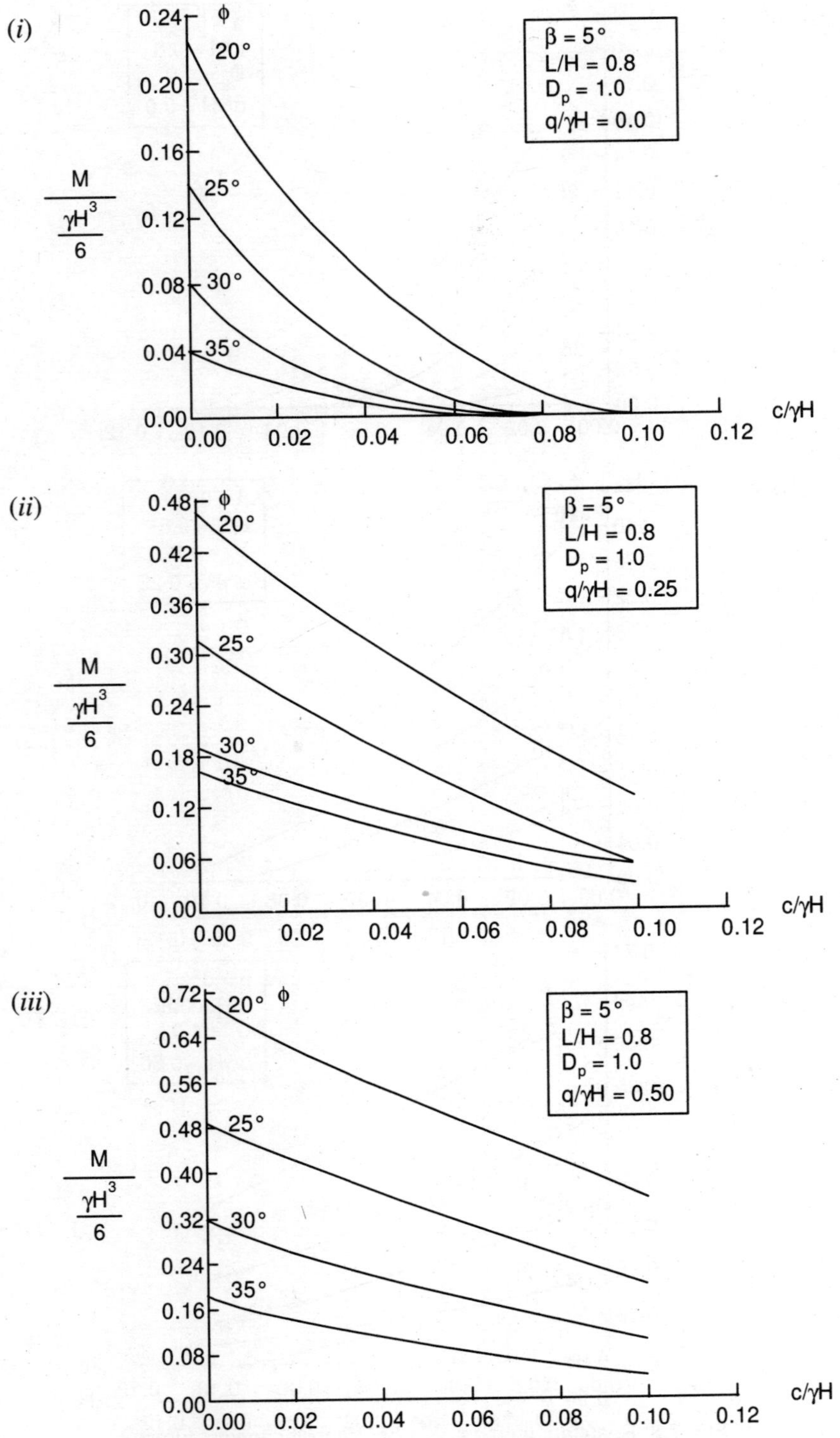

Fig. 7.9: Resultant moment *vs.* c/γH (Stability against turning)

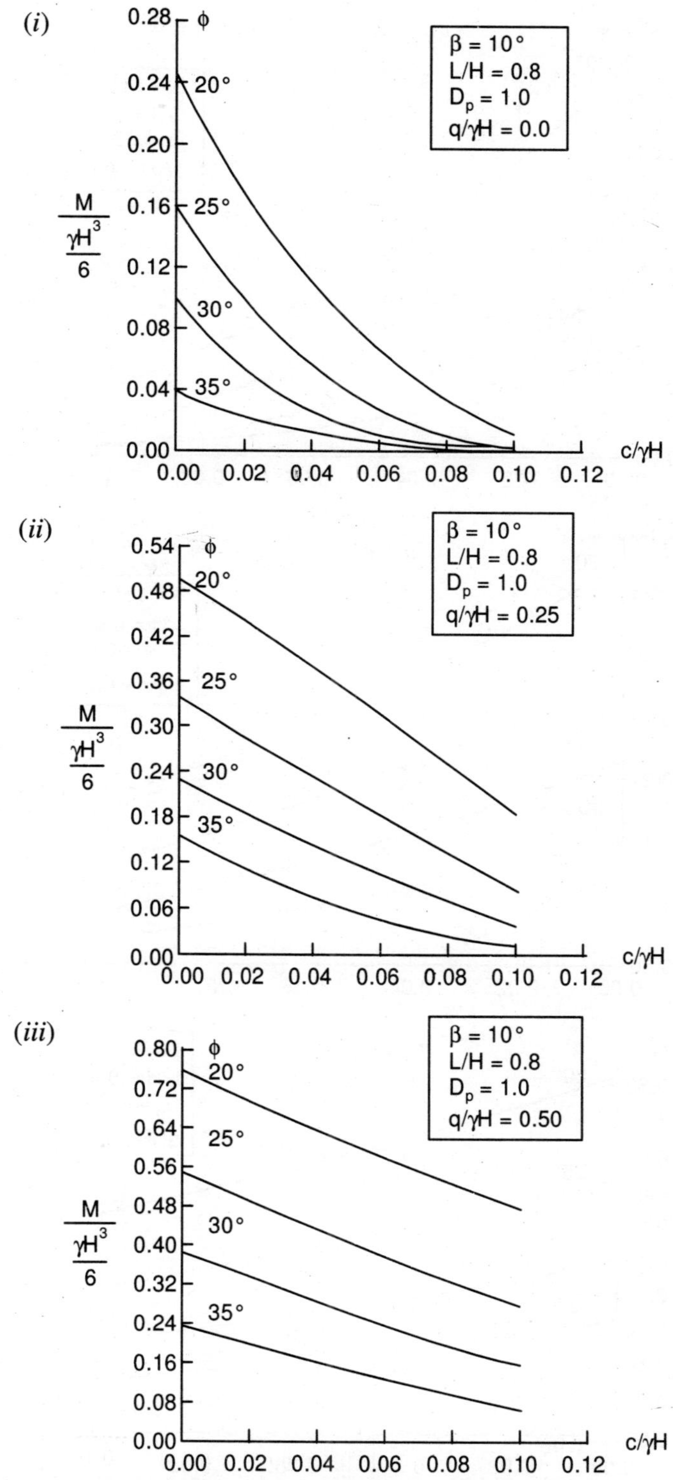

Fig. 7.10: Resultant moment *vs.* c/γH (Stability against turning)

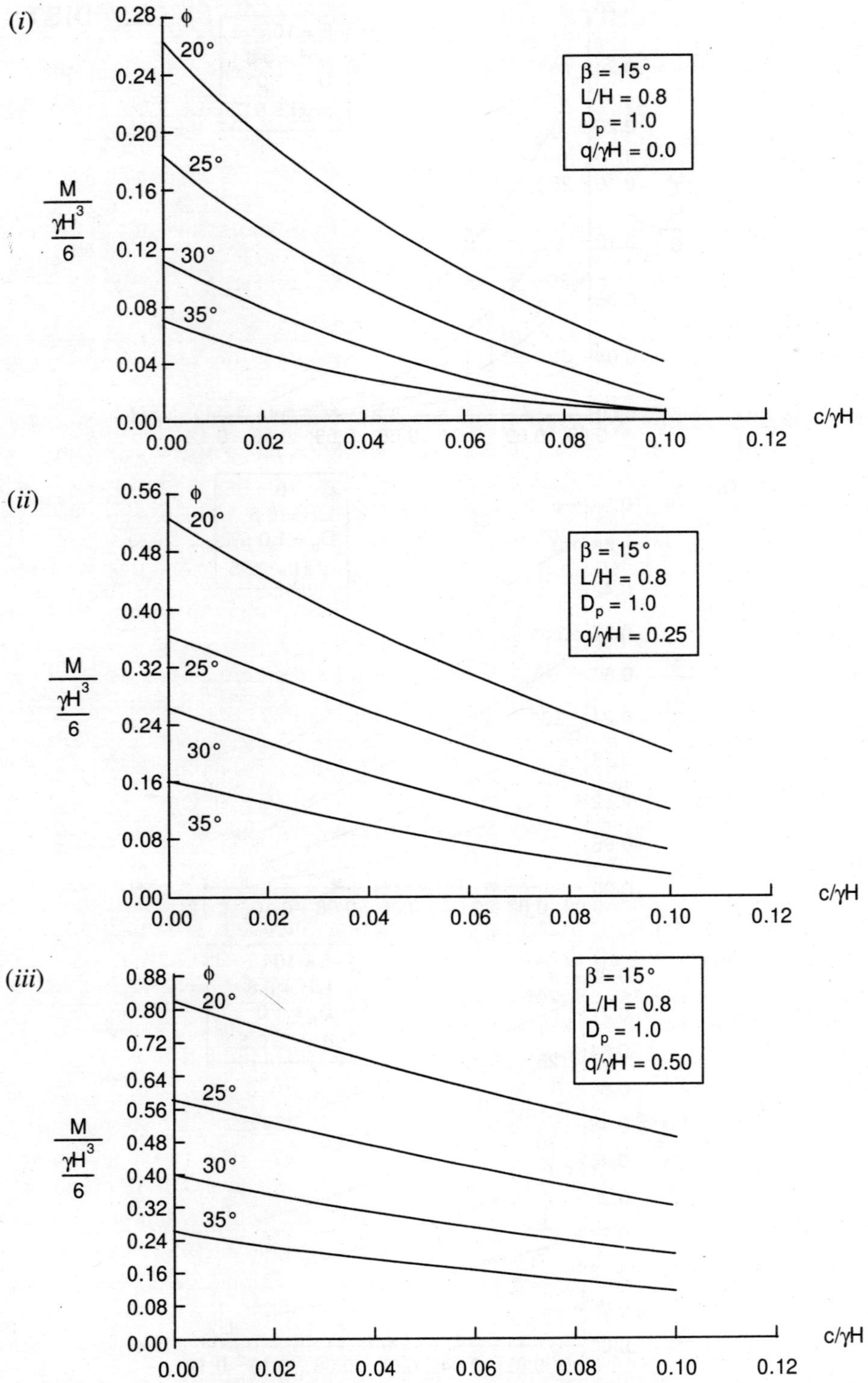

Fig. 7.11: Resultant moment *vs.* c/γH (Stability against turning)

7.6 RETAINING WALL WITH INCLINED BACK FACE HAVING UNREINFORCED FRICTIONAL COHESIVE BACKFILL WITH UNIFORMLY DISTRIBUTED SURCHARGE AND THE LOAD

7.6.1 Assumptions

The assumptions used in the analysis are as follows:

(*i*) The soil behind the retaining wall is considered to be dry isotropic and cohesive frictional material.

(*ii*) Rupture plane is assumed to be linear passing through the heel of the wall.

(*iii*) The wall yields to the extent that the rupture of the backfill soil takes place and a soil wedge is torn off from rest of the soil mass.

(*iv*) Resultant earth pressure acts at an angle δ to the normal of the wall where δ is the angle of wall friction.

7.6.2 Analysis

Theoretical analysis has been developed for retaining wall with inclined backface having horizontal unreinforced frictional cohesive backfill subjected to line load and uniformly distributed surcharge together by considering the equilibrium of the Coulomb wedge.

Consider a retaining wall of height H with inclined back making an angle β with vertical, retaining a horizontal cohesive-frictional backfill of dry density γ, internal friction angle φ and cohesion, c_s. The backfill supports uniformly distributed surcharge load of intensity q per square metre and line load of intensity Q per metre length of wall acting at a distance of x from back face of wall (Fig. 7.12).

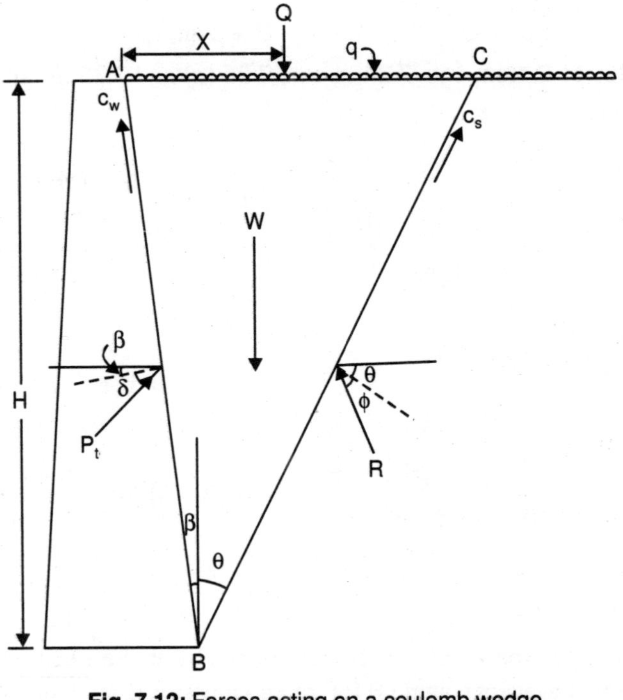

Fig. 7.12: Forces acting on a coulomb wedge

Following forces act on the wedge ABC:

Q : Line load per unit length of wall

q : Uniformly distributed surcharge load per square unit length of wall

c_w : Adhesion for wall-soil interface

c_s : Cohesion between soil particles

R : Reaction acting at an angle of ϕ with the normal to BC

P_t : Total pressure acting on AB acting at an angle with the normal to AB

W : Weight of wedge ABC

(*i*) Balancing all the forces acting on the wedge in the horizontal direction.

$$= 0$$

$$P \cos (+) = R \cos (+) + c_w H \tan - c_s H \tan$$

or,

$$R = \frac{P \cos(\delta + \beta) - c_w \cdot H \tan \beta + c_s \cdot H \tan \theta}{\cos(\theta + \phi)} \qquad \text{...(7.24)}$$

(*ii*) Balancing all the forces acting on the wedge in the vertical direction:

$$\Sigma V = 0$$

$$W + Q + qH (\tan \theta + \tan \beta) = c_w \cdot H \tan \beta - c_s \cdot H + p \sin (\beta + \delta) + R \sin (\theta + \phi)$$

If $H(\tan \theta + \tan \beta) < x$, then

$$W + qH (\tan \theta + \tan \beta) = c_w \cdot H + p \sin (\beta + \delta) + R \sin (\theta + \phi) \qquad \text{...[7.25 (a)]}$$

If $H(\tan \theta + \tan \beta) \geq x$, then

$$W + qH (\tan \theta + \tan \beta) + Q = c_w \cdot H + c_s \cdot H + p \sin (\beta + \delta) + R \sin (\theta + \phi) \qquad \text{...[7.25 (b)]}$$

Substituting the value of R from eqn. 7.24 in eqn. 7.25 (*a*) and 7.25 (*b*) and solving it, we get

(*a*) If H $(\tan \theta + \tan \beta) < x$, then

$$\frac{P_t}{1/2\,\gamma H^2} = \frac{\left[(\tan\theta + \tan\beta)\cdot\left(1 + \frac{2q}{\gamma H}\right)\right]\cdot\cos(\theta+\phi)}{\sin(\beta+\delta+\theta+\phi)} \cdot \frac{-2(c_w/\gamma H)\sec\beta\cdot\cos(\theta+\phi+\beta) - 2(c_s/\gamma H)\sec\theta\cdot\cos\phi}{\sin(\beta+\delta+\theta+\phi)} \qquad \text{...[7.26 (a)]}$$

(*b*) If H $(\tan \theta + \tan \beta) \geq x$, then

$$\left[\frac{P_t}{1/2\,\gamma H^2} = \frac{\left[(\tan\theta + \tan\beta)\cdot\left[1 + \frac{2q}{\gamma H}\right] + \frac{2Q}{\gamma H^2}\right]}{\sin(\beta+\delta+\theta+\phi)}\right]\cdot\cos(\theta+\phi)\cdot\frac{-2(c_w/\gamma H)\sec\beta\cdot\cos(\theta+\phi+\beta) - 2(c_s/\gamma H)\sec\theta\cdot\cos\phi}{\sin(\beta+\delta+\theta+\phi)} \qquad \text{...[7.26 (b)]}$$

The total pressure acting on the wedge can be determined by above equations [7.26 (*a*) and 7.26 (*b*)] by optimising it with respect to wedge angle θ for given values of β, ϕ, θ, Q and x/H.

7.6.3 Point of action of resultant pressure

Total earth pressure P_t consists of two parts, e.g., (*i*) total earth pressure (P) due to backfill and uniformly distributed surcharge q (as obtained earlier) and (*ii*) earth pressure (P_1) due to line load only.

The magnitudes of (P) and its corresponding moment (M) can be obtained from Figs. (7.4 to 7.7) and Figs. (7.8 to 7.11) respectively or L/H = 0.8. The value of P_l can therefore be obtained by subtracting the value of P and P_t (i.e., $P_l = P_t - P$).

To determine the point of application of total earth pressure (P_t), it is assumed that the earth pressure due to line load only (P_l) acts at the location suggested by Spangler (1973). According to him, if a line load of intensity Q and length of y_1, is acting at a distance x from the wall (Fig. 7.13), then the horizontal unit pressure on the wall at any depth z is given by

$$P_s = Q\int_0^{y_1} \frac{x^2 z}{R^5}\,dy \qquad ...(7.27)$$

where, $R = \sqrt{x^2 + y^2 + z^2}$

Solution of eqn. (7.37) is given below.

Put $x^2 + y^2 + a^2$

Thus, $$p_s = Q\int_0^{y_1} \frac{x^2 z}{\left(a^2 + y^2\right)^{5/2}}\,dy$$

Put $y = a\tan\theta,\ dy = a\sec^2\theta \cdot d\theta$ ...(7.28)

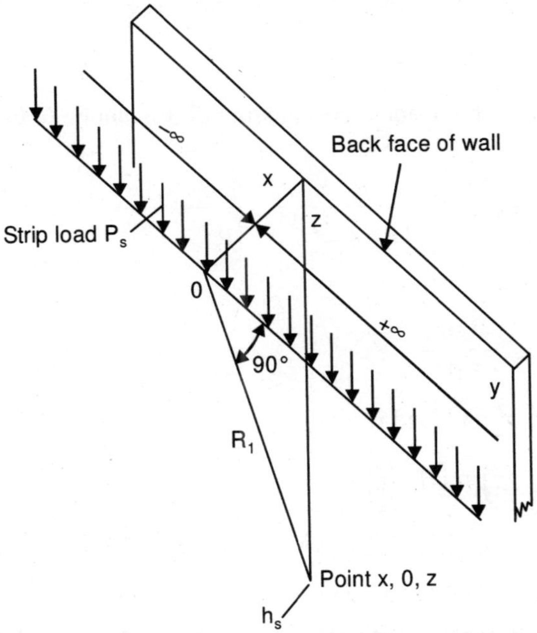

Fig. 7.13: Strip load of infinite length parallel to wall (After Spangler, 1973)

$$p_s = Q\int \frac{x^2 z\, a \sec^2\theta\, d\theta}{a^5\left(\sec^5\theta\right)} = \frac{Qx^2 z}{a^4}\int \cos^3\theta\, d\theta$$

$$= \frac{Qx^2 z}{a^4}\int\left[\cos\theta\, d\theta - \cos\theta\cdot\sin^2\theta\cdot d\theta\right]$$

Put $\sin\theta = t$

$$\cos\theta\cdot d\theta = dt$$

$$p_s = \frac{Qx^2 z}{a^4}\left[\frac{y}{\sqrt{a^2+y^2}} - \frac{1}{3}\frac{y^3}{(a^2+y^2)^{3/2}}\right]_0^{y_1} \quad ...(7.29)$$

[by substituting the value of sin θ and cos θ from eqn. (7.38)]

Now, substituting the value of $a^2 = x^2 + z^2$ in eqn. (7.29)

$$p_s = \frac{Qx^2 z}{\left(x^2+z^2\right)^2}\left[\frac{y}{\sqrt{x^2+y^2+z^2}} - \frac{y^3}{3\left[x^2+y^2+z^2\right]_0^{3/2}}\right]^{y_1}$$

On solving it further,

$$p_s = \frac{Qx^2 z}{3_{R_1}^4}\left[\frac{R^2 y_1}{\left[R_1^2+y_1^2\right]^{1.5}} - \frac{2y_1}{\left[R_1^2+y_1^2\right]^{0.5}}\right] \quad ...(7.30)$$

where, $R_1 = \sqrt{x^2+z^2}$.

If wall is infinitely long on both sides, the equation (7.30) can be written as below:

$$p_s = \frac{Qx^2 z}{3_{R_1}^4}\left[\frac{R^2 y_1}{\left[R_1^2+y_1^2\right]^{1.5}} - \frac{2y_1}{\left[R_1^2+y_1^2\right]^{0.5}}\right]_{-\infty}^{+\infty}$$

which yields,

$$p_s = \frac{4Qx^2 z}{3R_1^4} \quad ...(7.31)$$

The point of application of line load is given by

$$H_1 = \frac{\int p_s[H-z]\sec\beta\cdot dz}{\int p_s\cdot\sec\beta\cdot dz} \quad ...(7.32)$$

Substituting the value of p_s from eqn. (7.31) in eqn. (7.32) and solving it like an differential equation.

We get, the height of point of application (H_t) of total pressure P_t as follows, which is obtained using the following equation:

$$H_t = \frac{M + P_1\cdot\cos\delta\cdot\sec\beta}{P_t\cos\delta\sec\beta} \quad ...(7.33)$$

The moment of total earth pressure p_t about the heel has been designated as M_t and is given by

$$M_t = P_t \cdot H_t \cdot \cos\delta \cdot \sec\beta \qquad ...(7.34)$$

The total earth pressure P_t and the corresponding moment M_t have been presented in the charts in non-dimensional form as $\frac{p_t}{(1/2)\gamma H^2}$ and $\frac{M_t}{(1/6)\gamma H^3}$ vide Figs. (7.14) to (7.29).

7.6.4 Parametric study

Computation of total earth pressure and moment have been done for the practical range of parameters given in Table 7.3.

Table 7.3: Parameters studied (Unreinforced backfill)

Parameter	*Range*	*Interval*
ϕ	20°–35°	5°
δ	2/3 ϕ	–
β	0°, 10°	–
$q/\gamma H$	0–0.5	0.25
$Q/\gamma H^2$	0–0.5	0.1
x/H	0–1.6	0.2
c_s, c_w	0.0–0.10	0.05

7.6.5 Empirical approach for reinforced cohesive frictional backfill having line load surcharge

The analysis of obtaining the total earth pressure and the corresponding moment for walls having cohesive frictional backfill with uniformly distributed surcharge load has been discussed in section 7.2. For determination of total earth pressure and moment for the walls having reinforced cohesive-frictional backfill with line load surcharge, an empirical approach is suggested below. Following steps are adopted for the same:

(*a*) Firstly the backfill is considered unreinforced. The pressure P_t for a given line load can either be determined by using charts Figs. 7.14 (*a*) to Fig. 7.29 (*a*).

(*b*) Equivalent value of uniformly distributed surcharge was then obtained using Figs. 7.14 (*b*) to Fig. 7.29 (*b*) so that the earth pressure becomes equal to P_t for L/H = 0. For convenience, curves of $\frac{p_t}{(1/2)\gamma H^2}$ *vs.* $\frac{q_e}{\gamma H}$ have been plotted in Fig. 7.30 (*a*) which represent values of each pressure for L/H = 0. This equivalent uniformly distributed surcharge q_e will be used for checking factor of safety against sliding.

(*c*) The value of equivalent surcharge q_e are obtained by equalising the moment M_t and M for L/H = 0, same as discussed in (*a*) and (*b*) above. Thereafter curves of $\frac{M_t}{(1/6)\gamma H^3}$ *vs.* $\frac{q_e}{\gamma H}$ are plotted as shown in Fig. 7.30 (*b*) for L/H = 0. This equivalent q_e will be used for checking of safety against overturning.

The details of these steps are provided in the design example, discussed below.

Design Example

Design an 8.0 m high retaining wall with inclined back retaining horizontal backfill carrying a uniformly distributed surcharge load of 64 kN/m^2 and a line load surcharge of 102.4 kN/m at a distance of 3.0 m from wall. Inclination of wall back is 10° from vertical. Other data are as under.

(*i*) density of backfill soil = 16 kN/m^3
(*ii*) angle of internal friction (ϕ) = 25°
(*iii*) angle of wall friction (δ) = 17°
(*iv*) cohesion between soil particles = 6.4 kN/m^2
(*v*) coefficient of sliding for fondation soil interface (μ) = 0.50
(*vi*) allowable soil bearing pressure = 300 kN/m^2
(*vii*) factor of safety against sliding = 1.5
(*viii*) factor of safety against overturning = 2.0
(*ix*) unit weight of masonry = 22 kN/m^3

[A] Wall with unreinforced backfill

Solution (per metre length of wall)

Given data:

$$H = 8.00 \text{ m}, q = 64 \text{ kN/m}^2, Q = 102.4 \text{ kN/m}$$
$$x = 3.0 \text{ m}, \beta = 10°, \gamma = 16 \text{ kN/m}^3, \phi = 25°$$
$$\delta = 17°, c = 6.4 \text{ kN/m}^2, \mu = 0.5, \sigma_a = 300 \text{ kNm}^2$$

Step 1:

$$\frac{q}{\gamma H} = \frac{64}{(16 \times 8)} = 0.5$$

$$\frac{x}{H} = \frac{3.0}{8} = 0.375$$

$$\frac{c}{\gamma H} = \frac{6.4}{(16 \times 8)} = 0.05$$

$$\frac{Q}{\gamma H^2} = \frac{102.4}{(16 \times 8 \times 8)} = 0.10$$

Step 2:

A conventional masonry retaining wall with unreinforced backfill will be designed as follows:

$$\frac{P_t}{\frac{1}{2}\gamma H^2} = 0.841 \text{ [from Fig. 7.23 (a) (iii)]}$$

$$\frac{M_t}{\frac{1}{6}\gamma H^2} = 1.21 \text{ [from Fig. 7.23 (b) (iii)]}$$

and $\dfrac{\overline{H}}{H} = \dfrac{1.21}{3 \times 0.841 \times \cos 17° \times \sec 10°}$ (from eqn. 7.33)

$= 0.493$

Thus, $\overline{H} = 0.493 \times 8 = 3.94$ m

$P_t = 0.841 \times 0.5 \times 16 \times 64 = 430.59$ kN

$P_h = P_t \cos(\delta + \beta)$

$= 430.59 \cos(17° + 10°) = 383.66$ kN

$P_v = P_t \sin(\delta + \beta)$

$= 430.59 \sin(17° + 10°) = 195$ kN

Step 3: Design against sliding failure

Choose section of wall as shown in Fig. 7.31 (*a*)

$$w_1 = \frac{1}{2} \times 1.41 \times 8 \times 22$$

$$= 124.10 \text{ kN}$$

$$w_2 = 3.52 \times 8 \times 22$$

$$= 619.52 \text{ kN}$$

$$w_3 = \frac{1}{2} \times 3.10 \times 8 \times 22$$

Total weight $W = w_1 + w_2 + w_3 = 1016.42$ kN

F.O.S. against sliding $F_s = \dfrac{(W + P_v) \times \mu}{P_h}$

$$= \frac{(1016.42 + 195.48) \times 0.5}{383.66} = 1.57 > 1.50 \quad \text{(O.K.)}$$

Step 4: Design against overturning failure

Moment of resistance about A, (M_{RA}), *refer* Fig. 7.31.

$$M_{RA} = w_3 \times \frac{2}{3} \times 3.10 + w_2 \times 4.86 + w_1 \times \left(6.62 + \frac{1.41}{3}\right) + P_v \times (6.62 + 1.41 - 3.94 \times \cot 80°)$$

$$= 272.8 \times 2.06 + 619.52 \times 4.86 + 124.10 \times 7 \times 0.9 + 195.48 \times 7.335$$

$$= 5886.55 \text{ kN-m/m}$$

Overturning moment about A, (M_{OA})

$$M_{OA} = P_h \times \overline{H}$$

$$= 383.66 \times 3.94 = 1511.62 \text{ kN-m/m}$$

F.O.S. against overturning F_0

$$= \frac{M_{RA}}{M_{OA}}$$

$$= \frac{5886.55}{1511.62} = 3.86 > 2.0 \quad \text{(O.K.)}$$

Step 5: Check against bearing failure

(*a*) Distance $\bar{x}$ of point of application of resultant force at the base from toe of wall,

$$\bar{x} = \frac{M_{RA} - M_{OA}}{w + P_v} = \frac{5886.55 - 1511.62}{1016.42 + 195.48} = 3.62 \text{ m}$$

(*b*) eccentricity, $e = \frac{b}{2} - \bar{x}$

where, b = total base width of wall

$$= \frac{8.03}{2} - 3.62$$

or, $e = 0.398$ m

$$\frac{b}{6} = \frac{8.03}{6} = 1.33 > e \qquad \text{(O.K.)}$$

(*c*) Check against maximum and minimum pressure intensities:

$$\sigma_{max} = \frac{V}{b}\left[1 + \frac{6 \cdot e}{b}\right]$$

where, V = sum of total vertical forces

$$\sigma_{max} = \frac{(1016.42 + 195.48)}{8.03}\left[1 + \frac{6 \times 0.395}{8.03}\right]$$

$$= 195.46 \text{ kN/m}^2 < 300 \qquad \text{(O.K.)}$$

Similarly,

$$\sigma_{min} = \frac{V}{b}\left(1 - \frac{6 \cdot e}{b}\right)$$

$$= \frac{1211.9}{8.03}\left(1 + \frac{6 \times 0.395}{8.03}\right) = 124.42 \text{ kN / m}^2 > 0 \qquad \text{(O.K.)}$$

Hence, the wall section as chosen above is O.K. for unreinforced backfill. The volume of the wall section per metre comes out as 46.20 m^3.

[B] Wall with reinforced backfill

Step 1: Adopting

(*a*) 256 cm wide strips of geogrid (CE-121, having tensile strength of 7.68 kN/m for 20% strain) as reinforcing element.

(*b*) $\frac{L}{H} = 0.7$, thus L = 0.70 × 8 = 5.60 m

Since, $D_p = \frac{wf * H}{S_x S_z}$

Therefore, $S_x S_z = \frac{0.25 \times 0.5 \times 1}{1}$

Keeping $S_x = S_z = 1.0$ m.

Step 2: Design against sliding failure

Firstly the total pressure caused by line load and u.d.l. shall be calculated, considering the backfill as unreinforced. For this purpose, curves of Fig. [7.22 (*a*) (*iii*)] shall be used. Thereafter equivalent surcharge corresponding to this pressure shall be determined using Fig. 7.30 (*a*). Now corresponding to this (equivalent), pressure shall be calculated by linear interpolation.

Further steps are as follows:

(*a*) $\dfrac{P_t}{\frac{1}{2}\gamma H^2} = 0.841$ for unreinforced backfill [from Fig. 7.22 (*a*) (*iii*)]

(*b*) Equivalent surcharge $\dfrac{q_e}{\gamma H}\left[\text{for } \dfrac{P_t}{\frac{1}{2}\gamma H^2} = 0.841\right] = 0.75$ [from Fig. 7.30 (*a*)]

(*c*) $\dfrac{P}{\frac{1}{2}\gamma H^2} = 0.486$ for reinforced backfill for $\dfrac{c}{\gamma H} = 0.05,\ \dfrac{L}{H} = 0.7,$

$D_p = 1.0$ and $\dfrac{q}{yH} = 0.75$ (by linear interpolation)

(*d*)

$$P = 0.486 \times 0.5 \times 16 \times 4 = 248.83 \text{ kN}$$
$$P_h = 218.11 \times \cos(17° + 10°) = 221.71 \text{ kN}$$
$$P_v = 218.11 \times \sin(17° + 10°) = 112.96 \text{ kN}$$

Choose the wall section as shown in Fig. 7.31 (*b*).

$$w_1 = \frac{1}{2} \times 1.41 \times 8 \times 22 = 124.08 \text{ kN}$$
$$w_2 = 1.40 \times 8 \times 22 = 246.40 \text{ kN}$$
$$w_3 = \frac{1}{2} \times 2.50 \times 8 \times 22 = 220.00 \text{ kN}$$

Total weight $W = w_1 + w_2 + w_3 = 590.48$ kN

F.O.S. against sliding $F_s = \dfrac{(W + P_v) \times \mu}{P_h}$

$$= \frac{(590.48 + 112.96) \times 0.5}{221.71} = 1.58 > 1.50 \qquad \text{(O.K.)}$$

Step 3: Design against overturning failure

(*a*) $\dfrac{M_t}{\frac{1}{6}\gamma H^3} = 1.21$ for reinforced backfill [from Fig. 7.22 (*b*) (*iii*)]

(*b*) Equivalent u.d.l. $q_e/\gamma H = 0.7$ [from Fig. 7.30 (*b*)]

(*c*) $\dfrac{M_t}{\frac{1}{6}\gamma H^3} = 0.654$ for reinforced backfill for $c/\gamma H = 0.05$, $L/H = 0.7$

$D_p = 1.0$ and $q/\gamma H = 0.7$ (by linear interpolation)

Similarly,

(*d*) $$\frac{P}{\frac{1}{2}\gamma H^2} = 0.325 \quad \text{(from set of Fig. 7.5)}$$

(*e*) $$\frac{\overline{H}}{H} = \frac{0.654}{3 \times 0.325 \cos 17° \times \sec 10°} = 0.69$$

Thus, $$\overline{H} = 0.69 \times 8 = 5.52 \text{ m}$$

$$P = 0.325 \times 0.5 \times 16 \times 64 = 166.4 \text{ kN}$$

$$P_h = 166.4 \times \cos (17° + 10°) = 148.26 \text{ kN}$$

$$P_v = 166.4 \times \sin (17° + 10°) = 75.54 \text{ kN}$$

F.O.S. of wall section against overturning shall be determined as follows:

Overturning moment about A, M_{OA}

$$M_{OA} = P_h \times \overline{H} = 148.26 \times 5.52 = 819.29 \text{ kN-m/m}$$

Moment of resistance about A, M_{RA}

$$= W_3 \times \frac{2}{3} \times 2.50 + w_2 \times 3.20 + w_1 \times \left(3.90 \times \frac{1.41}{3}\right) + 75.54 \times (3.90 + 1.41 - 5.52 \cot 80°)$$

$$= 220 \times \frac{5}{3} + 246.40 \times 3.20 + 124.08 \times \left(3.90 + \frac{1.41}{3}\right) + 75.54 \times 4.33$$

or, $M_{RA} = 2024.46$ kN-m/m

F.O.S. against overturning F_0

$$= \frac{M_{RA}}{M_{OA}} = \frac{2024.46}{819.29} = 2.47 > 2.0 \qquad \text{(O.K.)}$$

Step 4: Check against bearing failure

(*a*) Distance $\bar{x}$ of point of application of resultant force at the base from toe,

$$\bar{x} = \frac{2024.46 - 819.29}{590.48 + 75.54} = 1.80$$

(*b*) Eccentricity, $$e = \frac{b}{2} - \bar{x} = \frac{5.31}{2} - 1.80 = 0.85$$

$$\frac{b}{6} = \frac{5.31}{6} = 0.88 > e \qquad \text{(O.K.)}$$

(*c*) $$\sigma_{max} = \frac{V}{b}\left(1 + \frac{6 \cdot e}{b}\right) = \frac{590.48 + 75.54}{5.31}\left(1 + \frac{6 \times 0.85}{5.31}\right)$$

$$\sigma_{min} = \frac{V}{b}\left(1 - \frac{6 \cdot e}{b}\right) = \frac{666.02}{5.31}\left(1 - \frac{6 \times 0.85}{5.31}\right)$$

$$= 4.96 \text{ kN/m}^2 > 0.0 \qquad \text{(O.K.)}$$

Thus, a wall section with 5.31 m base width and 1.40 m top width can be used for the given loading conditions. The section of the wall is shown in Fig. 7.31 (*b*). The volume of wall per metre length comes out as 26.84 m^3.

It is clearly evident that there is a considerable amount of saving of horizontal space on ground and thereby the overall volume of wall per m length. Even after adding of cost of geogrid, the overall cost of wall comes out as 60% of conventional gravity wall, besides saving in precious ground space. Hence this wall should be constructed where skilled labours are not available and ground space is also less.

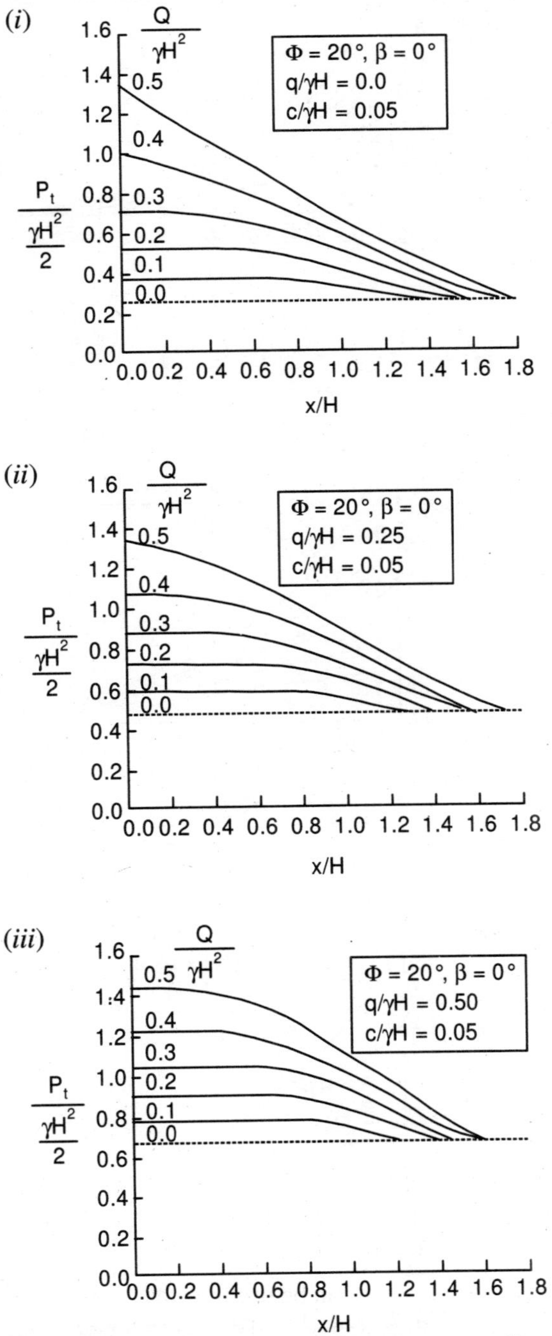

Fig. 7.14: (a) Resultant earth pressure *vs.* x/H

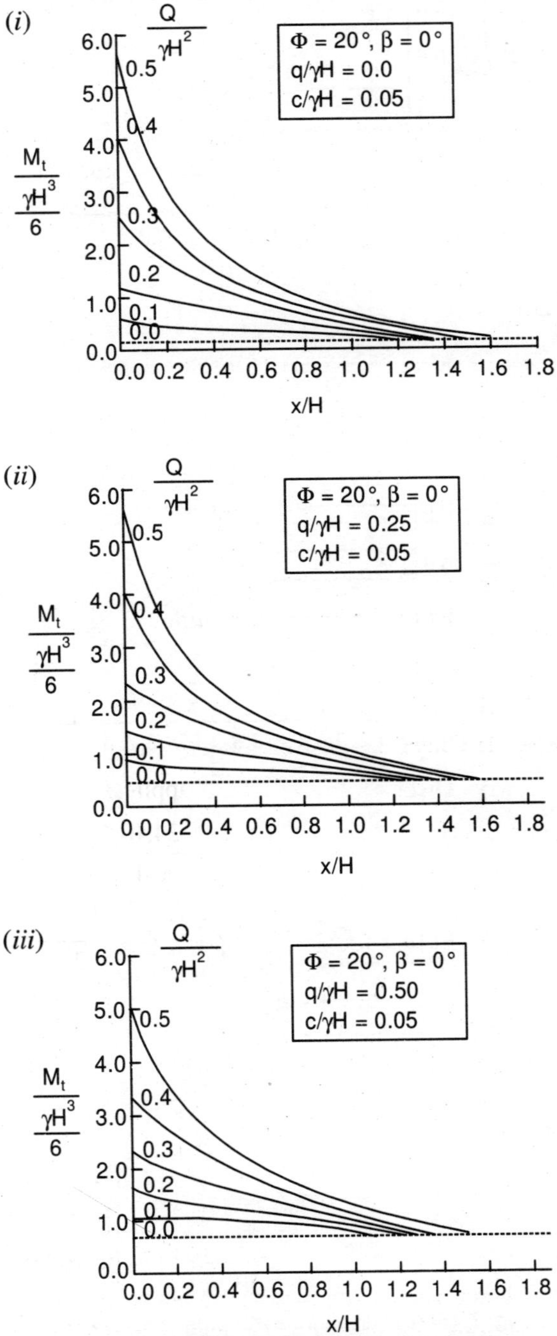

Fig. 7.14: (b) Resultant moment *vs.* x/H

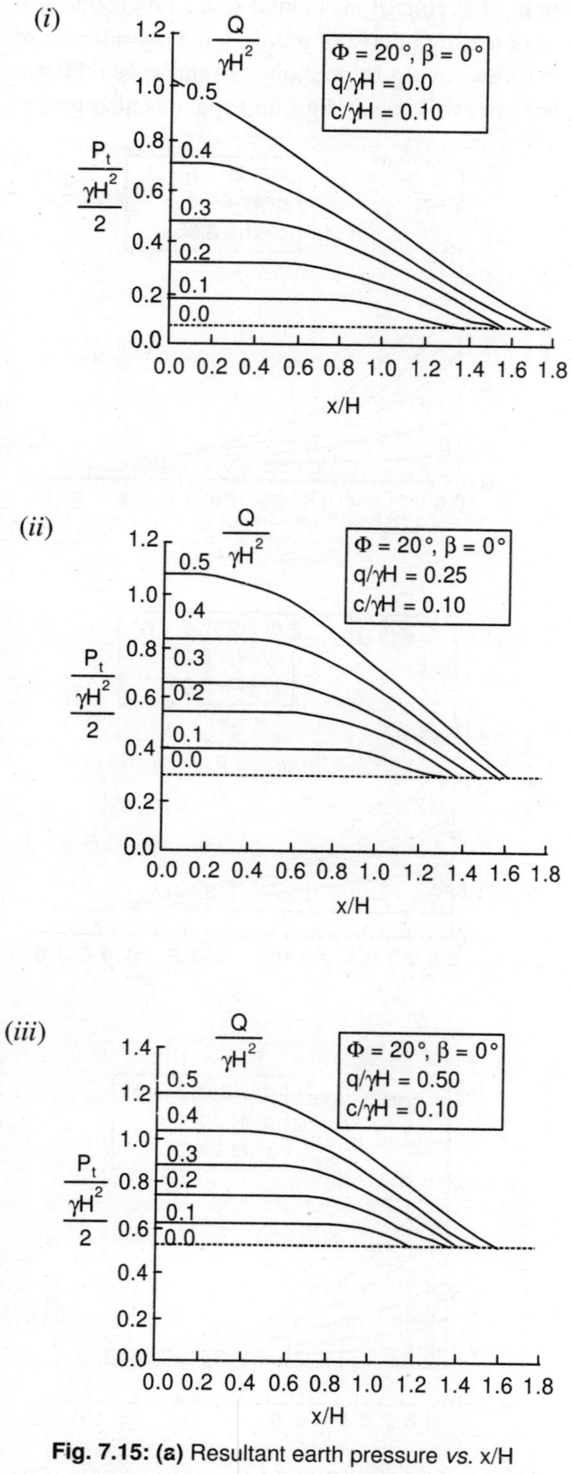

Fig. 7.15: (a) Resultant earth pressure *vs.* x/H

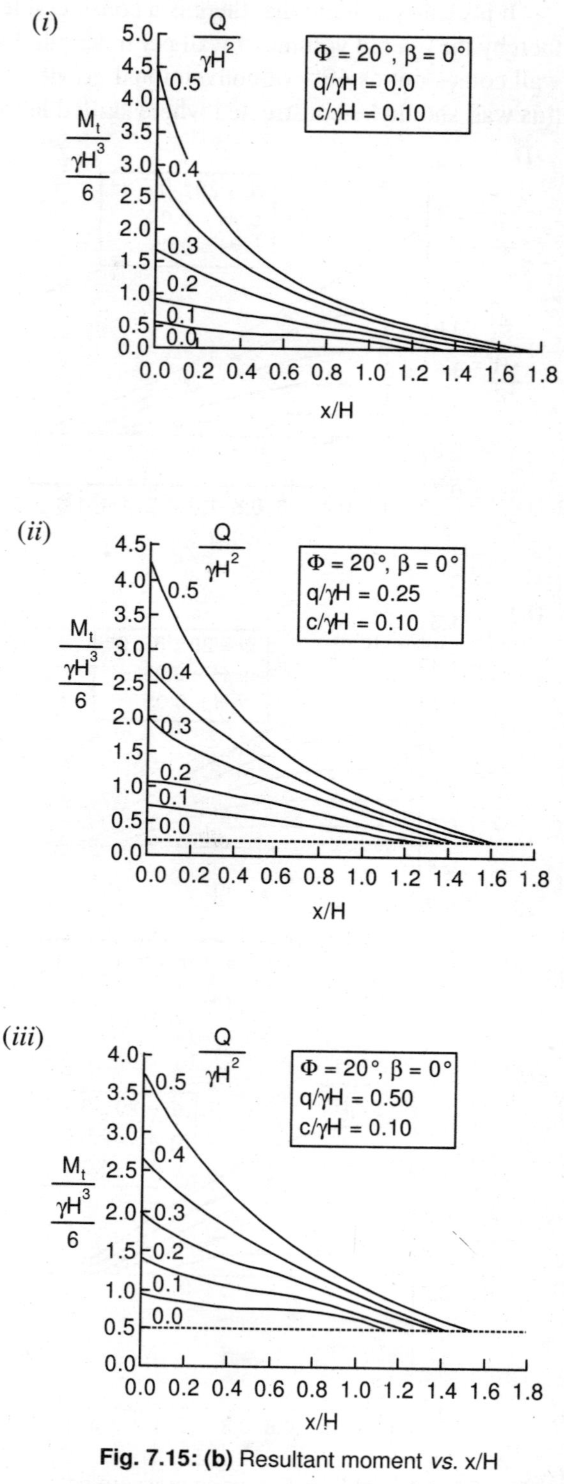

Fig. 7.15: (b) Resultant moment *vs.* x/H

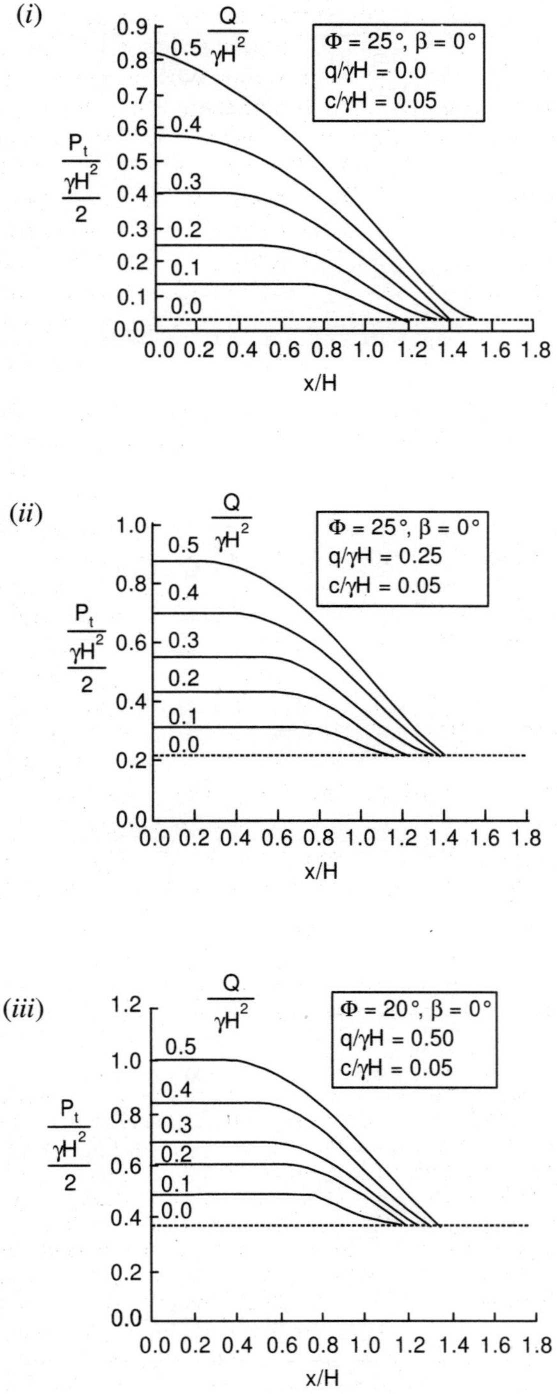

Fig. 7.16: (a) Resultant earth pressure *vs.* x/H

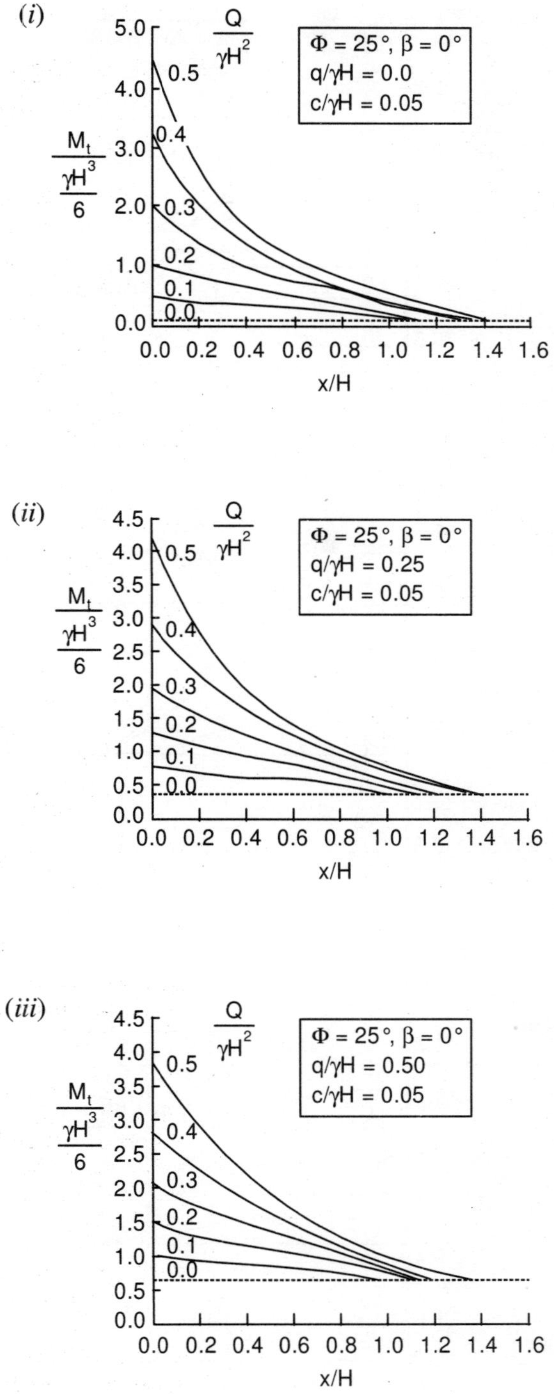

Fig. 7.16: (b) Resultant moment *vs.* x/H

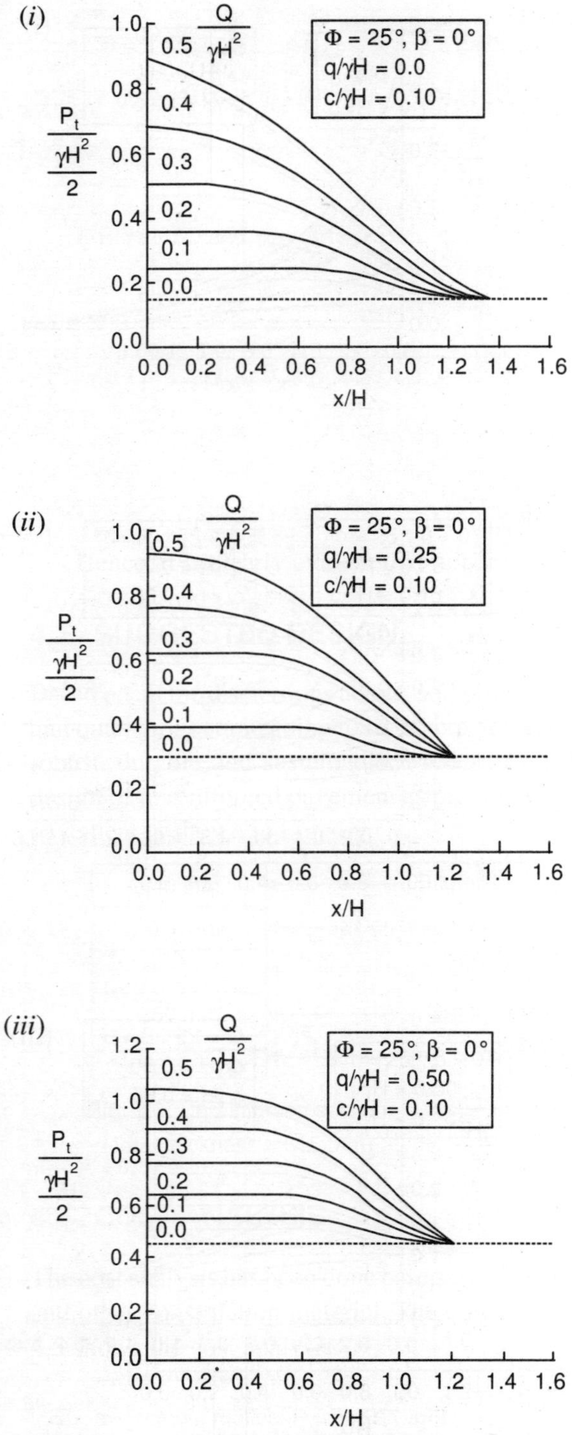

Fig. 7.17: (a) Resultant earth pressure *vs.* x/H

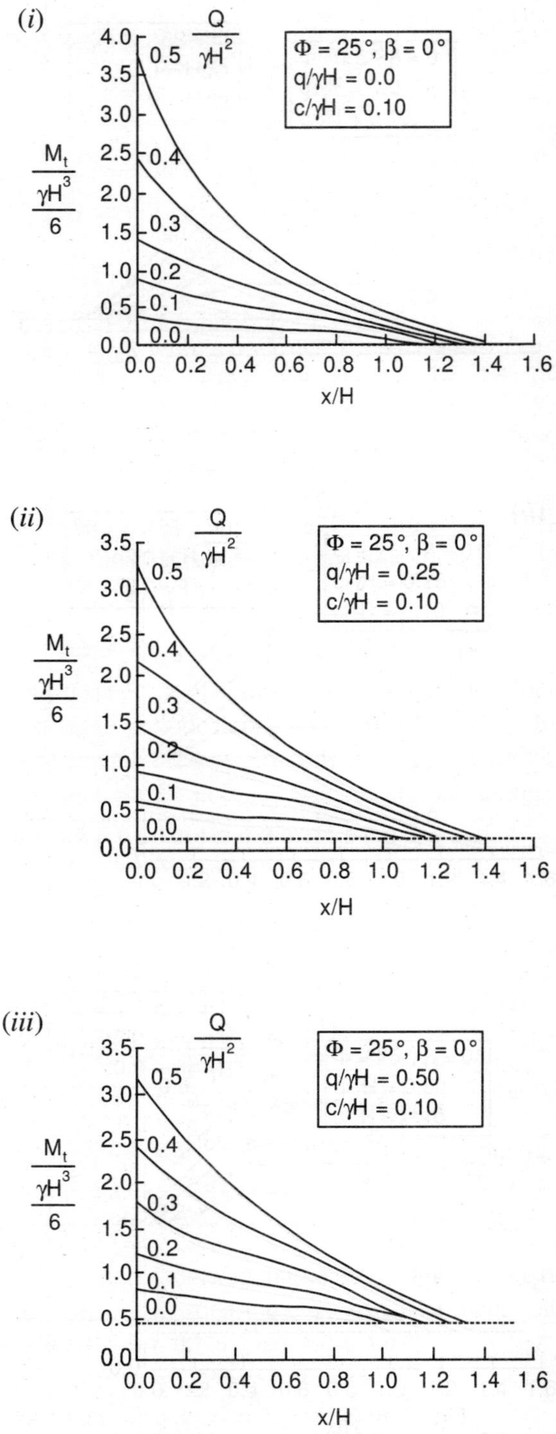

Fig. 7.17: (b) Resultant moment *vs.* x/H

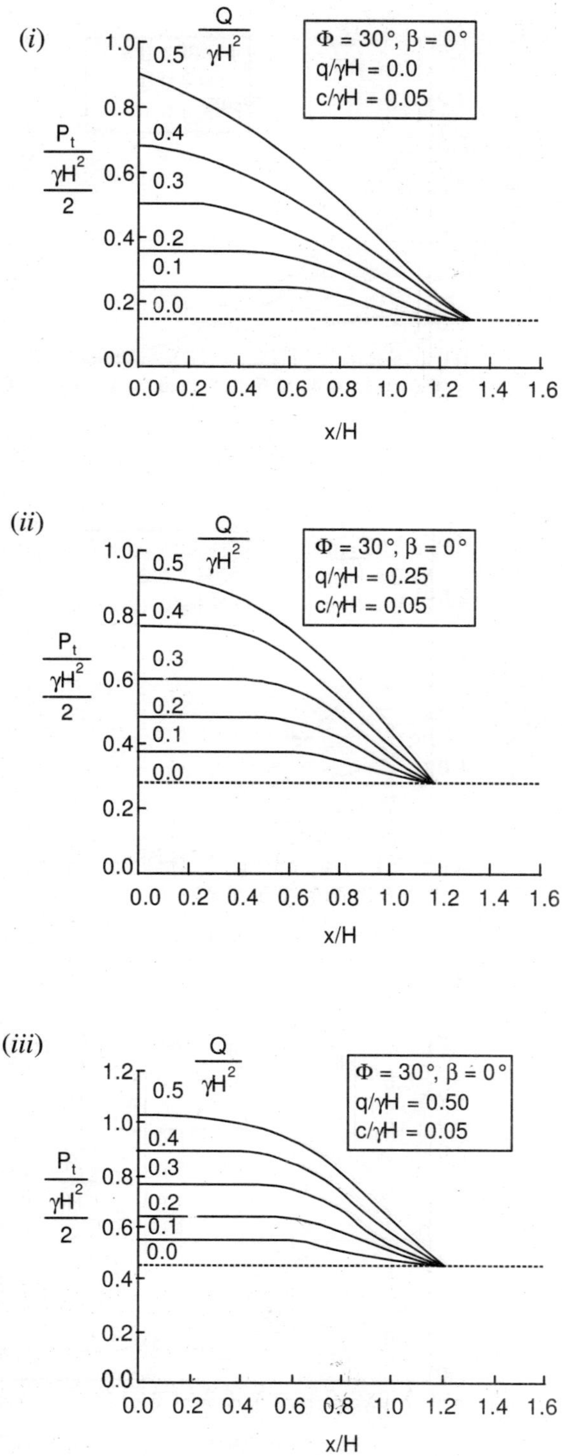

Fig. 7.18: (a) Resultant earth pressure *vs.* x/H

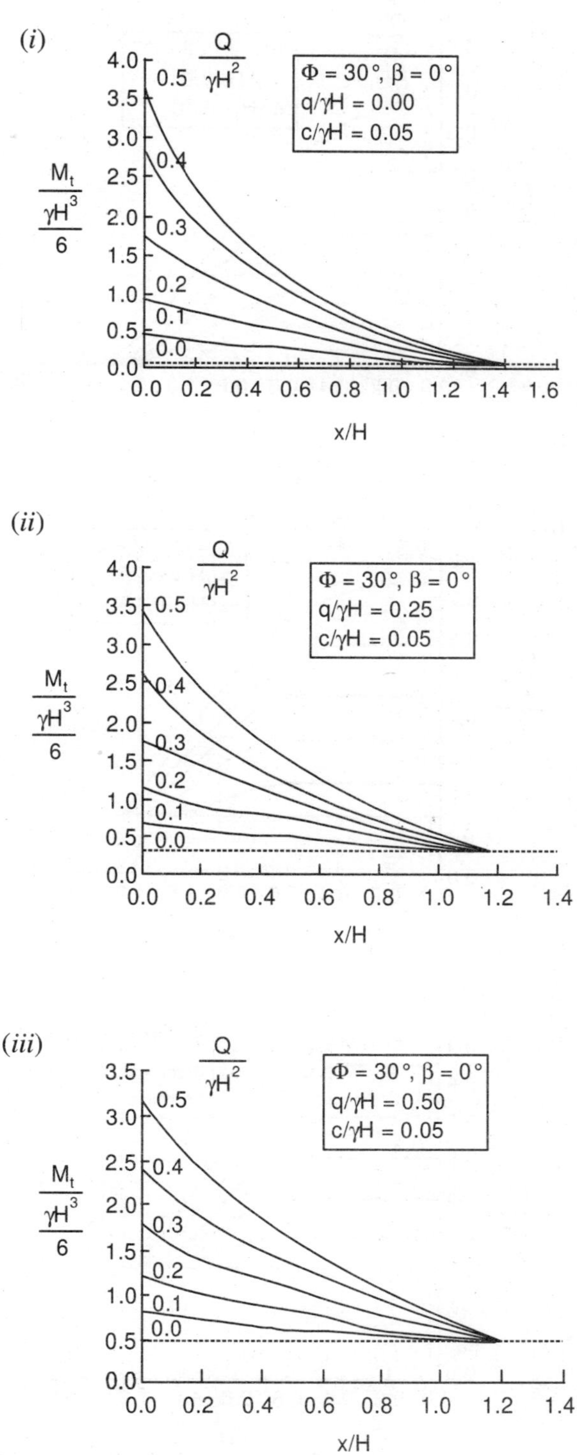

Fig. 7.18: (b) Resultant moment *vs.* x/H

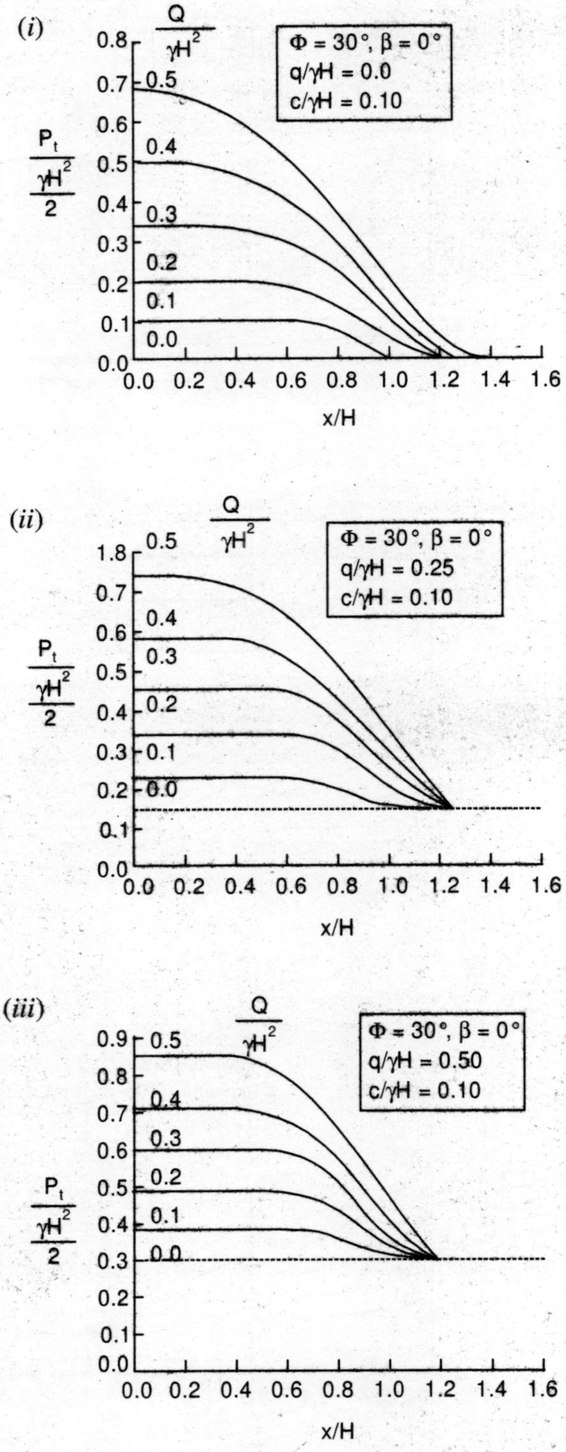

Fig. 7.19: (a) Resultant earth pressure *vs.* x/H

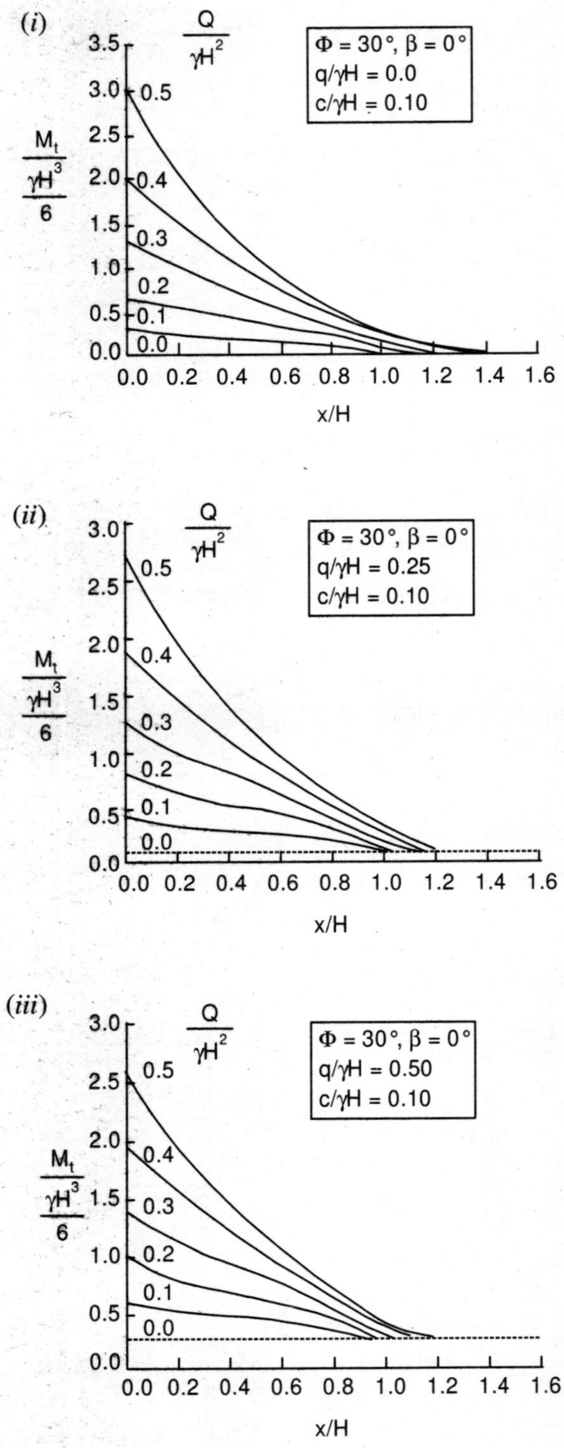

Fig. 7.19: (b) Resultant moment *vs.* x/H

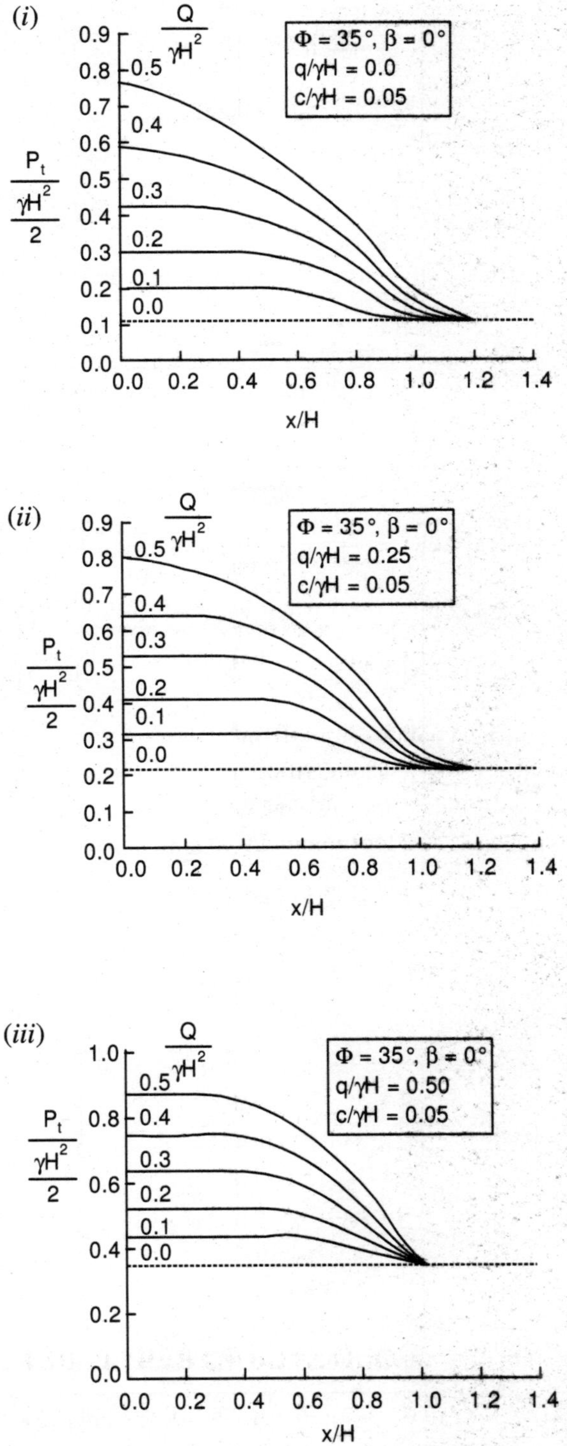

Fig. 7.20: (a) Resultant earth pressure *vs* x/H

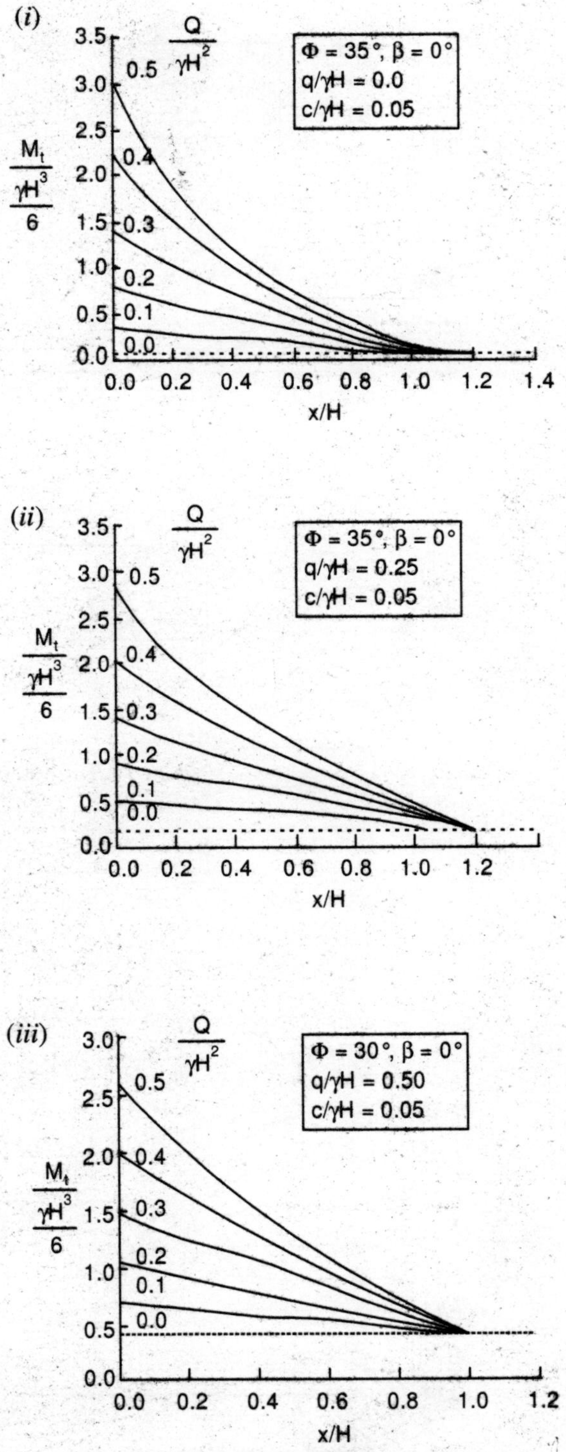

Fig. 7.20: (b) Resultant moment *vs.* x/H

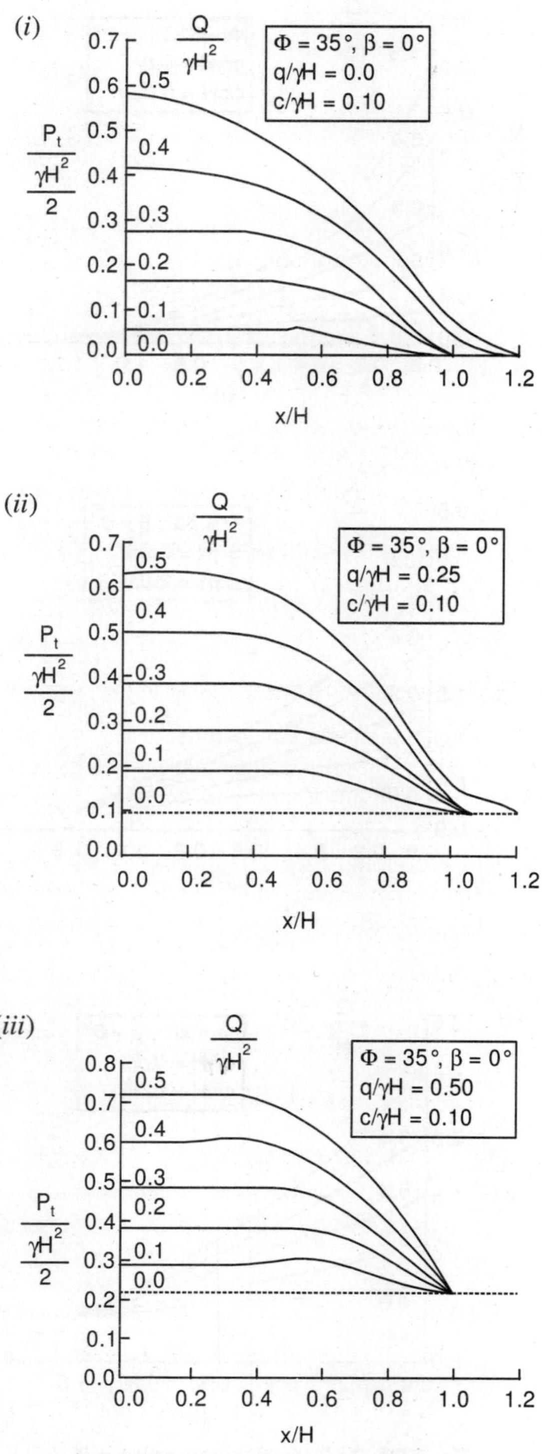

Fig. 7.21: (a) Resultant earth pressure *vs.* x/H

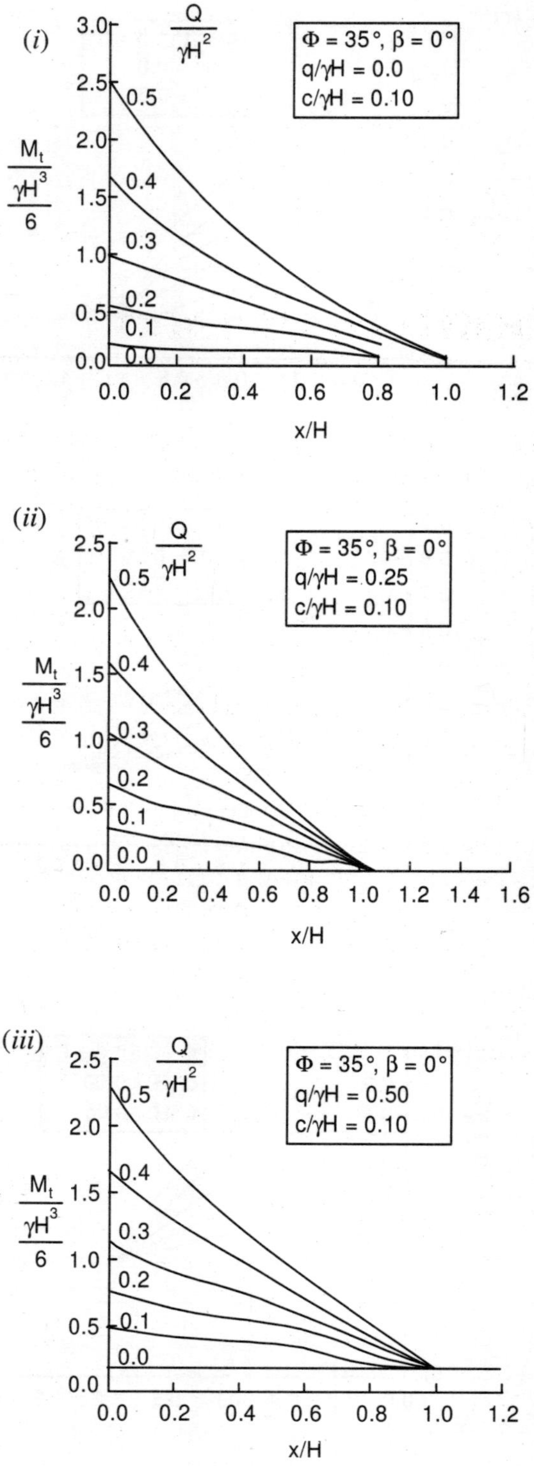

Fig. 7.21: (b) Resultant moment *vs.* x/H

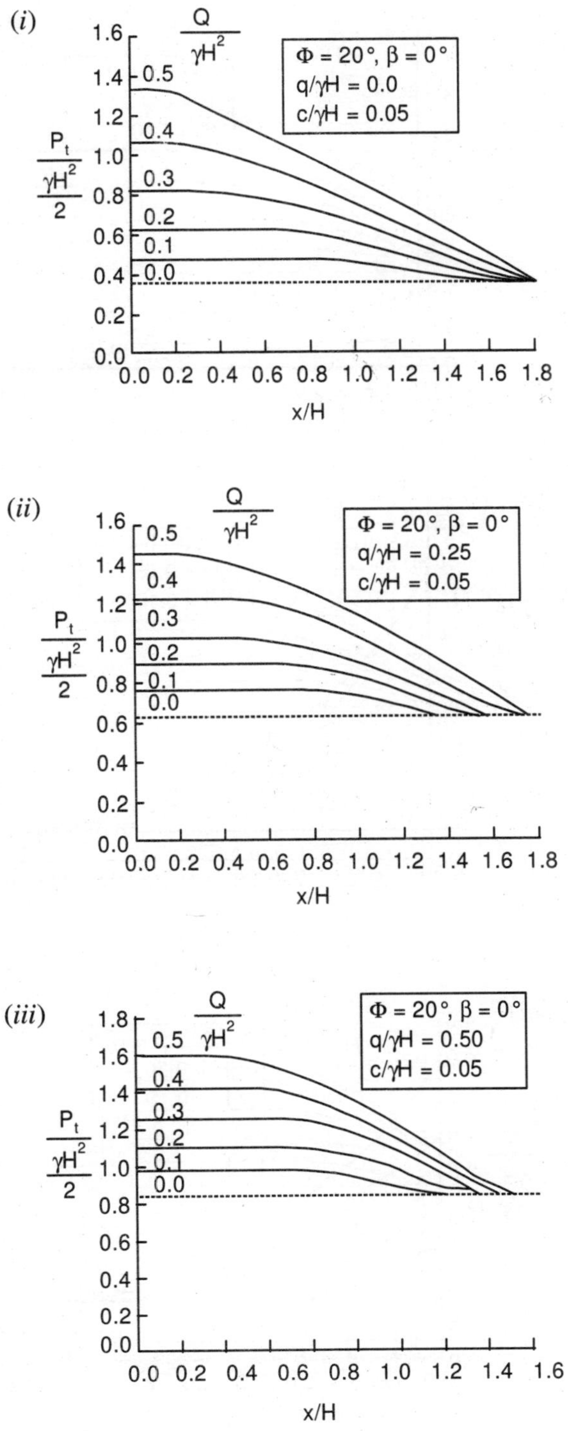

Fig. 7.22: (a) Resultant earth pressure *vs.* x/H

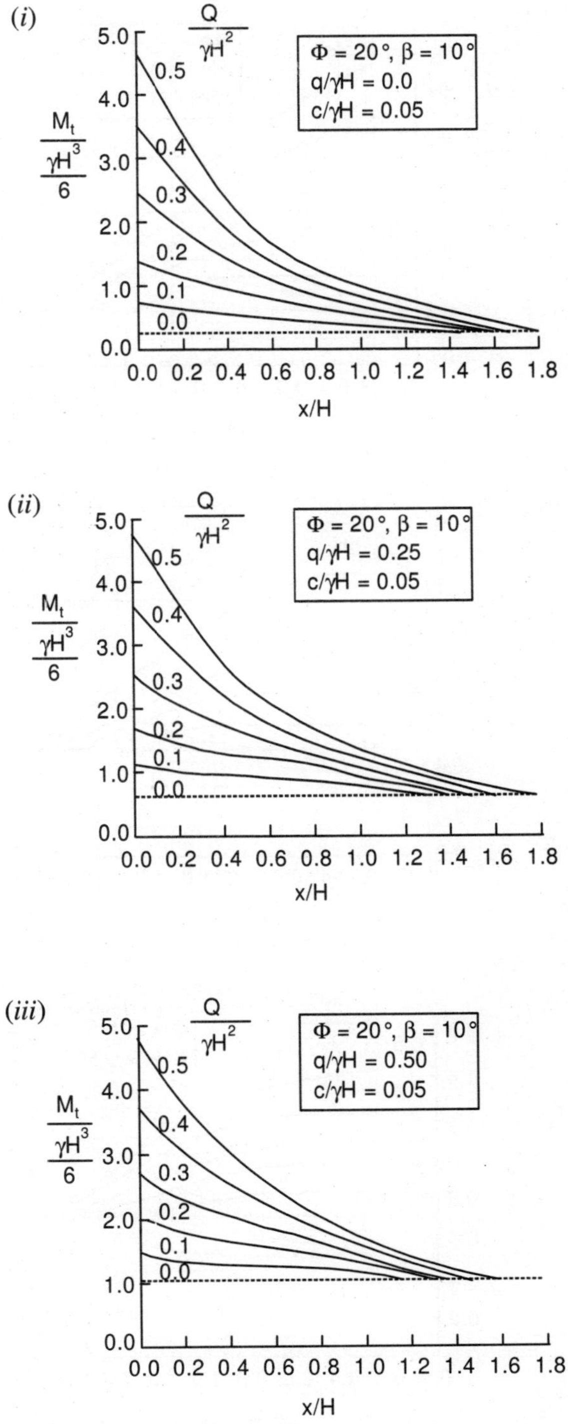

Fig. 7.22: (b) Resultant moment *vs.* x/H

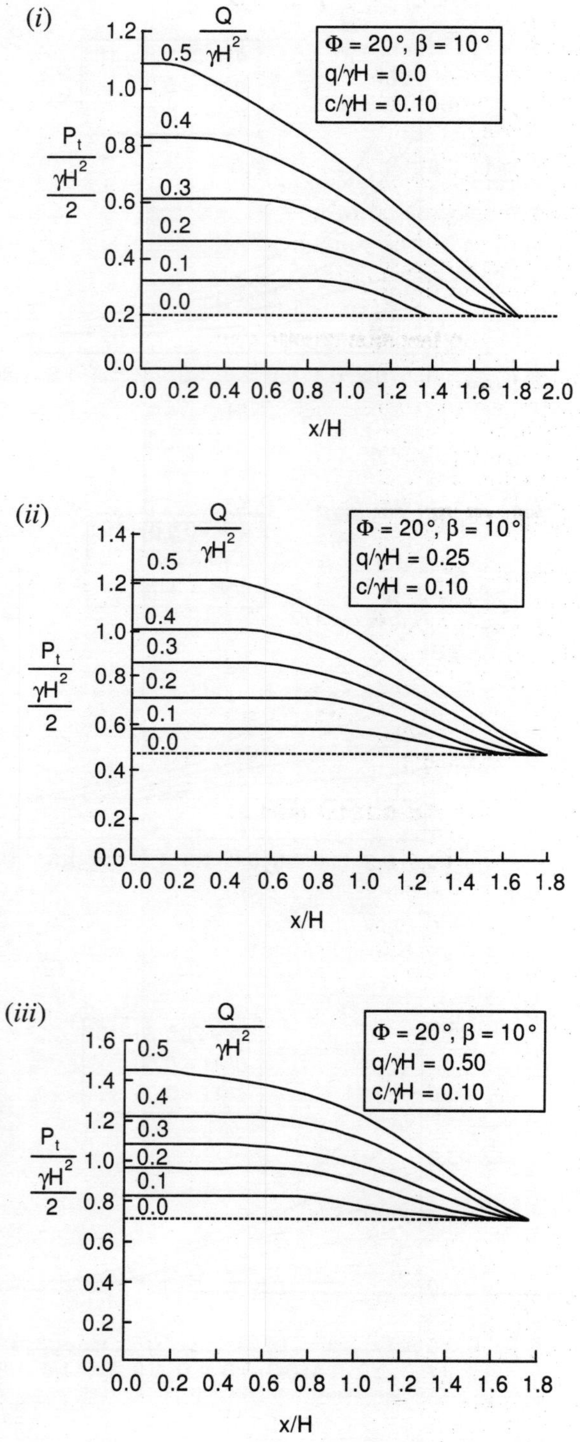

Fig. 7.23: (a) Resultant earth pressure *vs.* x/H

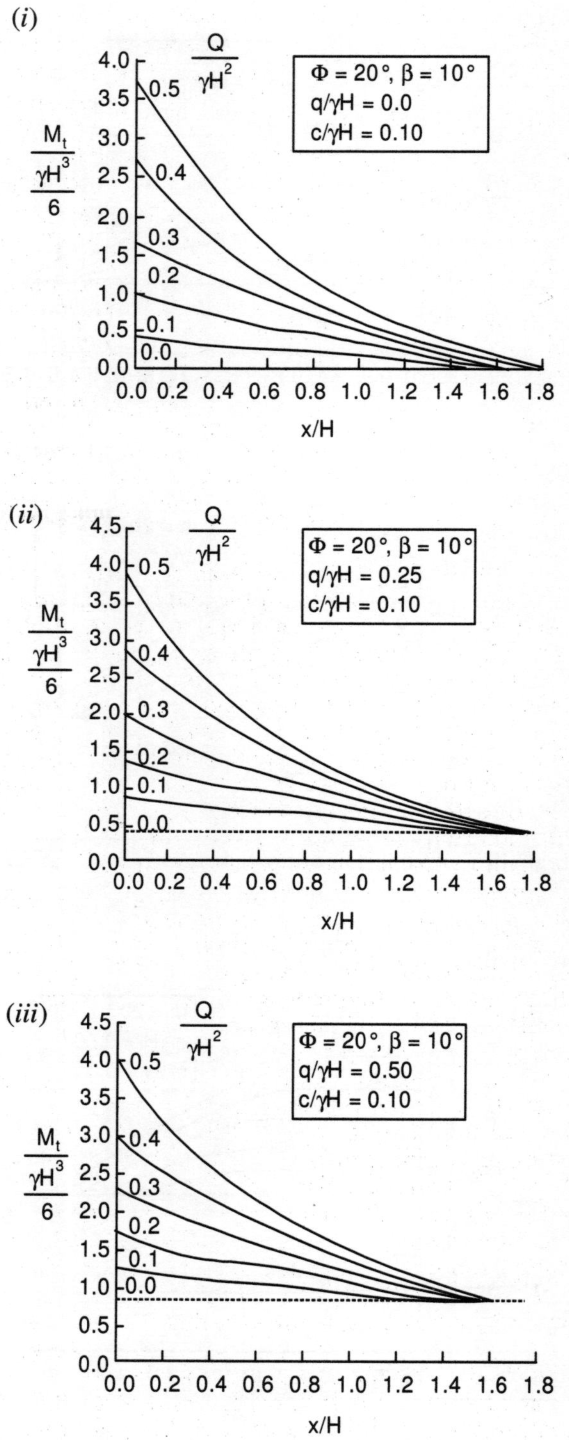

Fig. 7.23: (b) Resultant moment *vs.* x/H

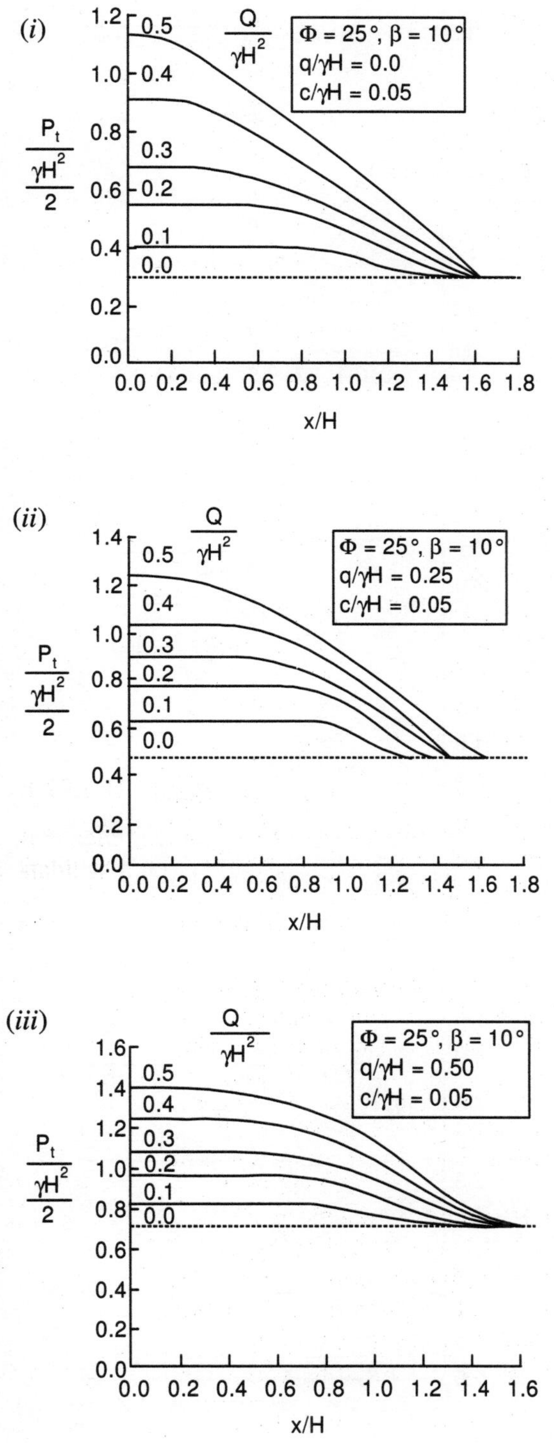

Fig. 7.24: (a) Resultant earth pressure *vs. x*/H

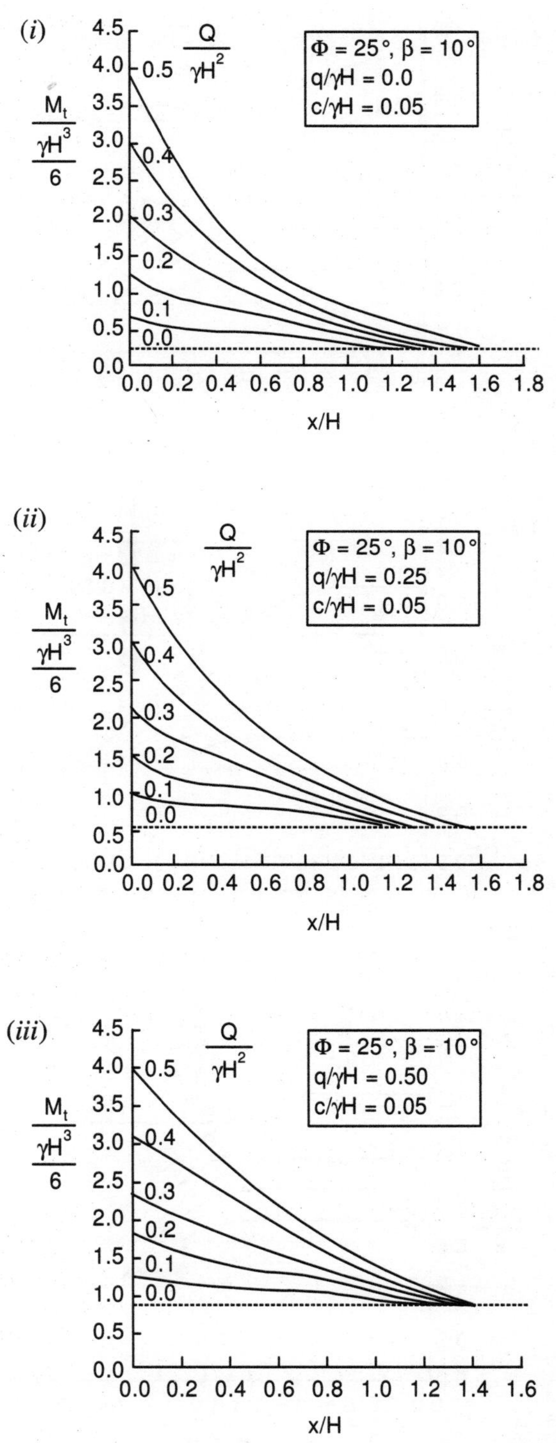

Fig. 7.24: (b) Resultant moment *vs. x*/H

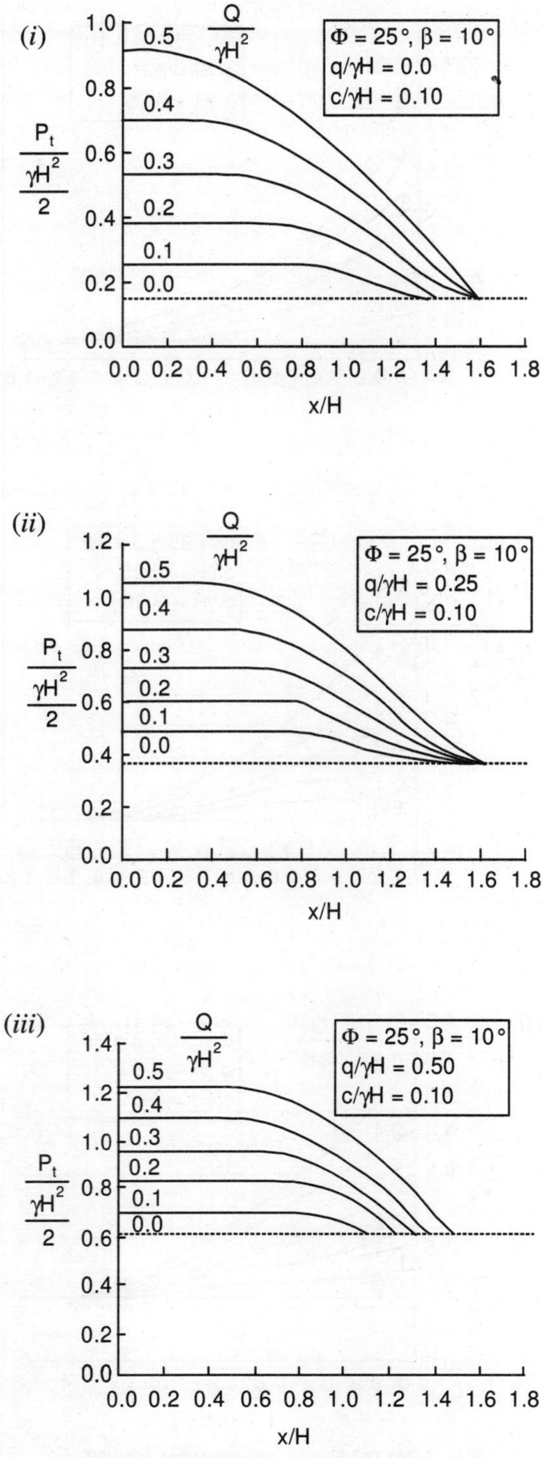

Fig. 7.25: (a) Resultant earth pressure *vs.* x/H

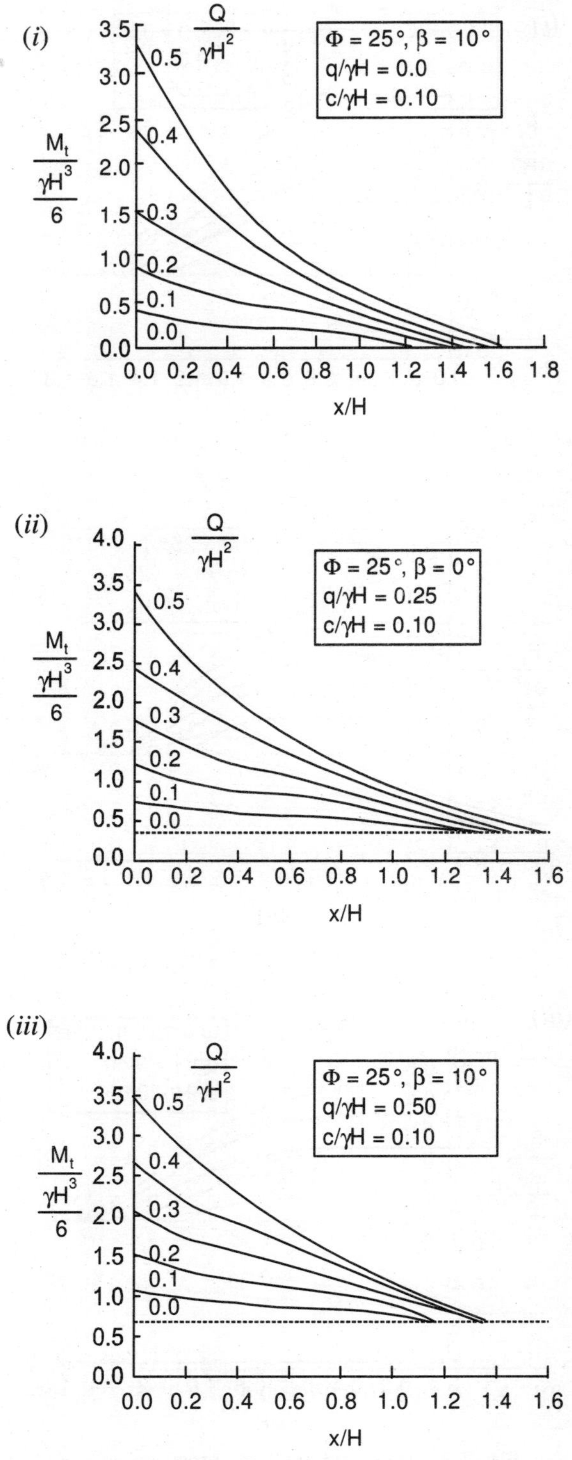

Fig. 7.25: (b) Resultant moment *vs.* x/H

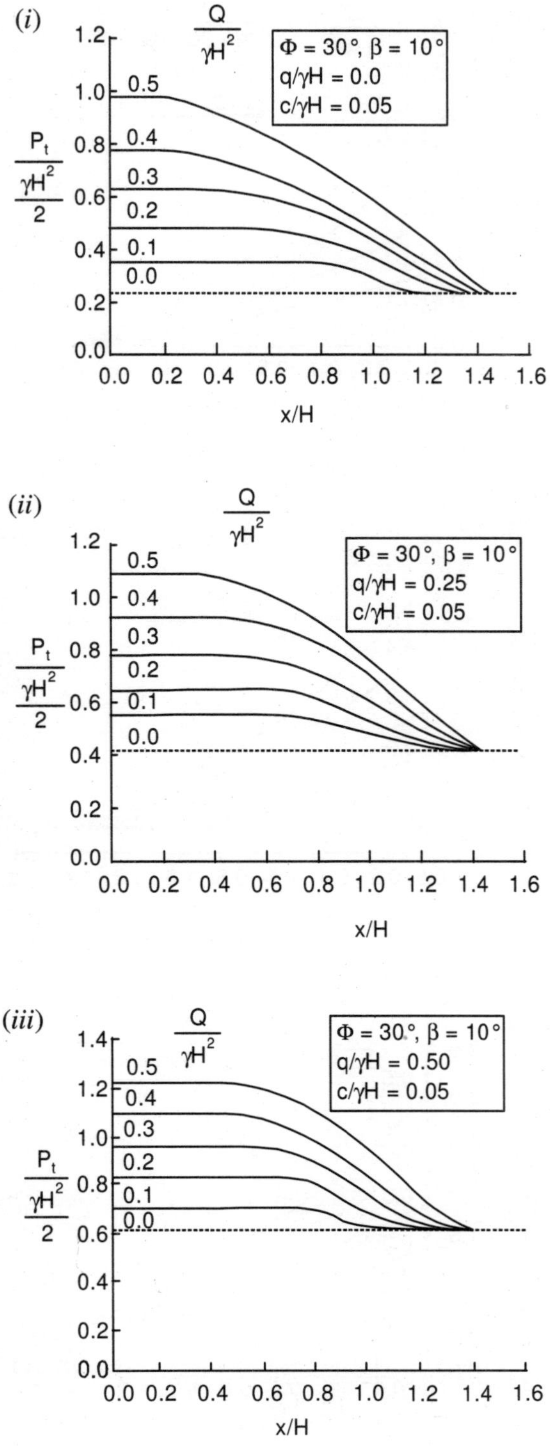

Fig. 7.26: (a) Resultant earth pressure *vs.* x/H

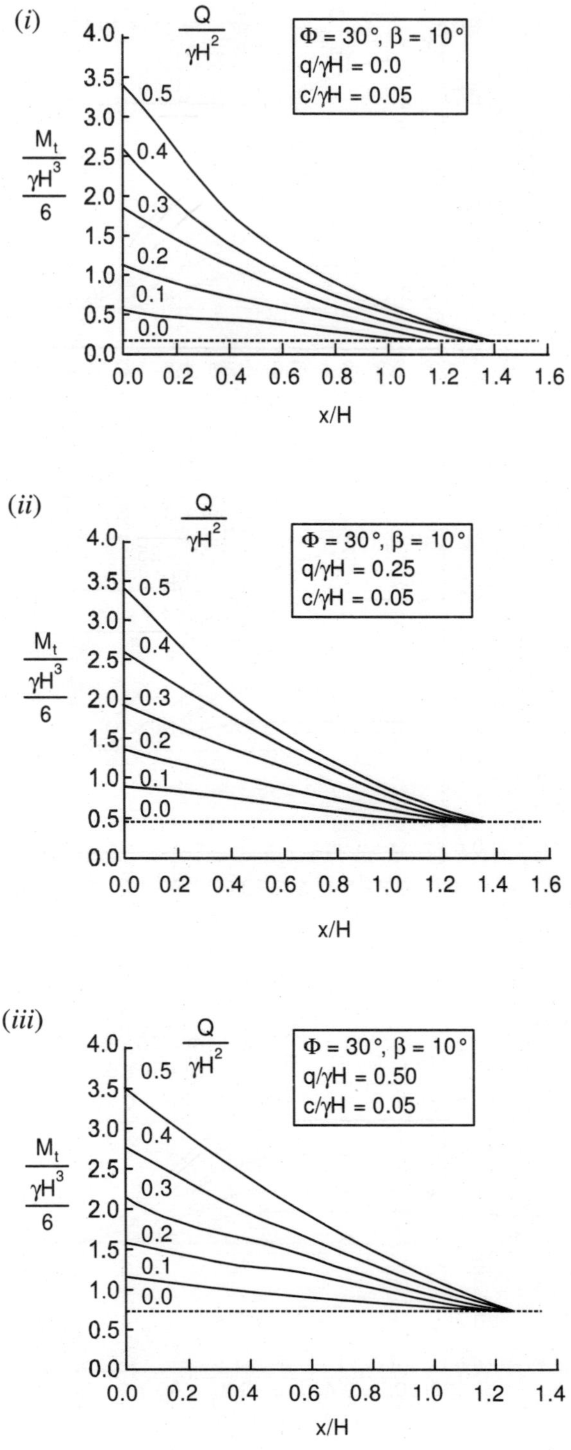

Fig. 7.26: (b) Resultant moment *vs.* x/H

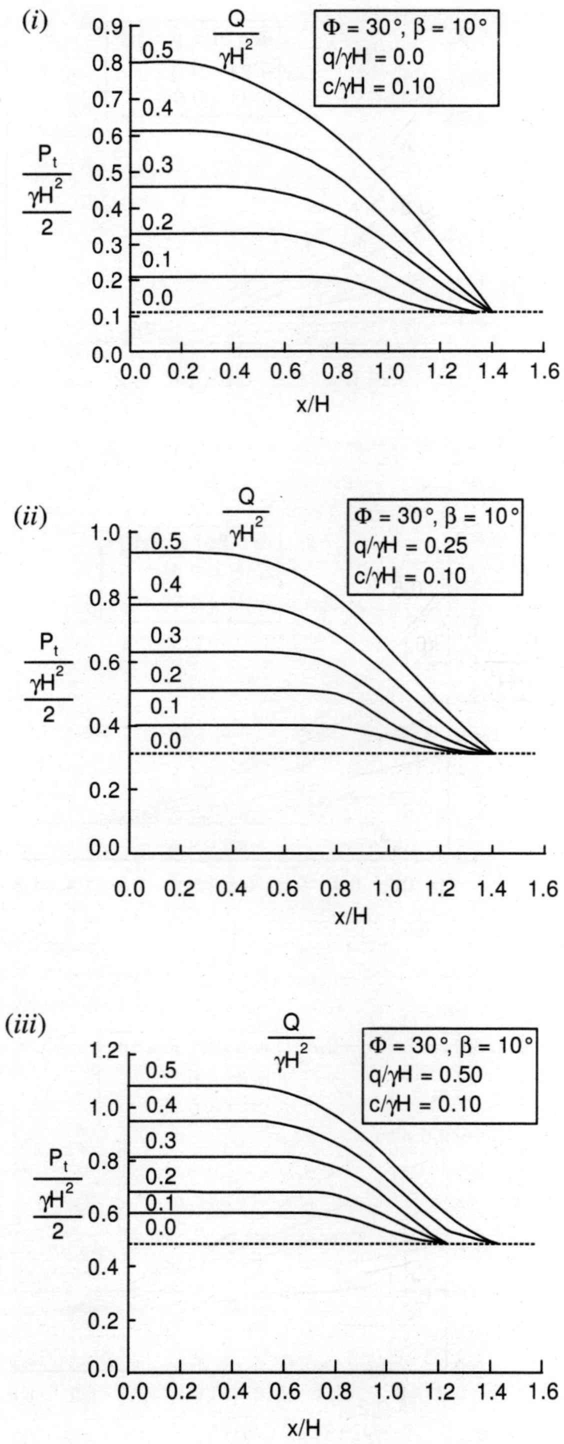

Fig. 7.27: (a) Resultant earth pressure *vs.* x/H

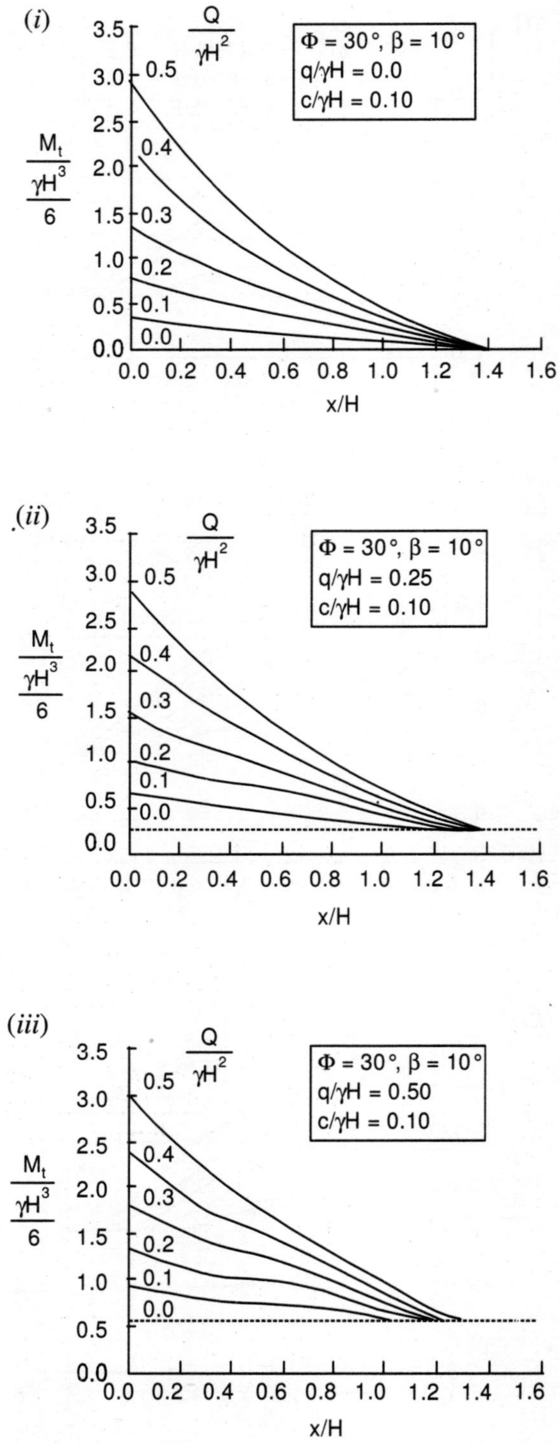

Fig. 7.27: (b) Resultant moment *vs.* x/H

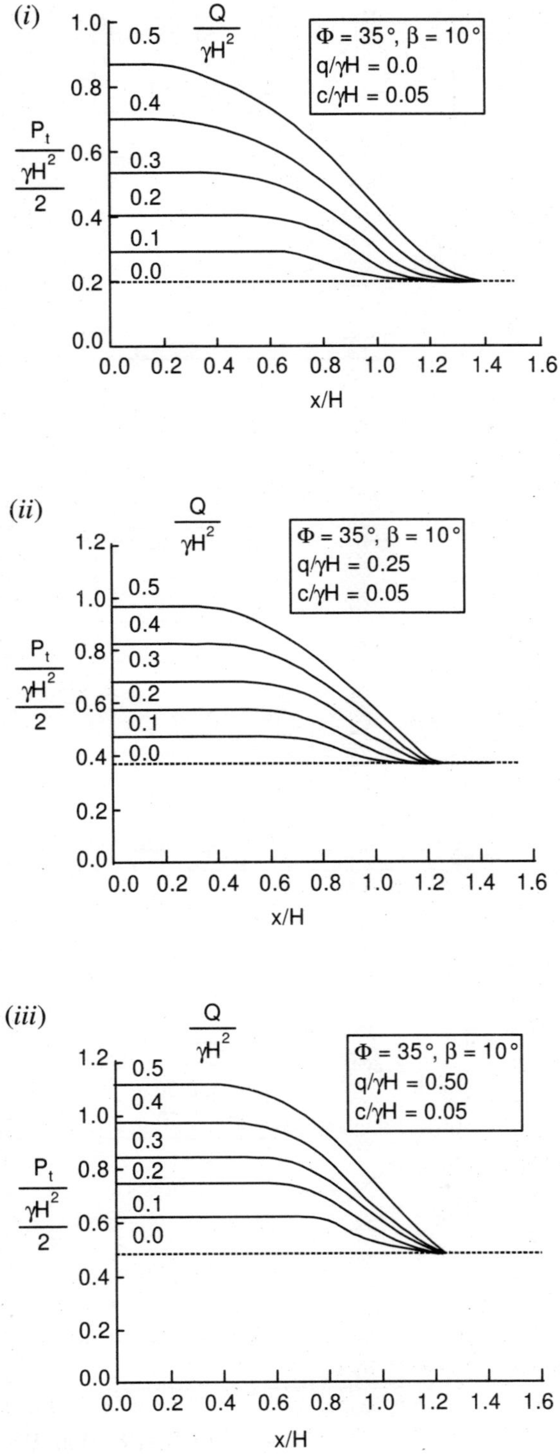

Fig. 7.28: (a) Resultant earth pressure *vs.* x/H

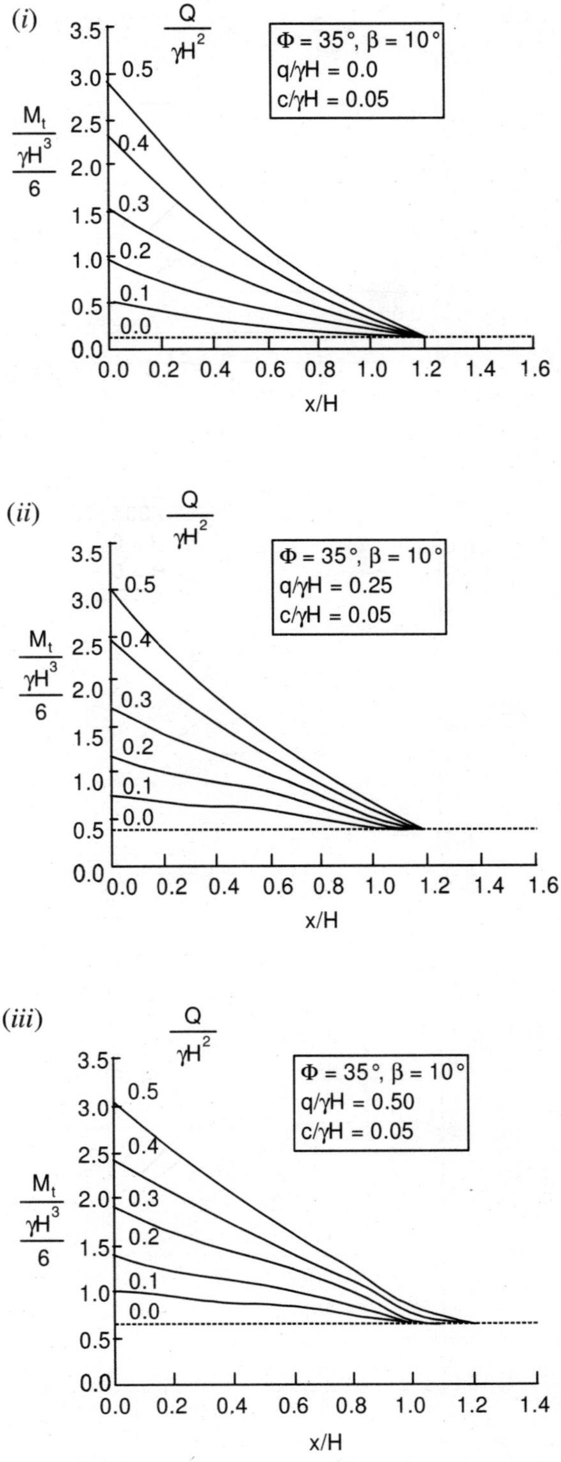

Fig. 7.28: (b) Resultant moment *vs.* x/H

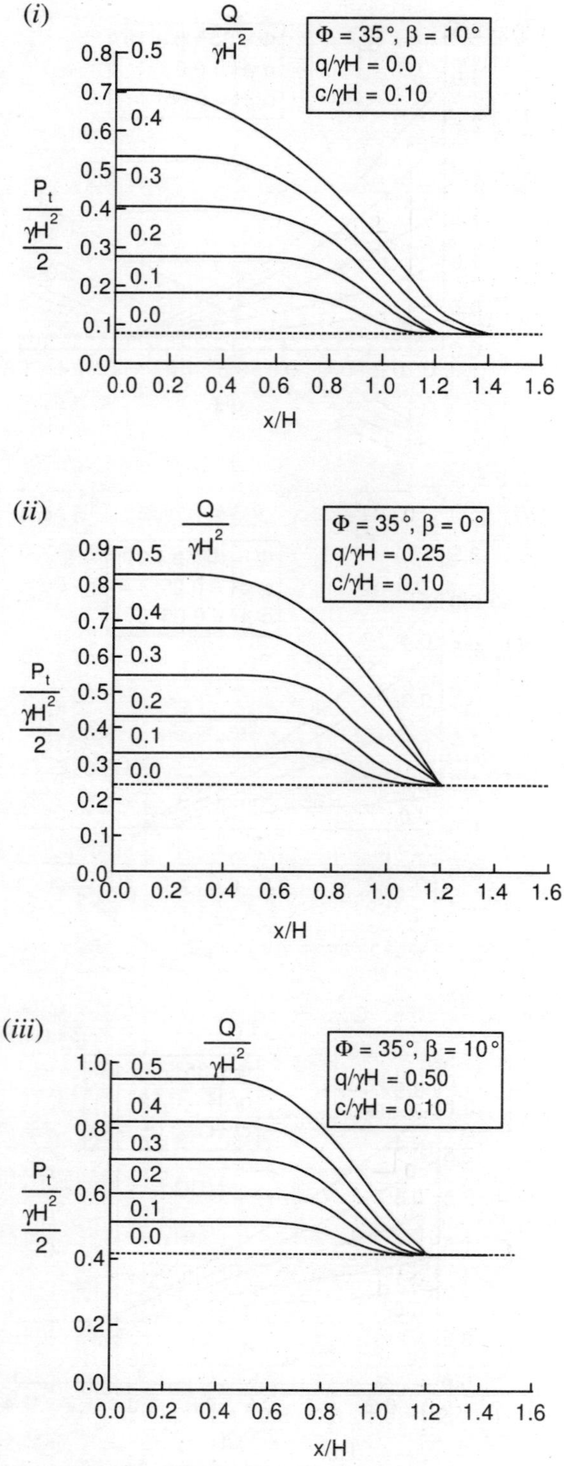

Fig. 7.29: (a) Resultant earth pressure *vs.* x/H

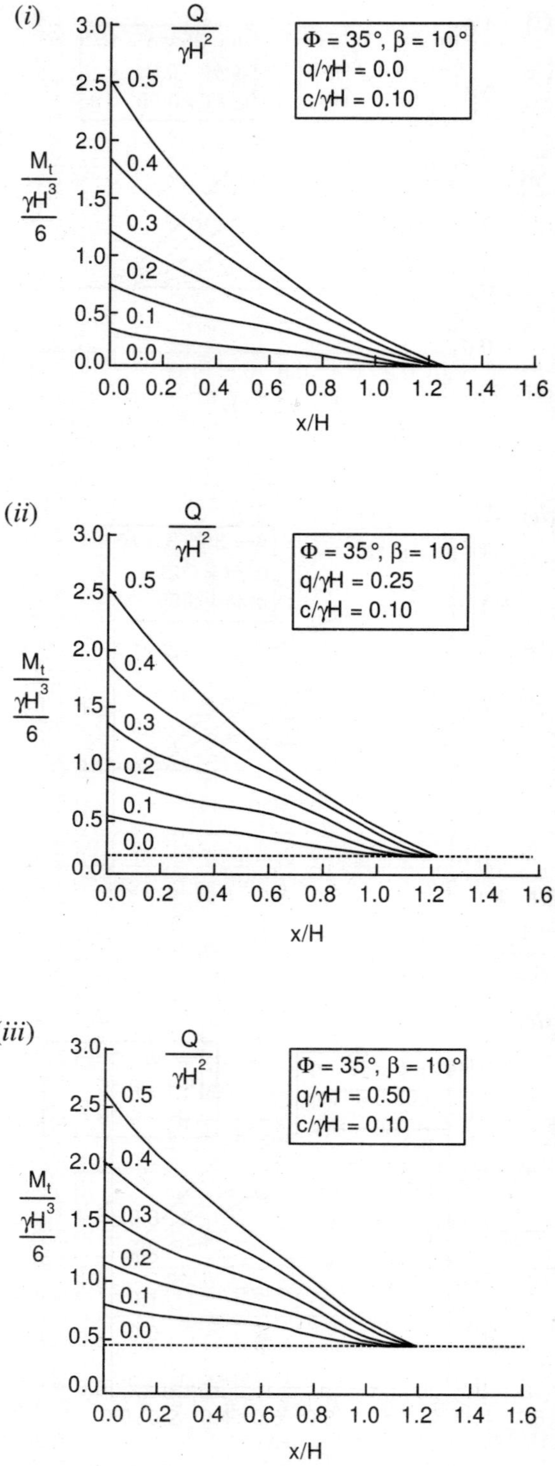

Fig. 7.29: (b) Resultant moment *vs.* x/H

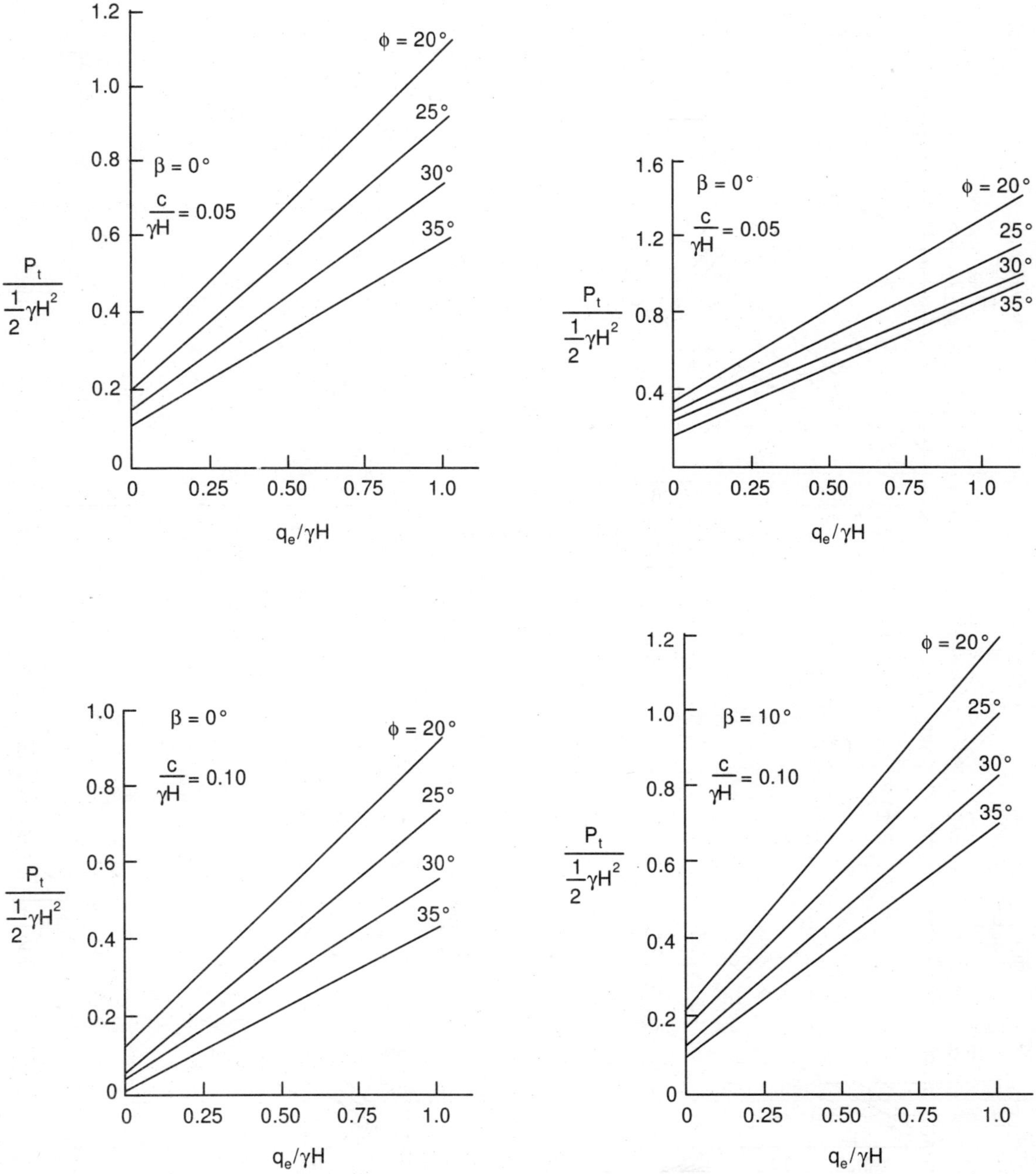

Fig. 7.30: (a) Resultant pressure *vs.* equivalent surcharge (u.d.l.) (Unreinforced case)

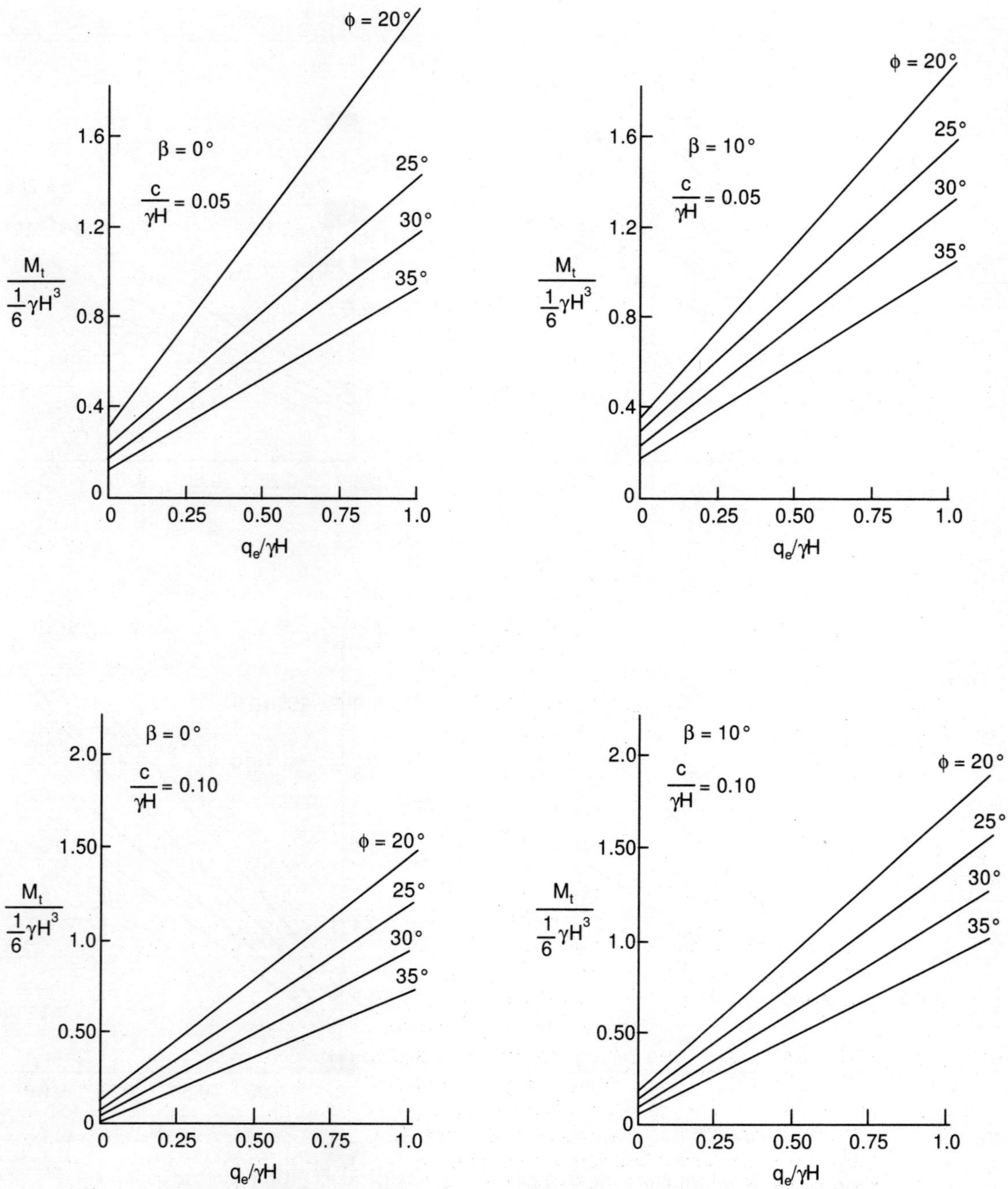

Fig. 7.30: (b) Resultant moment *vs.* equivalent surcharge (u.d.l.) (Unreinforced case)

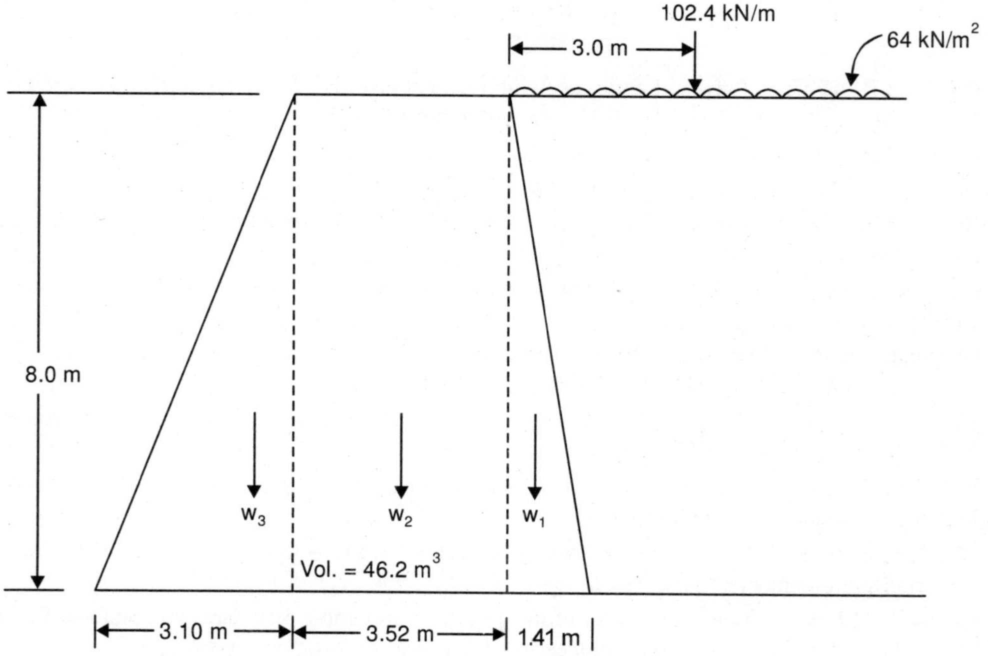

Fig. 7.31: (a) Wall with reinforced backfill

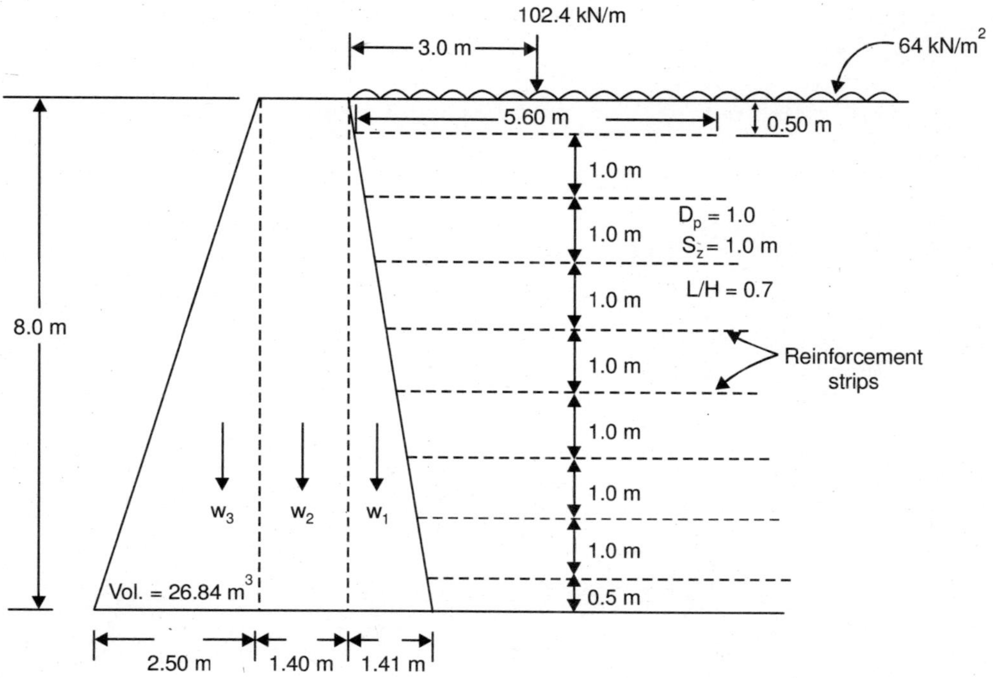

Fig. 7.31: (b) Wall with reinforced backfill

References

1. Broms, B.B. (1977a). "Polyster Fabrics as Reinforcement in Soils", *Proc. International Conference on the Use of Fabrics in Geotechnics,* Paris, France, Vol. I, pp. 129-135.
2. Broms, B.B. (1977b). "Triaxial Tests with Fabric-Reinforced Soil", *Proc. International Conference on the Use of Fabrics in Geotechnics,* Paris, France, Vol. III, pp. 129-133.
3. Broms, B.B. (1987). "Fabric Reinforced Soils", *Proc. International Symposium on Geosynthetics-Geotextiles and Geomembranes*–Kyoto, Japan, pp. 13–54.
4. Garg, K.G. (1988). "Earth Pressure Behind Retaining Wall with Reinforced Backfill", Ph.D. Thesis, University of Roorkee.
5. Hausmann, M.R. and Lee, K.L. (1978). "Rigid Model Wall with Soil Reinforcement", *Proc. Symposium on Earth Reinforcement,* ASCE, Pittsburgh, pp. 400-427.
6. Khan, I.N. (1991). "A Study of Reinforced Earth Wall and Retaining Wall with Reinforced Backfill", Ph.D. Thesis, University of Roorkee.
7. Mittal, S. (1998). "Behaviour of Retaining Walls with Reinforced c-ϕ" Soil Backfill, Ph.D. Thesis, Indian Institute of Technology, Roorkee.
8. Mittal, S., Garg. K.G., Saran, S. and Pathak, A.N. (1998). "Rigid Wall retaining Bottom Ash Backfill with Geogrid Reinforcement", *Indian Geotechnical Journal*, Vol. 31, No. 1, pp. 88–102.
9. Talwar, D.V. (1981). "Behaviour of Reinforced Earth in Retaining Structures and Shallow Foundation", Ph.D. Thesis, Department of Civil Engineering, University of Roorkee, Roorkee, India.
10. Vidal, H. (1966). "La Terre Armee", Annales de 1 Institute Technique du Batiments et des Travaux Publics, Paris, France.
11. Vidal, H. (1969). "The Principle of Reinforced Earth", Highway Research Record No. 282, USA.

Micropiling

CHAPTER 8

8.1 INTRODUCTION

Micropiles are small diameter piles which can either be drilled/grouted or simply driven pile. The majority of micropiles are between 100 and 250 mm in diameter, 5–30 m in length and 300–1000 kN in compressive or tensile service load, (although still more greater depths and much higher loads are common in the United States, Bruce et al., 1995). The grout is either placed or injected under pressure (with grouting pressures about 0.8–1 MPa). It consists of a continuously threaded hollow reinforcing tendon as a load carrying steel member together with a grout body of furnace (Portland) cement which allows transfer of tension and compression forces mainly from the friction of

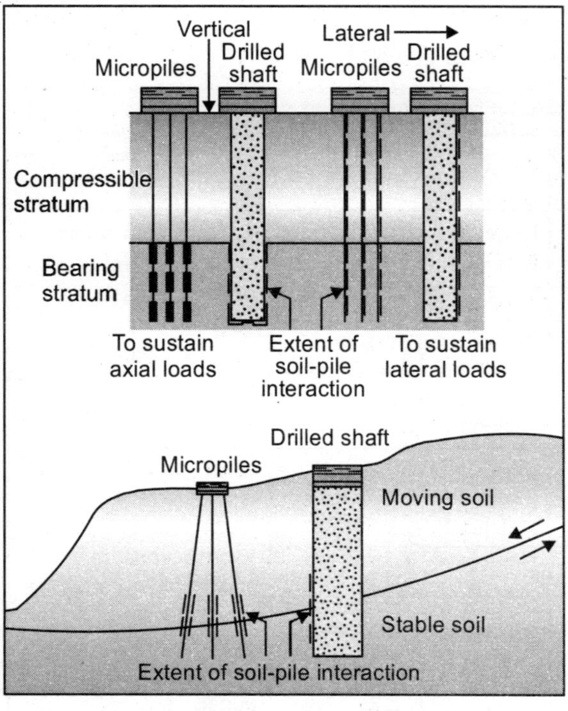

Fig. 8.1: Micropiles (Directly loaded)

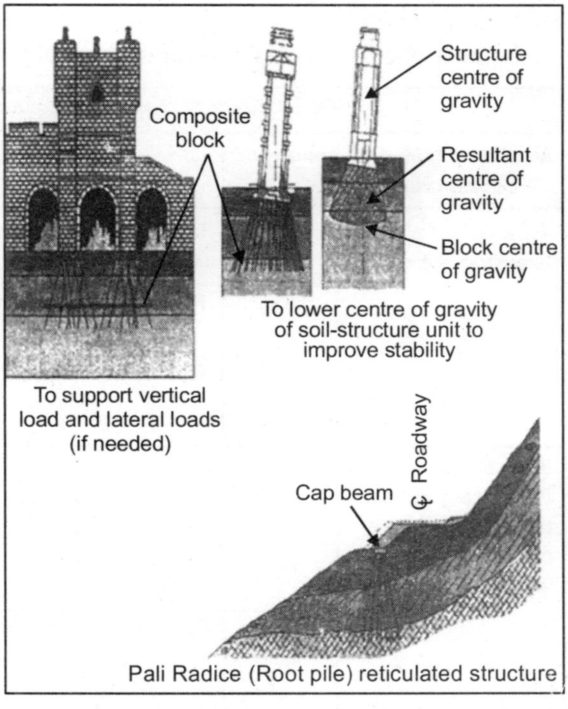

Fig. 8.2: Micropiles—Reticulated pile network with reinforced soil mass loaded or engaged (Directly loaded)

the threaded tendon via the grout body into the surrounding soil. Due to high pressures grouting there is insignificant shrinkage between the pile and the soil. The penetration of the fluid part of the cement mix into the surrounding soil creates a transitory zone between the body of the micropile and the soil leading to a strong grout/ground bond. Due to this reason ultimate load carrying capacity of micropile is based on conventional bearing capacity theory. Micropiles are typically used for structural support. It is usually installed for bridge and building foundations supports and seismic retrofits. Another main application of micropile is soil reinforcement for slope stabilisation/earth retention projects.

Micropiling technology was first started in Italy in the early 1950s, when the search was on for innovative restoration work and reliable method of underpinning historic buildings and monuments. It was in 1952 when Lizzi, F., the then technical director of Italian firm Fondedile developed a new type of pile and piling system called pali radice (root piles) and reticolo-di-radice (reticulated root piles) respectively. The pali radice developed by Lizzi (1952) is recognised as the original micropiles used in those days for variety of geotechnical problems (Figs. 8.1 and 8.2). Fondedile introduced root piles in U.K. in 1962 and the same was introduced into the United States after two decades by Bruce, 1989.

Since the middle 1980s in particular, there has been a rapid growth in its use, mainly as foundation support elements in static and seismic applications and as *in situ* reinforcement for slope and excavation stabilisation (Sharma, 2011).

Strengthening of foundations of existing buildings in earthquake prone megacities and addition of new floors to existing buildings due to very high land values in big cities, usually need excavation and temporary support system below the foundation level. This creates problem due to limited head room and access in congested area. Also there is risk of collapse of the structure during the excavation process itself. Moreover, underground spaces need to be utilised for alternate modes of transport. Distresses historical monuments need to be saved. Micropiles may be used in such situations as the economic solution.

Some of the past works done in the area of micropiling are given in Table 8.1.

Table 8.1: Details of micropile classification based on type of grouting

Micropile type and Grouting method	*Sub-type*	*Drill casing*	*Reinforcement*	*Grout*
Type A: Gravity grout only	A1	Temporary or unlined (open hole or auger)	None, monobar, cage, tube or structural section	Sand/cement mortar or neat cement grout, tremied to base of neat cement grout, hole (or casing), no excess pressure applied.
	A2	Permanent, full length	Drill casing itself	
	A3	Permanent, upper shaft only	Drill casing in upper shaft, bar (s) or tube in lower shaft (may extend full length)	
Type B: Pressure-grouted through the casing or auger during withdrawal	B1	Temporary or unlined (open hole or auger)	Monobar(s) or tube (cages rare due to lower structural capacity)	Neat cement grout is first tremied into drill casing/auger. Excess pressure (up to 1 MPa typically) is applied to additional grout injected during withdrawal of casing auger.
	B2	Permanent, partial length	Drill casing itself	
	B3	Permanent, upper shaft	Drill casing in upper shaft, bar (s) or tube in lower shaft (may extend full length)	

Contd...

Type C: Primary grout placed under gravity head, then one phase of secondary 'global' pressure grouting	C1	Temporary or unlined (open hole or auger)	Monobar(s) or tube in lower shaft (cages rare due to lower structural capacity)	Neat cement grout is first tremied into hole (or casing/auger). Between 15 to 25 minutes later, similar grout injected through tube (or reinforcing pipe) from head, once pressure is greater than 1 MPa.
	C2	Not conducted	–	
	C3	Not conducted	–	
Type D: Primary grout placed under gravity head (Type A) or (Type B). Then one or more phases of secondary 'global' pressure grouting	D1	Temporary or unlined (open hole or auger)	Monobar(s) or tube (cages are rare due to lower structural capacity)	Neat cement grout is first tremied (Type A) and/or pressurised (Type B) into hole or casing/auger. Some hours later, similar grout injected through sleeved pipe (or sleeved reinforcement) via packers, as many times as necessary to achieve bond.
	D2	Possible, only if regrout	Drill casing itself tube placed full-length outside casing	
	D3	Permanent, upper shaft only	Drill casing in upper shaft, bar(s) or tube in lower shaft (may extend full length)	

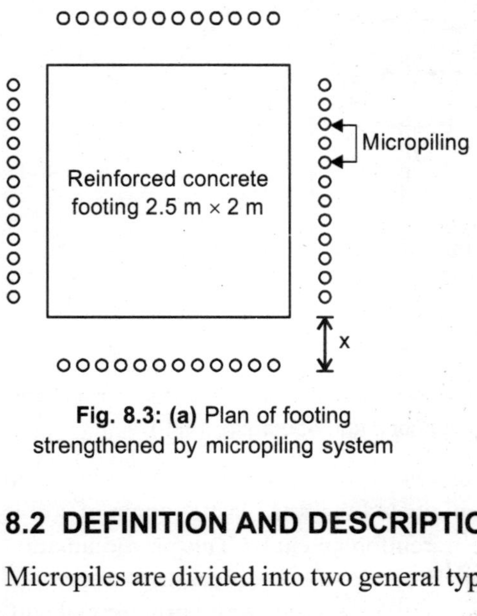

Fig. 8.3: (a) Plan of footing strengthened by micropiling system

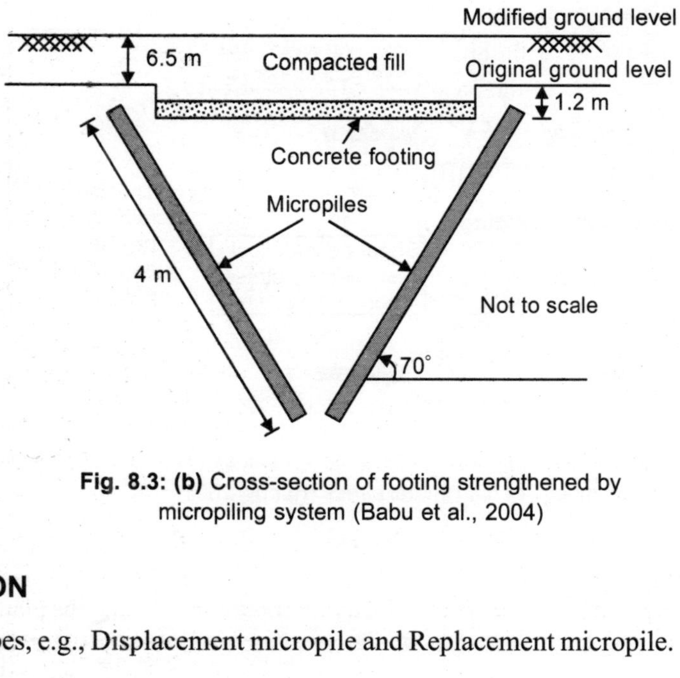

Fig. 8.3: (b) Cross-section of footing strengthened by micropiling system (Babu et al., 2004)

8.2 DEFINITION AND DESCRIPTION

Micropiles are divided into two general types, e.g., Displacement micropile and Replacement micropile.

8.2.1 Displacement micropiles

These are the members that are driven or vibrated into the ground, thereby displacing the surrounding soil laterally during installation.

8.2.2 Replacement micropiles

These are placed or constructed within a previously drilled bore hole, thus replacing the excavated ground.

A typical plan and cross-section of micropiles used as by Babu et al., (2004) for strengthening of foundation of a two storeyed building is shown in Fig. 8.3 (*a*) and (*b*).

8.3 REINFORCING MECHANISM

A simple model illustrating the effect of lateral confinement in soil is shown in Fig. 8.4 (Handy, 2001), where top diagram shows shear directions and contact friction in a normally consolidated soil. Here friction acts in support of the higher vertical load. With the application of horizontal stress shear directions are reversed, and as shown in middle diagram contact friction acts to oppose the increased horizontal stress. In the bottom diagram when a foundation load is applied, shear direction again are reversed, and as in case of the normally consolidated soil, consolidation can begin only when the vertical stress is sufficient to overcome the frictional resistance.

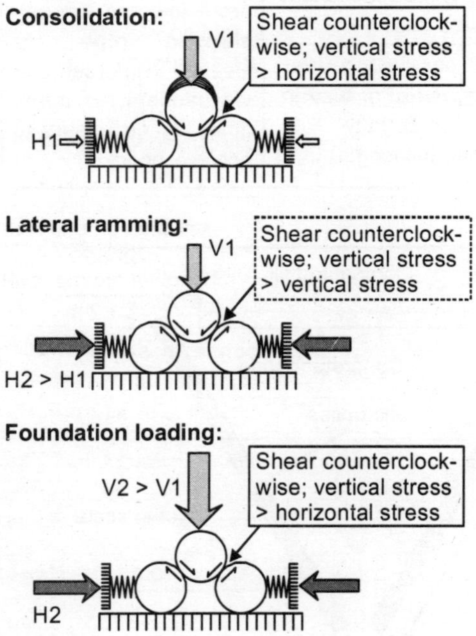

Fig. 8.4: Increase in the allowable vertical stress due to impose of high lateral stress (Handy, 2001)

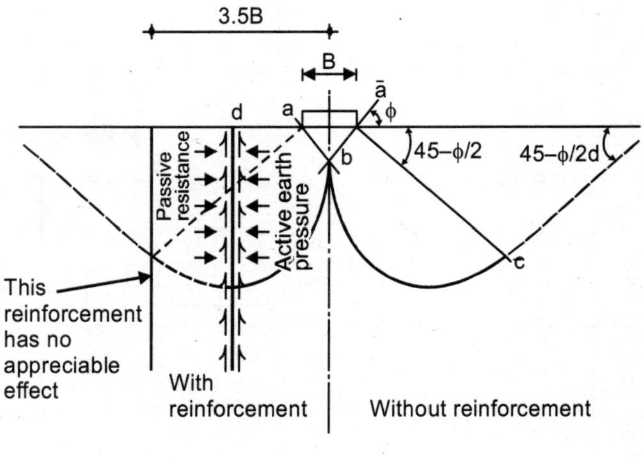

Fig. 8.5: Anticipated failure surface for vertical confinement (Mahmoud and Abdrabbo, 1989)

The vertical reinforcing element resists the lateral displacement of soil underneath the footing and creates the sand confinement situation. From Fig. 8.5, it is anticipated that the plastic region of soil is limited by the plane of velocity discontinuity *abc* and the plane of reinforcement *cd*. This plastic situation is smaller than that in similar situation without reinforcement. Thus, a similar vertical displacement of footing resting on reinforced soil is expected compared with a similar case but with unreinforced soil. The lateral passive resistance of soil developed along the top part (*cd*) of the reinforcing elements confined the plastic flow of soil within the region *abcd*, i.e., the reinforcing elements create some sort of lateral confinement (Mahmoud and Abdrabbo, 1989).

It can be concluded that high lateral stress can differ consolidation settlement to a substantially higher foundation load, without any influence from densification of the soil and without necessarily increasing the pore water pressure. It should be emphasised that the vertical reinforcement factor is fully developed only in soil that has been horizontally stressed to the passive limit, as where the reinforcing element are closely spaced in lines on in groups.

Many investigators have done some research work on micropiling. Their work in brief is presented in Table 8.2.

Table 8.2: Brief literature review

Researcher	***Research conducted***
Verma & Char (1986)	Tests on model footings on sand subgrades reinforced with multiple rows of galvanised rods placed vertically in the subgrade to show beneficial effects of reinforcement.
Mahmoud and Abdrabbo (1989)	Experimental study on strip footing resting on subgrade utilising vertical reinforcement made of aluminium strips 2 mm thick and 20 mm width.
Verma and Jha (1992)	Experimental study on three dimensional model footing resting on sand subgrade reinforced with multiple rows galvanised rods on all four sides of footing base.
Mandal and Manjunath (1995)	Experimental study to improve the bearing capacity of strip footing resting on subgrades, utilising a Georgia and bamboo as vertical reinforcement.
Researcher	***Their findings***
Hari Krishna and Ramana Murthy (2000)	Experimental study to determine the bearing capacity improvement by using of vertical and inclined GI strips driven in sand bed outside the footing.
Murthy et al. (2002)	The improvement of load-deformation response in sands due to densification of soil surrounding the inclined micropiles and interaction between the soil and micropiles by using finite difference software FLAC.
Bahloul et al. (2004)	Uses vertical reinforcement to increase the bearing capacity of the subgrade layer under shallow strip footing.
Babu Sivakumar et al. (2004)	Uses inclined micropiles for strengthening the existing foundation of a two storied building and the bearing pressure increases from 66.8 kPa to 106.3 kPa.
Bhattacharjee (2011)	Developed non-dimensional design charts for design of micropiles for different loading conditions, including eccentric loads. Verified results with FLAC 3D software.
Sharma, Binu (2011)	Conducted model study of micropiles subjected to lateral loading and oblique pull.

8.4 STUDY CONDUCTED AT IIT ROORKEE, INDIA (Mittal, 2012)

8.4.1 Salient points of study

- Micropiles were used as the vertical reinforcement.

- Study done in respect of improvement in the load carrying capacity and reduction in settlement characteristics of sand subgrades with micropiles placed at the side of the footing subjected to central and eccentric load.
- Parametric study was done to see the effect of different variables such as distance of the micropile from edge of the footing, mutual spacing of micropile, diameter and length of the micropile in bearing capacity improvement of the footing.

8.4.2 Test programme

Material used: Solani river sand was used in the study. The properties of sand are given in the following table: (Table 8.3)

Table 8.3: Properties of soil used in the investigation

Sl. No.	*Characteristics*	*Value*
1	Soil type/classification	SP
2	Coefficient of uniformity (C_u)	1.82
3	Coefficient of curvature (C_c)	1.11
4	Specific gravity of solids (G_s) at 27°C	2.58
5	Minimum void ratio (e_{min})	0.5
6	Maximum void ratio (e_{max})	0.91
7	Average density under test condition (γ_d)	16.1 kN/m^3
8	Relative density under test condition (D_R)	75%
9	Angle of shearing resistance by triaxial test in UU condition	38°

8.4.3 Material of micropile

Mild steel bars of 3.0 mm, 6.0 mm and 10.0 mm diameter coated with sand-fevicol (adhesive) mixture were used to simulate as RCC micropile.

8.4.4 Test footing

A square mild steel plate of size 150 mm × 150 mm × 10 mm was used as model footing (Fig. 8.6).

8.4.5 Test tank

A steel tank of size 115 cm × 115 cm × 50 cm was used to conduct model tests (Fig. 8.6). Firstly sand was filled up in tank at design relative density. After that footing was kept on top surface of sand at the centre of tank. Load was applied on footing in different increments similar to a plate load test and corresponding settlement of footing was measured with the help of 2 nos. dial gauges. The tests were conducted without any micropile and also with micropiles around footing to notice the effect of micropiles. Full care was taken that footing settles uniformly and does not get tilted. This was ensured by the help of a tilt meter (Fig. 8.6).

8.4.6 Loading device

Load were applied by means of a mechanical jack, attached with a calibrated proving ring (Fig. 8.7). Load was applied centrally over the footing and also in other tests, load was applied eccentrically

(Fig. 8.8) to see the effect of micropile under the action of eccentric load on footings. Every load applied was left till the settlement of footing becomes constant. For every load settlement was recorded by dial gauges till failure occurred.

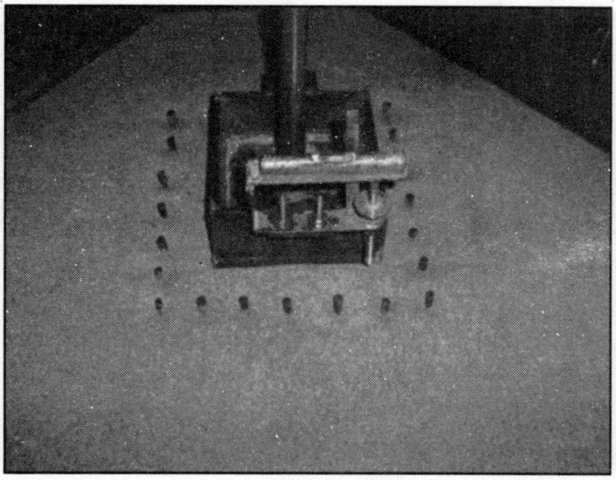

Fig. 8.6: Test set up showing micropile and tilt meter

Fig. 8.7: Test set up and loading device

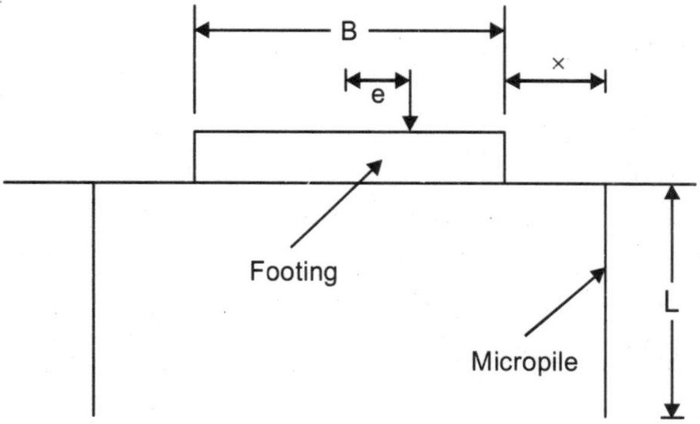

(a) Cross-section of reinforced footing

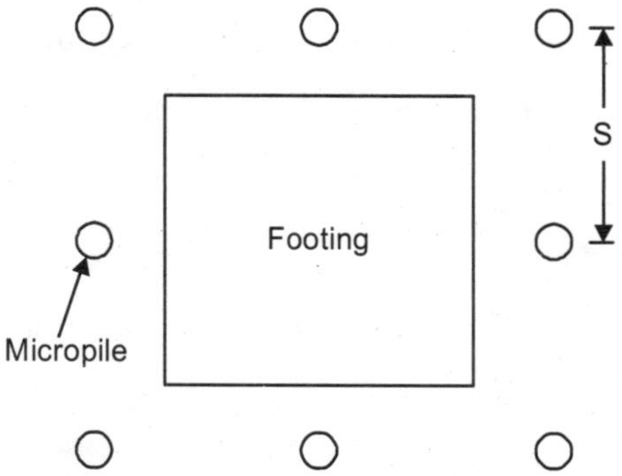

(b) Plan of reinforced footing

Fig. 8.8: Arrangement of micropile

8.5 PARAMETRIC STUDY

The following were the parameters, studied in model tests:

(*i*) Ratio of length of the micropile to width of footing (L/B).

(*ii*) Ratio of diameter of the micropile to width of footing (*d*/B).

(*iii*) Ratio of distance of micropile from edge of footing to the width of footing (*x*/B).

(*iv*) Ratio of spacing between the micropiles to width of footing (*s*/B).

(*v*) Ratio of eccentricity of load to width of footing (*e*/B).

Table 8.4: Different parameters and their values	
Parameters	***Values***
d/B	0.02, 0.04, 0.067
e/B	0.0, 0.1, 0.2
L/B	1.0, 2.0, 3.0
s/B	0.25, 0.5
x/B	0.1, 0.5, 1.0

The values of different parameters are given in the following Table 8.4.

A total of one hundred sixty-five tests were performed. The results were presented in the form of bearing capacity ratio (BCR) and settlement ratio (SR).

The bearing capacity ratio, BCR is defined as

$$\text{BCR} = q/q_0$$

where,

q = ultimate bearing capacity for soil reinforced with micropile

q_0= ultimate bearing capacity for unreinforced soil (i.e., no micropile)

The settlement ratio, SR for a particular bearing pressure is defined as

$$\text{SR} = \text{S}/\text{S}_0$$

where,

S = settlement of footing on reinforced soil at a particular bearing pressure

S_0 = settlement of footing on unreinforced soil at that bearing pressure

8.6 NUMERICAL SIMULATION USING FLAC 3D SOFTWARE

A numerical model was also developed to verify the results obtained from model tests. For the numerical simulation, following soil properties were used. (Table 8.5).

Table 8.5: Properties of Solani river sand	
Properties	***Values***
Modulus of elasticity for test condition	23.08 MPa
Poisson's ratio	0.3
Bulk modulus for test condition	19.23 MPa
Shear modulus for test condition	8.87 MPa
Angle of internal friction for test condition	38°

8.7 SIMULATION OF MICROPILE

The micropiles were modelled as pile element.

For micropile element, following input properties were used in numerical model

- Diameter of micropile
- Modulus of elasticity
- Poisson's ratio
- Perimeter (πd)
- Area of pile
- Moment of inertia (I)
- Coupling spring shear cohesion

= coupling spring normal cohesion

= cohesion of soil × perimeter of pile

- Coupling spring shear friction

= coupling spring normal friction

Coupling spring shear stiffness

= coupling spring normal stiffness

$$= \frac{(K + \frac{4}{3}G)}{(\Delta Z_{min})}$$

where,

K = bulk modulus of soil

G = shear modulus of soil

ΔZ_{min} = smallest width of an adjoining zone in the normal direction (in the mesh modelled to represent the soil).

The figures related to numerical simulation are given below (Figs. 8.9 to 8.12)

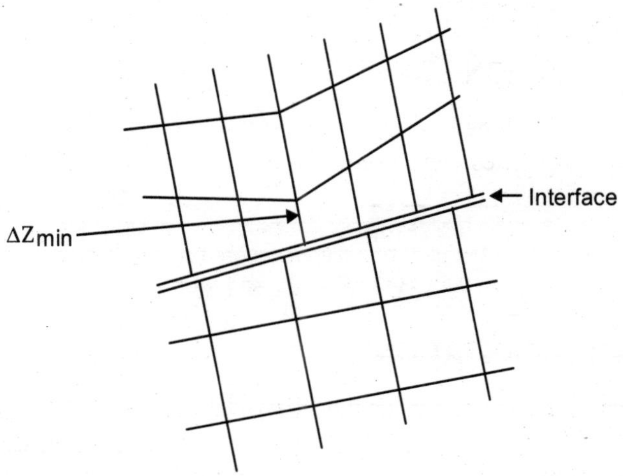

Fig. 8.9: Zone dimension used in stiffness calculation

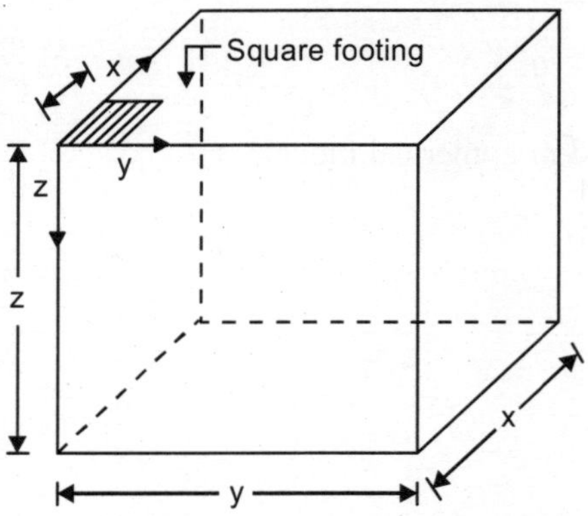

Fig. 8.10: Domain for FLAC 3D analysis —quarter symmetry

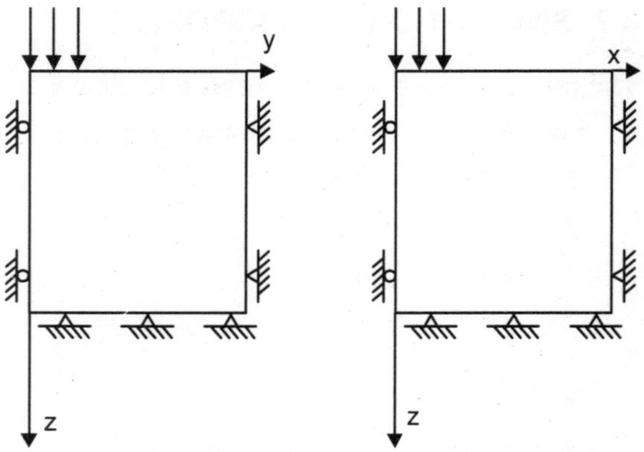

Fig. 8.11: Boundary condition—quarter symmetry

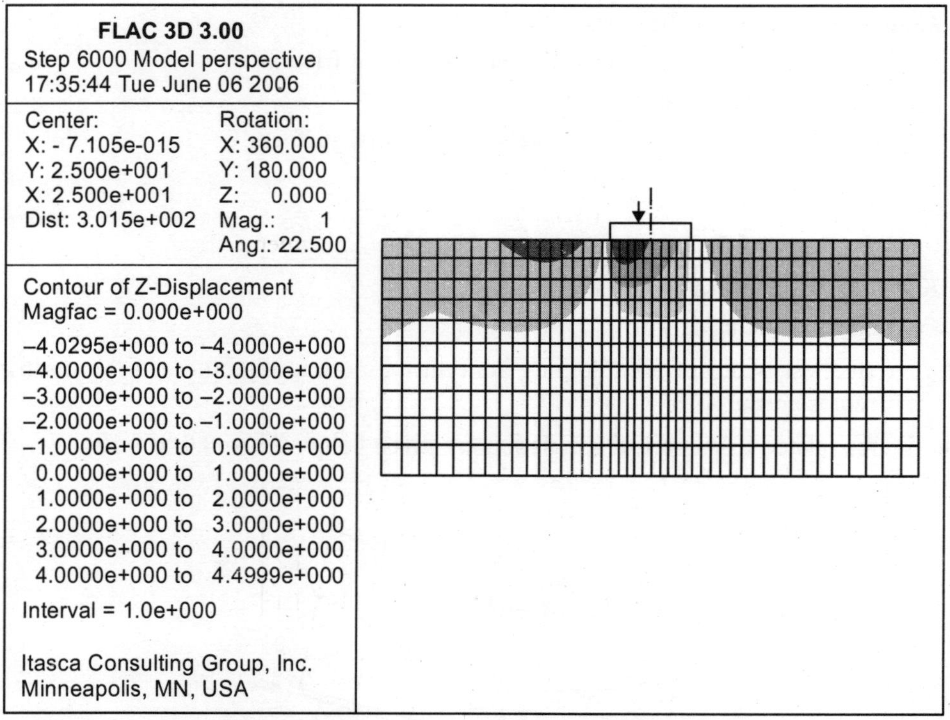

Fig. 8.12: FLAC 3D contour plot of z-displacement for reinforced soil (d/B=0.02, e/B=0.1,L/B=3,s/B=0.25 and x/B=0.1)

8.8 RESULTS AND INTERPRETATION

The pressure v/s settlement curves were drawn on the basis of laboratory model tests and also the numerical study. Some typical results are shown vide Fig 8.13 which are for virgin soil (no micropile). The numerical show that the observations from model tests and numerical methods are in close agreement. Fig. 8.14 clearly show that there is a significant improvement in bearing capacity of soil when micropiles were used, at any edge distance (x) from footing and at any mutual spacing between micropiles.

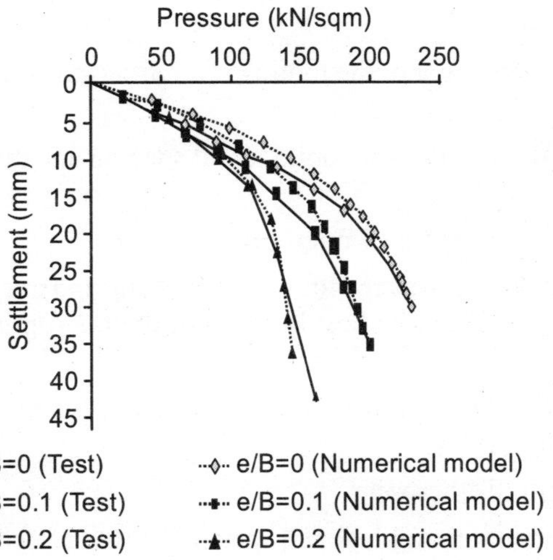

Fig. 8.13: Pressure *vs.* settlement curves

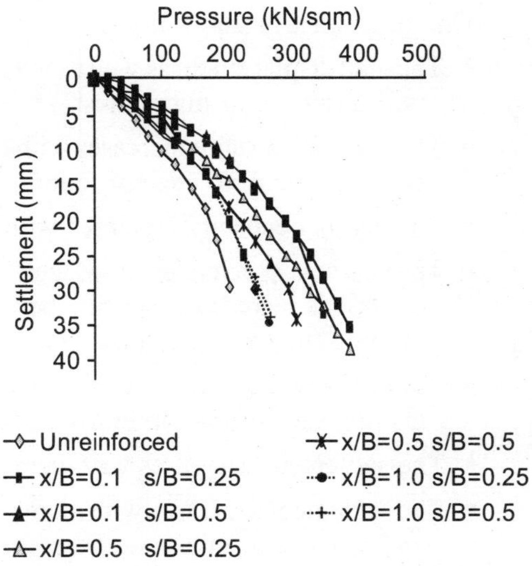

Fig. 8.14: Pressure-settlement curve for d/B = 0.067, e/B = 0 and L/B = 3 (Test)

Fig. 8.15 shows BCR v/s *d*/B plots. The curves indicate that bearing capacity ratio increases with increase in diameter of micropile. Fig. 8.16 shows that settlement of footing reduces drastically as the length of micropile increases. Model tests results are in close agreement with numerical model results.

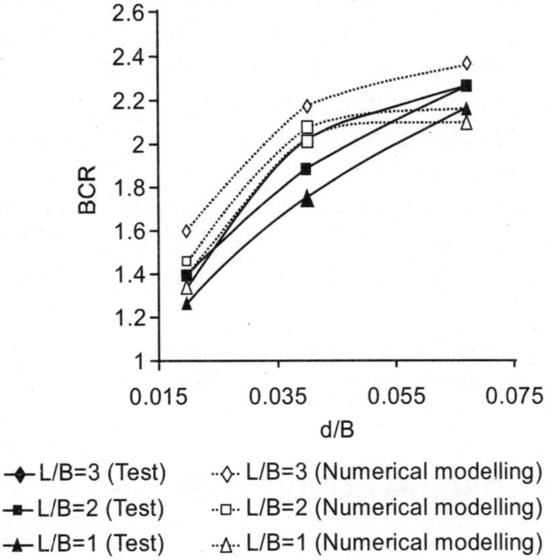

Fig. 8.15: BCR *vs.* d/B curve for e/B=0.2, x/B=0.1, s/B=0.25

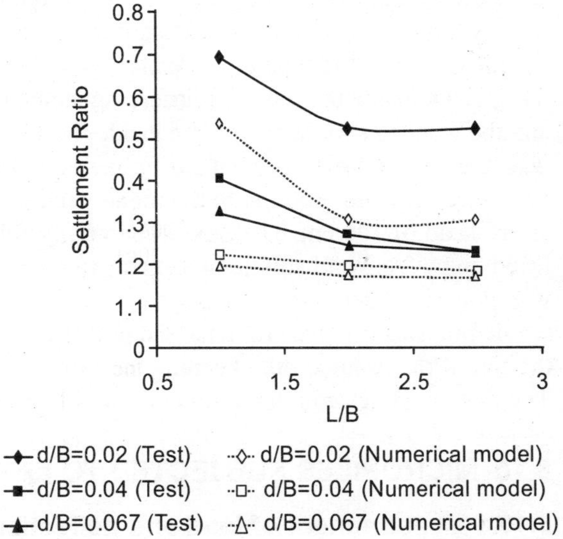

Fig. 8.16: SR *vs.* L/B for e/B=0.2, s/B=0.25, x/B=0.1 and Bearing pressure =150 kN/sqm

The above charts are some typical charts obtained from the study. More such charts are available elsewhere (Bhattacharjee, 2006).

This study clearly shows that:

- The load displacement as well as load settlement behaviour of the footing is modified considerably with insertion of micropiles.
- The bearing capacity increases with increase in diameter of micropile for all spacing and distance of micropile from edge of footing.
- The bearing capacity increases with the increase in length of micropile.
- The bearing capacity decreases with increasing distance between the edge of footing and micropile as well as increase in spacing between the micropiles. The best location of the micropile is as close to the footing as possible.
- The settlement decreases with increase in diameter of micropile for all spacing between micropiles and distance of micropile from edge of footing.
- The settlement decreases with increase in length of micropile.
- The settlement decreases with increase in diameter of micropile for all spacing between micropiles and distance of micropile from edge of footing.
- The bearing capacity improvement is more for eccentrically loaded footing than centrally loaded footing.
- The reduction of settlement is lesser for eccentrically loaded footing than centrally loaded footing.

8.9 INSTALLATION OF MICROPILES

Sharma (2011) conducted model study to understand behaviour of micropiles subjected to lateral loading and oblique pull. Dry sand was used in the model tests micropiles were cast in sand of 50% relative density through an aluminium casing pipe of external diameter 12 mm and internal diameter 10 mm. At the lower end of the pipe a 60° conical wooden shoe was attached. A 1.5 mm diameter mild steel rod was placed inside the pipe as reinforcing element. The aluminium casing pipe along with the steel rod and the wooden shoe were pushed inside the sand manually, keeping the pipe exactly vertical. The pipe was then grouted with cement slurry having a water cement ratio of 0.5.

Since pressure grouting in the model tank was not possible the piles were grouted under a constant slurry head of 100 cm. The pipe was initially filled up up to 100 cm by cement slurry, then gradually lifted up by 2 cm and additional cement slurry equivalent to 2 cm height was poured from the top. This was done in order to maintain a constant slurry head of 100 cm. This process was repeated till the whole pile was grouted. To make the piles free standing, a 50 mm long aluminium pile was attached to the top of the grouted pile, keeping the mild steel rod at the centre and again cast with cement slurry. The piles were tested after an interval of 4 to 5 days of grouting.

8.10 MICROPILES SUBJECTED TO LATERAL LOADING CONDITIONS

8.10.1 Measurement of load and deflection

Load deflection test was done according to IS: 2911 (Pt-4) 1985, code of practice for load test on pile except the gradual application of load was up to failure. Load applied to the pile cap by a mechanical jack, was measured by means of a proving ring and displacement measured by a dial gauge attached to the pile cap.

A mechanical jack was adopted for applying lateral load. The mechanical jack was mounted to a horizontal cross beam with bolts and nuts along which it was free to slide when required. One end of an extension rod of length 17 cm was fixed to the tip of the lever arm of the mechanical jack and the other end to a proving ring. The other end of the proving ring was pivoted to the pile cap, fixed at the top of the pile, by ball and socket arrangement. The pile cap was made of wood having size (6.4 cm × 6.2 cm × 2.8 cm). The cap consists of two halves with semi circular groove to accommodate the pile head. The two halves were screwed tightly to the pile top.

8.10.2 Test results

Single micropiles were cast for length/diameter (L/D) ratios of 8, 12, 15, 18, 21, 24, 27, 30, 35 and 55. Moreover for each L/D ratio three micropile were cast at three different relative densities of 30%, 50% and 80%. In total 30 numbers of micropiles were tested. Ultimate resistance of pile under lateral load has been taken as the point on the lateral load versus lateral displacement curve at which the curve maintains a continuous displacement increase with no further increase in lateral load.

Fig. 8.17 shows the variation of ultimate lateral load with length/diameter ratio of the micropiles at the three different relative densities. It is observed from the figure that at relative density 30%, the ultimate load increases at a high rate upto L/D 15.

Thereafter up to L/D 30, the rate of increase slows down and after L/D ratio 35 it becomes constant. At relative density of 50%, the ultimate lateral load increases at a high rate up to L/D 15, the rate slows down up to L/D 30, thereafter the rate comes out to be insignificant. At relative density of 80% the rate of increase is high up to L/D of 24, thereafter the increase becomes insignificant.

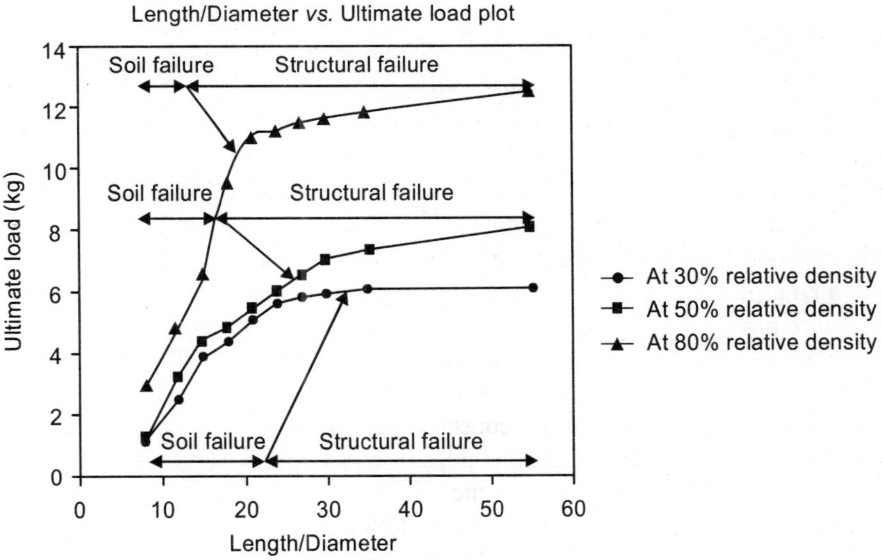

Fig. 8.17: Length/diameter ratio *vs.* Ultimate load plot at 30%, 50% and 80% relative density

8.11 DESIGN CRITERIA

1. The micropiles shall be designed to meet the specified loading and movement conditions.
2. The allowable working load on the pile shall not exceed the following values:
3. **For compression loads**

$$P_{allc} = (0.33 \times f_c \times A_{grout} + 0.4 \times f_{ycasing} \times A_{casing} + 0.4 \times f_{ybar} \times A_{bar}) \quad ...(8.1)$$

where,

P_{allc} = allowable working load (compression)

f_c = unconfined compressive strength (UCS) of grout

A_{grout} = area of grout

$f_{ycasing}$ = yield strength of casing up to 80 ksi

> **Note:** 1 ksi = 1000 psi = 7 MPa. Hence 80 ksi = 560 MPa which is equivalent to nearest Indian grade of 550 MPa strength of steel.

A_{casing} = area of steel casing (with allowance for corrosion if appropriate)

f_{ybar} = yield strength of rebar/core steel up to 80 ksi

A_{bar} = area of rebar/core steel

4. For tension loads

$$P_{allt} = 0.6 \times (f_{ycasing} \times A_{casingT} + f_{ybar} \times A_{bar}) \qquad \text{...(8.2)}$$

where,

P_{allt} = allowable working load (tension)

$f_{ycasing}$ = yield strength of casing

$A_{casingT}$ = area of steel casing (at threaded joints if appropriate)

f_{ybar} = yield strength of rebar/core steel

A_{bar} = area of rebar/core steel

Care must be taken in evaluating tension on threaded joints of drill casing, however, the load tests may verify threaded joint tension adequacy.

The ultimate structural capacity shall be determined as:

1. Compression

$$P_{uc} = (0.85 \times f_c \times A_{grout} + f_{ycasing} \times A_{casing} + f_{ybar} \times A_{bar}) \qquad \text{...(8.3)}$$

where,

P_{uc} = ultimate structural capacity (compression)

f_c = unconfined compression strength (UCS) of grout

A_{grout} = area of grout

$f_{ycasing}$ = yield strength of casing upto 80 ksi

A_{casing} = area of steel casing (corroded if appropriate)

f_{ybar} = yield strength of rebar/core steel upto 80 ksi

A_{bar} = area of rebar/core steel

The maximum usable strength of the steel of 80 ksi is based on the typical ultimate concrete strain of 0.003 (29000 ksi × 0.003 = 87 ksi). 80 ksi is also the maximum steel strength used.

2. Tension

$$P_{ut} = (f_{ycasing} \times A_{casing} + f_{ybar} \times A_{bar}) \qquad \text{...(8.4)}$$

where,

P_{ut} = ultimate structural capacity (tension)

$f_{ycasing}$ = yield strength of casing

A_{casing} = area of steel casing (at threaded joints if applicable)

f_{ybar} = yield strength of rebar

A_{bar} = area of rebar

8.12 PRODUCTS AND MATERIALS FOR MICROPILING

8.12.1 Water

Water for mixing grout shall be potable, clean and free from substances which may be in any way deleterious to grout or steel. If water is not potable, it shall be tested in accordance with AASHTO T26 for acceptability.

8.12.2 Admixtures

Admixtues shall conform to the requirements of ASTM C494 (AASHTO M194). Admixtures which control bleed, improve flowability, reduce water content, and retard set may be used in the grout subject to the review and acceptance of the Owner. Expansive admixtures shall be compatible with the grout and mixed in accordance with the manufacturer's recommendations. Their use will only be permitted after appropriate field tests on fluid and set grout properties. Admixtures with chlorides shall not be permitted.

8.12.3 Cement

All cement shall be Portland cement conforming to ASTM C150 (AASHTO M85) and shall be the product of one manufacturer. If the brand or type of cement is changed during a project, additional grout mix tests should be conducted to ensure consistency of quality and performance *in situ*.

8.12.4 Fillers

Inert fillers such as sand may be used in the grout in special situations (e.g., presence of large voids in the ground, when grout take and travel are to be limited).

8.12.5 Bar reinforcement

8.12.5.1 Reinforcing steel shall be deformed bars in accordance with ASTM A615 (AASHTO M31) Grade 60 or Grade 75 or 550 MPa tensile strength steel.

8.12.5.2 For cases of tensile loading, bar couplers, if required, shall develop the ultimate tensile stress of the bar, without any evidence of failure. For compressive loading, the coupler shall be compatible with efficient load transfer and overall reinforcement performance requirements.

8.12.6 Pipe/Casing

If the casing is to be relied upon to carry loads or reduce deflection, the permanent steel casing/pipe:

1. Shall meet the tensile requirements of ASTM A252, Grade 3, except the minimum yield strength shall be as used in the design submittal (typically 50,000 psi to 80,000 psi, psi = pounds per sq inch and minimum elongation shall be 15%.
2. May be new 'Structural Grade' steel pipe free from defects (dents, cracks, tears) and with two pipes tested per truckload.

It is recommended that if welding of the high strength steel casing is required, the welding procedure should be prepared by a welding specialist (Fig. 8.18).

Fig. 8.18: Typical casing threads-flush joint type (Deep Foundations Institute, 2004)

8.12.7 Centralisers

Centralisers are fabricated from plastic, steel or material that is non-detrimental to the reinforcing steel. Wood shall not be used. Fig. 8.19 shows a view of the centraliser. The central bulb of this centraliser is almost of size as diameter of borehole made for micropile. The top and bottom opening size of centraliser is same as that of micropile rebar diameter. The centraliser can be tightly fixed to the rebar with the help of small built in belt on both ends of centraliser.

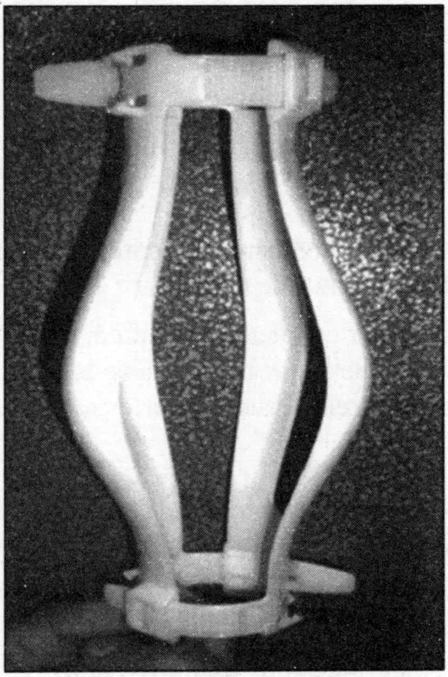

Fig. 8.19: A view of the centraliser

8.13 EXECUTION

8.13.1 Installation

(*i*) The micropile installation technique shall be such that it is consistent with the geotechnical, logistical, environmental, and load carrying conditions of the project.

(*ii*) The borehole must be open to the defined nominal diameter, full length, prior to placing grout and reinforcement. This condition may exist in stable, competent materials without the need for temporary hole support. Often however, some form of borehole wall support will be required for all or part of the pile length. Such methods include temporary steel casing, hollow stem augers, or the use of a hole-stabilising drilling fluid (provided it in no way has a deleterious effect on geotechnical bond development). Water, air, and polymer slurry foams are typical micropile hole flushing media. Air should be used with caution when in the vicinity of existing structures or services and granular soils below the water table.

(*iii*) All installation techniques shall be determined and scheduled such that there will be no interconnection or damage to piles in which grout has not achieved final set.

(*iv*) Centralisers shall be provided at 10 ft maximum vertical spacing on central reinforcement. The uppermost centralisers shall be located a maximum of 5 ft from the top of the micropile. Centralisers shall permit the free flow of grout without misalignment of the reinforcement.

(*v*) The central reinforcement steel with centralisers shall be lowered into the stabilised drill holes to the desired depth without difficulty. Partially inserted reinforcing bars shall not be driven or forced into the hole such that there will be no interconnection or damage to piles in which the grout has not achieved final set.

(*vi*) The contractor shall check pile top elevations and adjust all installed micropiles to the planned elevations.

8.13.2 Grouting

After drilling, the hole shall be flushed with water and/or air to remove drill cuttings and/or other loose debris. A stable, homogenous, neat cement grout or a sand cement grout with a minimum 28 days unconfined compressive strength of 4000 psi should be used. The grout shall not contain lumps or any other evidence of poor or incomplete mixing. Admixtures, if used, shall be mixed in accordance with manufacturer's recommendations. The pump shall be equipped with a pressure gauge to monitor grout pressure. The pressure gauge shall be capable of measuring pressures of at least 150 psi or twice the actual grout pressures used by the Contractor, whichever is greater. The grouting equipment shall be sized to enable the grout to be pumped in one continuous operation. The grout should be kept in constant agitation prior to pumping.

The grout shall be injected from the lowest point of the drill hole (by tremie methods) until clean, pure grout flows from the top of the micropile. The tremie grout may be pumped through grout tubes, hollow stem augers or drill rods. Subsequent to tremie grouting, all grouting operations associated with, for example, extraction of drill casing and pressure grouting, must ensure complete continuity of the grout column. The use of compressed air to directly pressurise the fluid grout is not permissible. The grout pressures and grout takes shall be controlled to prevent excessive heave in cohesive soils or fracturing of soil or rock formations. The entire pile shall be grouted to the design cut-off level. Grout within the micropiles shall be allowed to attain the minimum design strength prior to being loaded.

8.13.3 Pile splices

(*i*) Pile splices shall be constructed to develop the required design strength of the pile section.

(*ii*) Lengths of casing and reinforcing steel to be spliced shall be secured in proper alignments and in such a manner that no eccentricity between them results.

References

1. Babu Sivakumar, G.L., Murthy Srinivasa, B.R., Murthy, D.S.N and Nataraj, M.S. (2004). "Bearing Capacity Improvement Using Micropiles A Case Study", *Geosupport 2004*, pp. 692–699.
2. Bahloul, M.M., E.L. Shanawany, M.D., Azzam, W.R. (2004). "Strengthening of Loaded Soil System by Vertical Extensible Reinforcement", 5th International Conference on Ground Improvement Techniques: 22-23 March 2004, Kuala Lumpur, Malaysia, pp. 109–118.
3. Bhattacharjee, Arup. (2006). "Behaviour of Eccentrically Loaded Square Footing with Micropile", M.Tech. Thesis, Indian Institute of Technology Roorkee.
4. Bhattacharjee, Arup, Mittal, Satyendra & Krishna, Murali, A. (2011). "Bearing Capacity Improvement of Square Footing by Micropiles", *International Journal of Geotechnical Engg.,* USA, Vol. 5, pp. 113–118.
5. Bruce, D.A., DiMillio, A.F. and Juran, I. (1995). "Primer on Micropiles", *Civil Engineering*, Vol. 65, No. 12, December, pp. 51–54.
6. Guide to Drafting a Specification for Micropiles (2004). "Deep Foundations Institute", USA.
7. Handy, R.L. (2001). "Does Lateral Stress Really Influence Settlement?", *Journal of Geotechnical and Geoenvironmental Engineering,* ASCE, Vol. 127, No. 7, pp. 623–626.
8. Hari Krishna, P. and Ramana Murthy, V. (2000). "Effect of Orientation of Reinforcement on Bearing Capacity of Reinforced Sand Bed", IGC–2000, *The Millennium Conference*, pp. 325–326.
9. Lizzi, F. (1982). "The Pali Radice (Root Piles)", Symposium on Soil and Rock Improvement, AIT, Bankok, Nov. 29–Dec. 3, Paper D-1, p. 21.
10. Mahmoud, M.A. and Abdrabbo, F.M. (1989). "Bearing Capacity Tests on Strip Footing Resting on Reinforced Sand Subgrade", *Canadian Geotechnical Journal,* Vol. 26, No. 1, pp. 154–159.
11. Mandal, J.N. and Manjunath, V.R. (1995). "Bearing Capacity of Strip Footing Resting on Reinforced Sand Subgrade", *Construction and Building Materials*, Vol. 9, No. 1, pp. 35–38.
12. Micropile Design and Construction Guidelines (2000). "ADSC", Dallas, Texas.
13. Mittal, S. (2012). "Recent Experiences of Ground Improvement Techniques" Keynote Lecture Delivered in Indian Institute of Technology Guwahati.
14. Murthy Srinivasa, B.R., Babu Sivakumar, G.L. and Srinivas, A. (2002). "Analysis of Bearing Capacity Improvement using Micropiles", *Ground Improvement*, Vol. 6, No. 3, pp. 121–128.
15. Sharma, Binu. (2011). "A Model Study of Micropiles Subjected to Lateral Loading and Oblique Pull", *Indian Geotechnical Journal,* Vol. 41, No. 4, pp. 196–205.
16. Verma, B.P. and Char, A.N.R. (1986). "Bearing Capacity Tests on Reinforced Sand Subgrades", *Journal of Geotechnical Engineering,* Vol. 112, No. 7, pp. 701–706.
17. Verma, B.P. and Jha, J.N. (1992). "Three Dimensional Model Footing Tests for Improving Subgrades below Existing Footings", *Proceedings of International Symposium on Theory and Practice of Earth Reinforcement,* Balkema, Rotterdam, Vol. 1, pp. 707–711.

Stone Columns

CHAPTER 9

9.1 INTRODUCTION

It has always been a challenging task for Civil engineers to provide safe and sound foundations in poor sub-soil conditions, e.g., soft saturated clays and loose saturated sands etc. especially for structures with high design loads and permissible low settlements. The general practice is to improve the capacity of ground by various means, e.g., pre-consolidation with radial drainage with use of sand/band/wick drains, woven fabric reinforcement, dynamic compaction etc. Replacing existing weak cohesive soil with non-compressive soil is also an important technique of ground improvement. This granular material, if filled in boreholes and compacted properly, the resulting structure is called the granular pile or stone column. The stone columns or granular piles result in the modification of the modulus of compressibility of composite soil mass.

This is very effective method in cohesive soils or marine clays. In this method 25 to 33% of soft cohesive soil is replaced by compacted stone/crushed rock in drilled vertical boreholes to construct stone columns or granular piles in the soil. These stone columns (or granular piles) are more economical where gravel and sand are available in abundance nearby. The stone columns are not recommended if undrained shear strength of soil is less than 10 kPa and stone columns are not required if undrained shear strength is greater than 50 kPa (Madhav, 2012). The stone columns are very effective in liquefaction prone sites. In Kobe earthquake of Japan, wherever stone columns were made, almost negligible or no damages took place.

The improvement in the ground capacity is due to:

(*i*) the reinforcement effect provided by the stone columns

(*ii*) enhanced rate of sub-soil consolidation by vertical drainage below stone columns/piles.

Stone columns, being more stiffer than the surrounding soil, share major part of applied vertical load (Fig. 9.1).

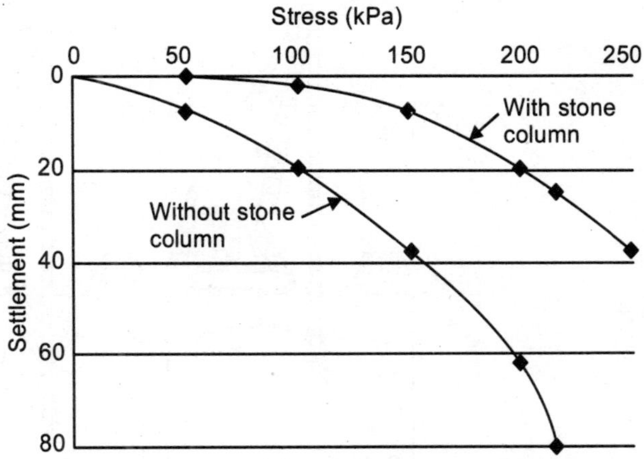

Fig. 9.1: Effectiveness of stone columns (Rao, 1983)

9.2 CONSTRUCTION METHODOLOGIES

Stone columns can be constructed by the following two methods:

(*i*) Vibroflotation process

(*ii*) Ramming process

(i) Vibroflotation process

The stone columns can be constructed using a vibroflot [Fig. 9.2 (*a*) and (*b*)]. In the case of non cohesive sub-soil stratum, the vibroflot sinks in the ground under its own weight (about 3–8 tonnes, depending on the total length) with the assistance of water jet and vibration. The length of the extension tube together with the vibrator and the lifting height of the crane is required to correspond with the total depth of penetration. After reaching the predetermined depth, the vibrator is then gradually withdrawn from the ground causing compaction. Three basic steps are involved in the construction [Fig. 9.2 (*b*)].

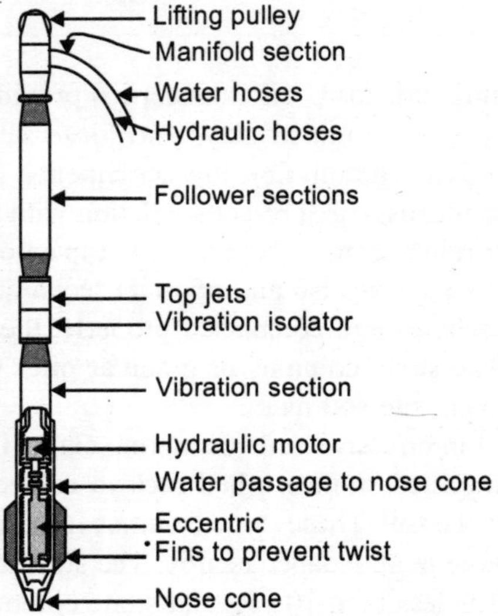

Fig. 9.2: (a) Essential features of a vibroflot

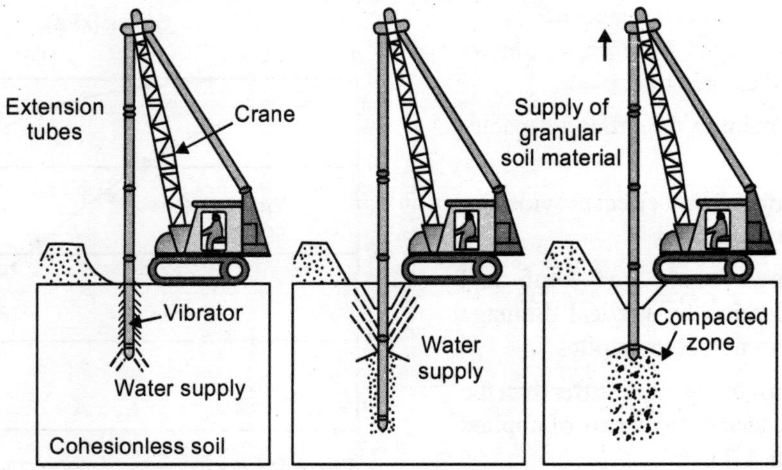

Fig. 9.2: (b) The vibro compaction process

The effectiveness of the compaction is dependent on the characteristics of the vibrator in terms of energy input, amplitude, frequency and its shape.

In well graded sand the vibrators require centre spacing of 3–3.5 m in equilateral triangular grids (Fig. 9.3) to produce 65–70 per cent of relative density in the centroid between three compaction points. But close spacings down to 1.5 m can produce minimum relative density of 90 per cent and even more.

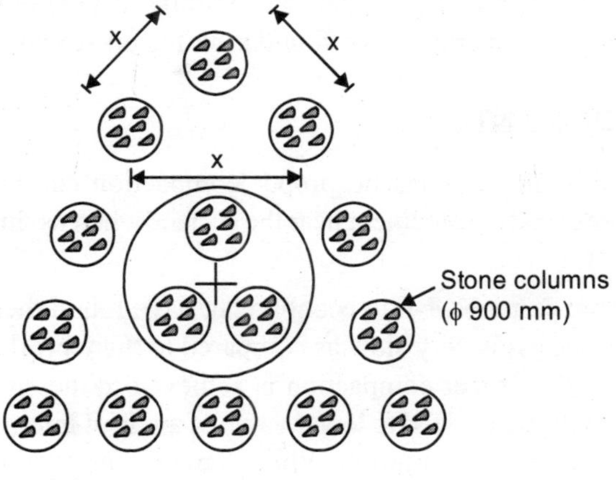

Fig. 9.3: Arrangement of columns

Note: '*x*' depends on soil property, loads on piles etc.

(ii) Ramming process

Datye and Nagaraju (1985) suggested the process of installation of stone columns/gravel piles by ramming of stone aggregate and sand within a cased borehole and also detailed the method of advancing borehole, placing of stone and sand and ramming procedure. In this method the granular fill is placed into a drilled hole and compacted by a heavy (1.5–2 Ton) rammer falling through the height of 1 m–1.5 m in the borehole (Fig. 9.4).

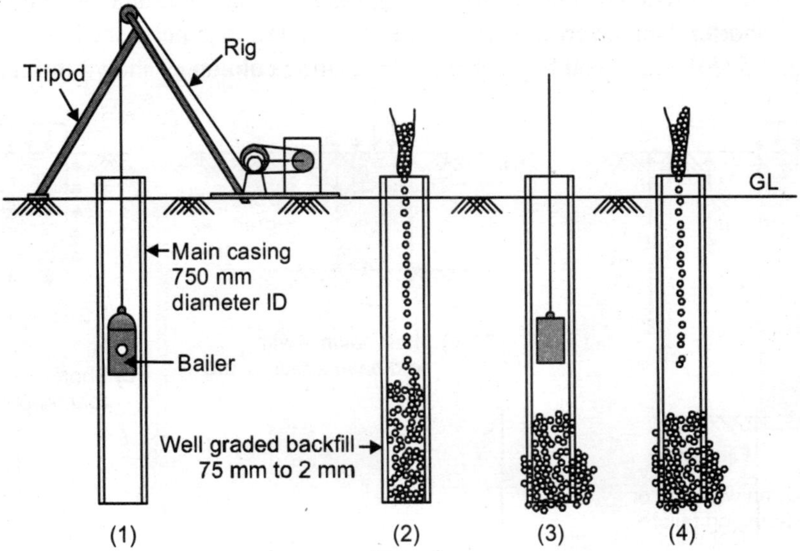

Fig. 9.4: Stone column installation by ramming method

Mixture of stone aggregate (12 mm–75 mm) and sand, in the ratio of 2 : 1 respectively, is used as backfill material so that it yields high angle of internal friction under applied compaction energy. Rao (1983), Ranjan (1989) also suggested making of stone column by making bore hole (by auger) and filling gravel + sand alternatively and thereafter compacting it. The gravel thickness in one layer is 30–50 cm and sand layer thickness is 8–10 cm. Filling of sand helps filling of voids in gravels. Granular piles of 15 m depth and 600 mm ϕ had been successfully installed by this technique by the authors in Haldia refinery (India) for construction of foundation for oil storage tanks.

9.3 GENERAL COMMENTS

In case of vibrofloted stone columns no proper compaction can be achieved in the top 1 m depth approximately due to reduced confinement near the surface whereas this part of stone column is subjected to higher load intensity.

1. Rammed stone columns, being economical, have found their wider application inspite of the fact that the process is very slow as compared to the vibroflotation. These are suited to almost all types of soils. Proper compaction is achieved in the entire length of stone column. High energy compaction may lead to lower value of angle of internal friction due to possible crushing of stone aggregate. The variation in pile diameter installed by vibroflot varies between 0.6 m (stiff clays) to 1.1 m (very soft cohesive soils).
2. The spacing of stone column is determined on the basis of settlement tolerance for the loads to be applied and the degree of improvement required. The spacing recommendations are same as that for sand piles. The furnace slag can also be used in place of gravel.

9.4 FAILURE MECHANISMS

Failure mechanism of a single stone column loaded over its area significantly depends upon the length of the column. For columns having length greater than its critical length (that is about 4 times the column diameter) and irrespective whether it is end bearing or floating, it fails by bulging [Fig. 9.5 (*a*)]. However, column shorter than the critical length are likely to fail in general shear if it is end bearing on a rigid base [Fig. 9.5 (*b*)] and in end bearing if it is a floating column as shown in Fig. 9.5 (*c*).

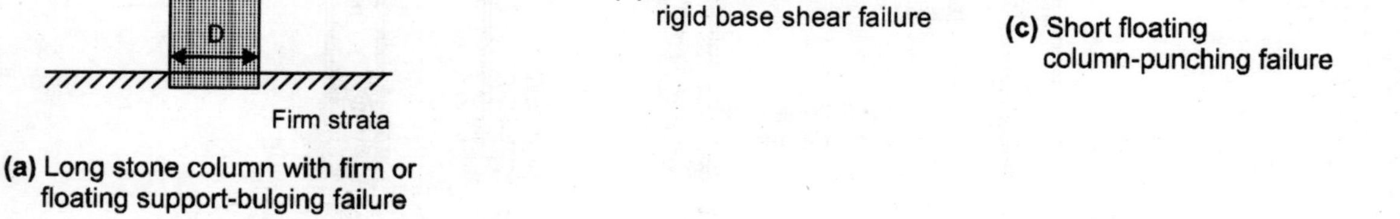

(a) Long stone column with firm or floating support-bulging failure

(b) Short column with rigid base shear failure

(c) Short floating column-punching failure

Fig. 9.5: Failure mechanisms of a single stone column in a homogeneous soft layer

In practice, however, a stone column is usually loaded over an area greater than its own (Fig. 9.6) in which case it experiences significantly less bulging leading to greater ultimate load capacity and reduced settlements since the load is carried by both the stone column and the surrounding soil.

> **Note:** The above failure mechanisms apply to stone columns installed in homogeneous soils. Practical situations may arise where isolated zones of very soft cohesive soils may result in significant bulging at both shallow and deep depths and hence, this should be duly considered wherever necessary.

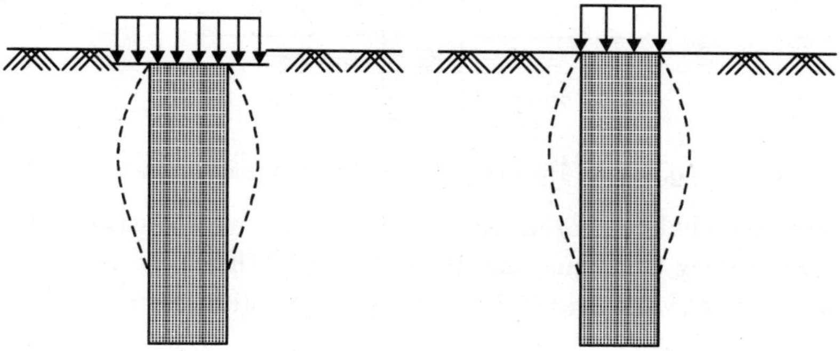

Fig. 9.6: Different type of loadings applied to stone columns

Wherever interlayering of sand and clay occurs, and if the sand layer is thick enough as compared to the size of the loaded area, the general compaction achieved by the action of the installation of the stone columns may provide adequate rigidity to effectively disperse the applied stresses thereby controlling the settlement of the weak layer. However, effective reduction in settlement may be brought about by carrying out the treatment of stone columns through the compressible layer. When clay is present in the form of lenses and if the ratio of the thickness of the lense to the stone columns diameter is less than or equal to 1, the settlement due to presence of lenses may be insignificant. In mixed soils, the failure of stone columns should be checked both for predominantly sandy soils as well as the clayey soil, the governing value being lower of the two calculated values.

The bearing capacity of an isolated stone column or that located within a group may be computed using the other established theories also. Besides the passive resistance mobilised by the soil, the increase in capacity of the column due to surcharge should be taken into consideration. In addition, capacity increase due to soil bearing should also be taken into account.

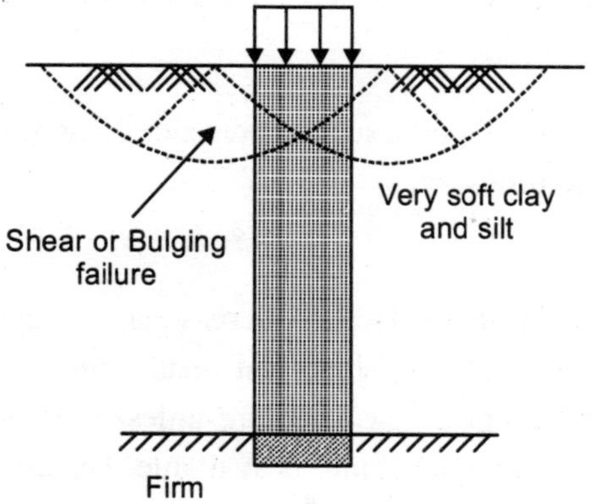

Fig. 9.7: (a) Soft layer at surface–bulging or shear failure

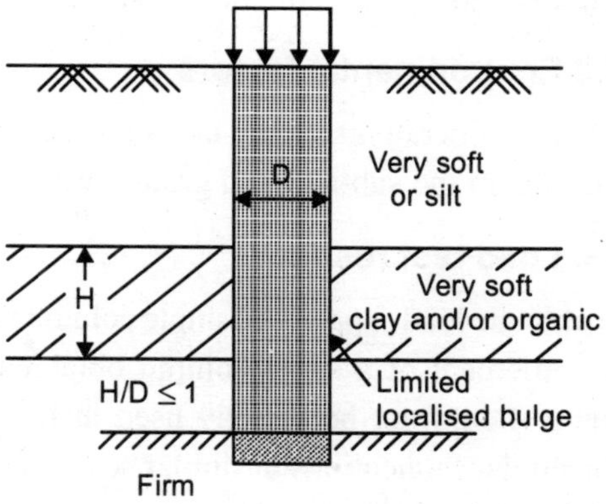

Fig. 9.7: (b) Thin very soft layer–contained local bulge

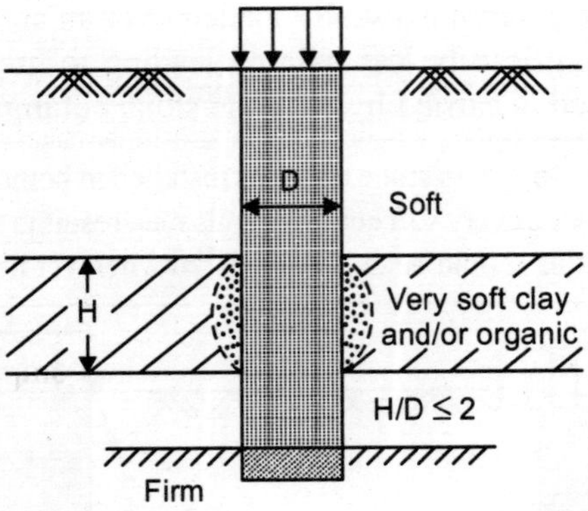

Fig. 9.7: (c) Thick very soft layer–local bulging failure

Particular attention should be paid to the presence of very weak organic clay layers of limited thickness where local bulging failure may take place (Fig. 9.7). Therefore, capacity of column in such weak clays should also be checked even if they are below the critical depth.

9.4.1 Adjacent structures

When working near existing structures, care should be taken to avoid damage to such structures by suitable measures.

In case of deep excavation adjacent to stone columns, prior shoring or other suitable arrangement should be done to guard against the lateral movement of soil or loss of confining soil pressure.

9.4.2 Ultimate load capacity and settlement

The ultimate load carrying capacity of stone column may be estimated approximately on the basis of soil investigation data or by test loading. However, it should be preferably determined by an initial load test on a test column specifically installed for the purpose and tested to its ultimate load particularly in a locality where no such previous experience exists.

Procedure for estimating the load capacity and settlement of a single column is given in section 9.5 and 9.7 respectively given on subsequent pages. Any other alternate formulae with substantially proven reliability depending upon the sub-soil characteristics and the method of installation, may also be used.

9.4.3 Environmental factors

Design considerations should take into account the environmental factors, such as presence of aggressive chemicals in the sub-soil and ground water, an artesian conditions, etc.

9.4.4 Load test results

The ultimate load capacity of single column may be determined from load tests with reasonable accuracy. The settlement of a stone column obtained at safe/working load from load test results on a single column should not be directly used in forecasting the settlement of the structure unless experience from similar foundations in similar soil conditions on its settlement behaviour is available. The average settlement may be assessed on the basis of sub-soil data and loading details of the structures as a whole using the principles of soil mechanics.

9.4.5 Granular blanket

Irrespective of the method used to construct the stone columns, the blanket laid over the top of the stone columns should consist of clean medium to coarse sand compacted in layers to a relative density of 75–80 per cent.

Minimum thickness of the compacted sand blanket should be 0.5 m. This blanket should be exposed to atmosphere at its periphery for pore water pressure dissipation.

After ensuring complete removal of slush deposited during boring operations, a minimum depth of 0.5 m, preferably 0.75 m below the granular blanket should be compacted by other suitable means, such as rolling/tamping to the specified densification criteria.

9.4.6 Field loading tests

Irrespective of the method used to construct the one columns, the initial load tests should be performed at a trial test site to evaluate the load settlement behavior of the soil-stone column system. The tests should be conducted on a single [Fig 9.8 (*a*)] and also on group of minimum three columns.

For the initial load tests, in order to simulate the field conditions of compaction of the intervening soil, minimum of seven column for a single test and twelve columns for three column group test may be constructed for triangular pattern as shown in Fig. 9.8 (*b*).

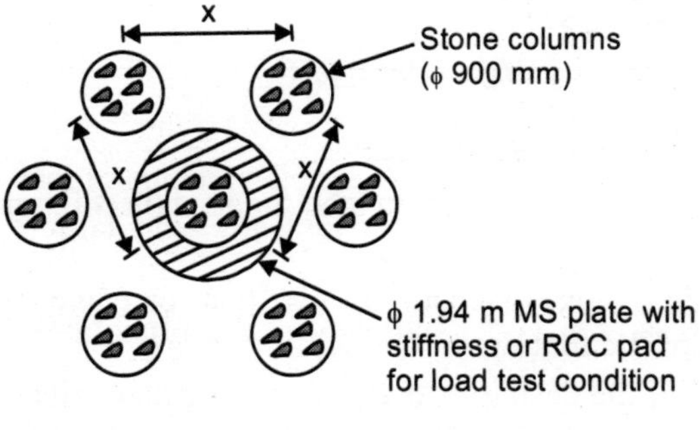

Fig. 9.8: (a) Single column test

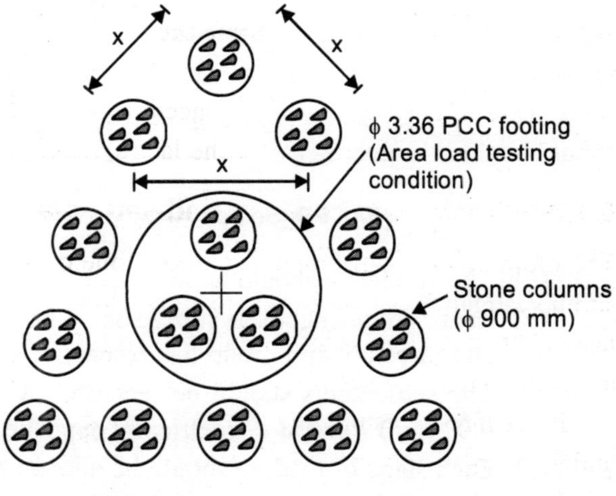

Fig. 9.8: (b) Three column group test

The diameter of the circular concrete footing or equivalent steel plate of adequate thickness and rigidity may be based on effective tributary soil area of stone column for a single column test and three times the effective area of single column for a three column group test. In each case, the footing may cover the equivalent circular effective area centrally.

The initial and final soil conditions at the trial site should be investigated by drilling at least one borehole and one static cone test/pressure meter test/dynamic cone test prior and subsequent to the installation of columns (IS: 15284, Part 1, 2003). All these tests including the standard penetration test, field vane shear tests and collection of undisturbed/disturbed samples and laboratory testing on the samples should be as per relevant Indian Standards.

A granular blanket of medium to coarse sand having thickness not less than 300 mm should be laid over the test column(s) as per section 9.4.5. Over the blanket, a properly designed footing should be laid. The footing may be cast away from the test site and transported to the test location so as to fix it properly over the sand blanket.

In case high water table conditions exist at site the water level during the tests should be maintained at the footing base level by dewatering.

Following procedure should be followed for application of load:

(*a*) The load should be applied to the footing by a suitable kentledge (see Fig. 9.9), taking care to avoid impact, fluctuations or eccentricity.

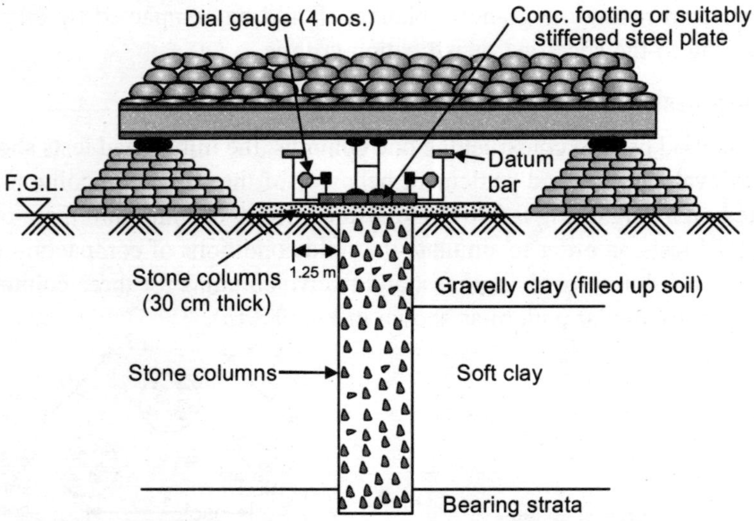

Fig. 9.9: Section showing loading arrangement for a single column test

(*b*) The kentledge should be minimum 1.30 times the minimum test load.

(*c*) Load settlement observations should be taken to 1.5 times the design load for a single column and three column group test respectively.

(*d*) The settlements should be recorded by four dial gauges (sensitivity less than or equal to 0.02 mm) fixed at diametrically opposite ends of the footing.

(*e*) Each stage of loading should be near about 1/5 of the design load and should be maintained till the rate of settlement is less than 0.05 mm/hr at which instant the next stage of loading should be applied.

(*f*) The design as well as the maximum test load should be maintained for a minimum period of 12 hr after stabilisation of settlement to the rate as given above point (*e*).

(*g*) Load settlement and time settlement relationships should be plotted from the settlements observed for each increment of load at intervals of 1 min, 2 min, 4 min, 8 min, 16 min, ½ hr, 1 hr, 1½ hrs, 2 hrs, 3 hrs, 4 hrs and so on till the desired rate of settlement has been achieved. The time intervals may be suitably modified if so desired.

(*h*) The test load should be unloaded in five stages. At each stage enough time should be allowed for settlements to stabilise.

(*i*) The load test should be considered acceptable if it meets the following settlement criteria:

(*a*) 10–12 mm settlement at design load for a single column test, and

(*b*) 25–30 mm settlement at the design load for a three column group test.

(*j*) For routine load test few job columns (say 1 test for 625 m^2 area) may be tested up to 1.1 times the design load intensity with minimum kentledge of 1.3 times the design load.

9.5 ESTIMATION OF LOAD CAPACITY OF A STONE COLUMN (IS : 15284, Part 1, 2003)

9.5.1 Stone columns in cohesive soils

Load capacity of the treated ground may be obtained by summing up the contribution of each of the following components for wide spread loads, such as large size tanks resting on ground and embankments:

(*a*) Capacity of the stone column resulting from the resistance offered by the surrounding soil against its lateral deformation (bulging) under axial load.

(*b*) Capacity of the stone column resulting from increase in resistance offered by the surrounding soil due to surcharge over it, and

(*c*) Bearing support provided by the intervening soil between the columns.

9.5.2 Capacity based on bulging of column

Considering that the foundation soil is at failure when stressed horizontally due to bulging of stone column, the limiting (yield) axial stress in the column is given by the sum of the following:

$$\sigma_v = \sigma_{rL} . K_{p_{col}}$$

or,

$$\sigma_v = (\sigma_{ro} + 4C_u)\, K_{p_{col}} \quad \ldots(9.1)$$

Here,

σ_v = limiting axial stress in the column when it approaches shear failure due to bulging, and

σ_{rL} = limiting radial stress

$= \sigma_{ro} + 4C_u$

Here,

C_u = undisturbed undrained shear strength of clay surrounding the column, and

σ_{ro} = initial effective radial stress

$= K_0 \sigma_{vo}$

where,

K_0 = average coefficients of lateral earth pressure for clays equal to 0.6 or alternatively, as determined from the relationship $K_0 = 1 - \sin\phi$, where ϕ is the effective angle of internal friction of soil, and

σ_{vo} = average initial effective vertical stress considering an average bulge depth as 2 times diameter of the column [see Fig. 9.5 (*a*)], that is $\sigma_{vo} = \gamma.2D$.

where,

γ = effective unit weight of soil within the influence zone

$$K_{p_{col}} = \tan^2\left(45° + \frac{\phi_c}{2}\right)$$

where

ϕ_c = angle of internal friction of the granular column material and it may vary depending upon angularity, surface characteristics and density of column material. Value applicable for the stones intended to be used as backfill material may be determined using large shear box tests or laboratory shear test. In absence of such tests, the design may be based on the best engineering judgement. As a broad guide, the ϕ_c may range from 38°–42° depending upon the compactness achieved during construction of stone columns.

$$\text{Yield load} = \sigma_v \left(\frac{\pi}{4}\right) D^2$$

Safe load on column alone:

$$Q_1 = \left(\sigma_v \frac{\pi}{4} D^2\right)/2 \qquad \text{...(9.2)}$$

where 2 is the factor of safety.

9.5.3 Surcharge effect

(*a*) Initially, the surcharge load is supported entirely by the rigid column. As the column dilates some load is shared by the intervening soil depending upon the relative rigidity of the column and the soil. Consolidation of soil under this load results in an increase in its strength which provides additional lateral resistance against bulging.

(*b*) The surcharge load may consist of sand blanket and sand pad (being applicable to tank foundations). If thicknesses of these elements are not known, the limiting thickness of the surcharge loading as represented by the safe bearing capacity of the soil may be considered.

(*c*) The increase in capacity of the column due to surcharge may be computed in terms of increase in mean radial stress of the soil as follows:

$$\Delta\sigma_{ro} = \frac{Q_{safe}}{3}(1 + 2K_0) \qquad \text{...(9.3)}$$

where $\Delta\sigma_{ro}$ is the increase in mean radial stress due to surcharge and Q_{safe} is the safe bearing pressure of soil with the factor of safety of 2.5 (see IS: 6403).

$$Q_{safe} = C_u N_c / 2.5$$

Increase in ultimate cavity expansion stress $\Delta\sigma_{ro} F_q'$

where,

F_q' = Vesic's dimensionless cylindrical cavity expansion factor = 1 for $\phi_g = 0$

Increase in yield stress of the column = $K_{p_{col}} \Delta\sigma_{ro}$

(*d*) Allowing a factor safety of 2, increase in safe load of column, Q_2 is given by the following formula:

$$Q_2 = \frac{K_{p_{col}} \Delta\sigma_{ro} A_s}{2} \qquad \text{...(9.4)}$$

The surcharge effect is minimum at edges and it should be compensated by installing additional columns in the peripheral region of the facility.

9.5.4 Bearing support provided by the intervening soil

This component consists of the intrinsic capacity of the virgin soil to support a vertical load which may be computed as follows:

Effective area of stone column including the intervening soil for triangular pattern = $0.866\ S^2$

Area of intervening soil for each column, A_g is given by the following formula:

$$A_g = 0.866\ S^2 - \frac{\pi D^2}{4}$$

Safe load taken by the intervening soil,

$$Q_3 = q_{safe} A_g \qquad \text{...(9.5)}$$

Overall safe load on each column and its tributory soil = $Q_1 + Q_2 + Q_3$.

Note: The number of columns to be provided under a structure may be obtained if the total load to which the structure is subjected to and the reduction in settlements required considering the permissible total and differential settlements for it are known.

9.5.5 Safe load capacity of stone column (Nayak, 1985)

Safe load capacity of a stone column is given as:

$$Q = Q_1 + Q_2 + Q_3$$

where,

Q_1 = load capacity resulting from resistance offered by the surrounding soil against lateral deformation (bulging) of stone column under axial load.

$$= \frac{1}{F.S.} [\ \gamma^2 .\ 4\ DK_0 + 4c)\ \tan^2 (45 + \phi_1 / 2]\ A$$

where,

A = cross-sectional area of stone column

γ^2 = density of soil

K_0 = 0.6 to 0.8

FS = Factor of safety usually taken as 1.5 to 2

c = cohesion of field value taking into consideration in conditions before filling or loading.

Q_2 = Bearing support provided by soil in between the stone column

Q_2 for (*i*) triangular grid pattern of stone column

$$= \left[S^2 \sin 60 - \frac{\pi D^2}{4}\right] \left\{\frac{CN_c}{FS}\right\}$$

For (*ii*) square grid pattern of stone column,

$$= \left[S^2 - \frac{\pi D^2}{4}\right] \left\{\frac{CN_c}{FS}\right\}$$

FS is tken as 4, and 'S' is taken as spacing of column

Q_3 = increase in resistance to lateral deformation due to superimposed loads i.e., surcharge effect on surrounding soil

$$= \frac{\pi D^2}{4} \left[\frac{Q_{safe}}{3} (1 + 2\ K_0)\ \tan^2 (45 + \phi_1 / 2\ )\right] \times \frac{1}{(FS)}$$

Here, $Q_{safe} = \frac{CN_c}{4}$, FS = 1.5 to 2.

The recent research conducted at some places (IIT Roorkee, IIT Madras, etc.) suggest that while compacting the stone columns, the granular material (ballast) etc. may penetrate into surrounding soil. Therefore, if it is encased in a geosynthetics casing, such migration of ballast material shall be mitigated. Monographs published by CBIP (1993) have given some ready tables to see the difference in load carrying capacity of stone columns without encasing and duly encased.

A typical table is shown below for ready reference:

Soil Data, $c = 1$ t/m², $\phi = 35°$ S = 1.25 m.

Load carrying capacity (t)	Stone column diameter (mm)						Remarks
	500	600	750	800	900	1000	
$Q = Q_1 + Q_2 + Q_3$ (t)	4.49	5.65	7.89	8.77	10.75	13.02	Without geosynthetics encasing
-do-	4.58	5.77	8.03	8.92	10.91	13.21	With geosynthetics encasing
$\phi = 35°$ spacing = 1.75 m, c = 1 t/m²							
Q	6.44	7.60	9.84	10.72	12.70	14.97	Without encasing
Q	6.53	7.71	9.98	10.37	12.86	15.16	With encasing
$\phi = 45°$ spacing = 1.25 m, c = 1 t/m²							
Q	6.08	8.0	11.67	13.11	16.35	20.08	Without encasing
Q	6.23	8.18	11.89	13.35	16.62	20.37	With encasing
$\phi = 45°$ spacing = 1.75 m, c = 1 t/m²							
Q	8.03	9.95	13.62	15.06	18.30	22.03	Without casing
Q	8.18	10.12	13.83	15.29	18.56	22.32	With casing

Interestingly, here it is seen that–

(*a*) when geosynthetic casing is provided the load carrying capacity is also increased.

(*b*) when spacing between columns is increased, the load carrying capcity is also increased.

9.6 STONE COLUMNS IN MIXED SOILS

In soils having both C and ϕ, the capacity may be estimated by using Bell's formula for passive pressure:

$$\sigma_{rL} = p_p = \gamma\, z\, K_p + 2C_u \sqrt{K_p} \qquad ...(9.6)$$

where,

p_p = passive pressure,

z = average bulge depth

= 2 times the column diameter, and

K_p = passive pressure coefficient of soil

= $(1 + \sin \phi_g)\,(1 - \sin \phi_g)$

where,

ϕ_g = angle of internal friction of soil.

Limiting axial stress in the column, when it approaches shear failure due to bulging σ_v is given by the following formula:

$$\sigma_v = \sigma_{rL} K_{p_{col}}$$
$$= \sigma_{rL} (1 + \sin \phi_c) / (1 - \sin \phi_c)$$

Safe load on column, $Q_1 = \sigma_v \dfrac{\pi/4D^2}{2}$ ...(9.7)

The surcharge effect (Q_2) and bearing support of the intervening soil (Q_3) may be obtained as per section 9.5.4.

9.6.1 Determination of column spacing and pattern

Stone columns should be installed preferably in an equilateral triangular pattern which gives the most dense packing although a square pattern may also be used. A typical layout in an equivalent triangular pattern is shown in Fig. 9.10.

The design of stone columns should be site specific and no precise guidelines can be given on the maximum and the minimum column spacing. However, the column spacing may broadly range from 2 to 3 depending upon the site conditions, loading pattern, column factors, the installation technique, settlement tolerances, etc.

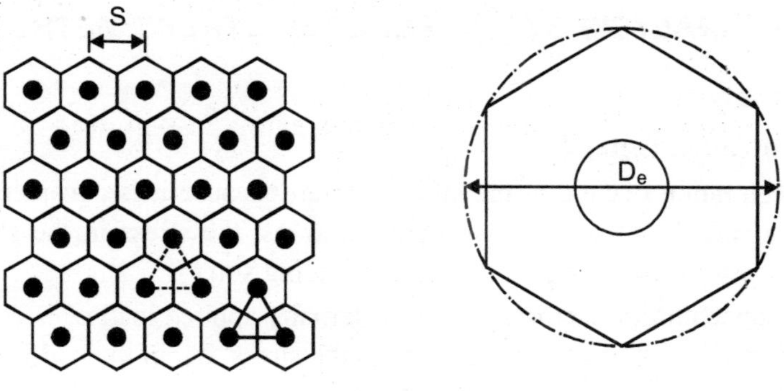

Triangular arrangement of stone columns

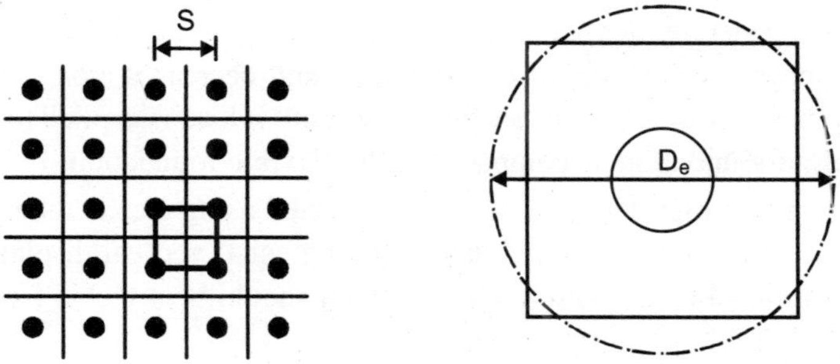

Square arrangement of stone columns

Fig. 9.10: Various pattern of stone columns

For large projects, it is desirable to carry out field trials to determine the most optimum spacing of stone columns taking into consideration the required bearing capacity of the soil and permissible settlement of the foundation.

From the plan area of the structure and the number of columns as assessed in section. 9.4.4, area per column is arrived at. This, in turn, will lead to effective spacing between the columns depending upon the pattern of columns as follows:

Pattern	Area for column
Triangular	0.866 S^2
Square	1.0 S^2

Design calculations should be repeated till there is convergence of the assumed and the calculated column spacing. One or two trials may be required to achieve an acceptable degree of convergence.

Additional stone columns may be required inside and outside the periphery of the loaded area considering pressure distribution, presence/absence of surcharge and permissible or expected settlement of the structure. These additional columns may be provided either as rings or at a closer spacing for an appropriate distance inside as well as outside the periphery of the loaded area.

The load capacity of the stone columns computed as in accordance with section 9.4.6 should be verified by load test as per Fig. 9.9.

9.7 SETTLEMENT ANALYSIS BY THE REDUCED STRESS METHOD

9.7.1 Settlement analysis

9.7.1.1 Stress concentration factor (*n*)

(*i*) Stress concentration occurs on the stone column because it is considerably stiffer than the surrounding soil. From equilibrium considerations, the stress in the stiffer stone columns should be greater than the stress in the surrounding soil.

(*ii*) The stress concentration factor, n, due to externally applied load σ_s is defined as the ratio of the average stress in the stone column, σ_s to the stress, σ_g, in the soil within the unit cell,

$$n = \frac{\sigma_s}{\sigma_g} \qquad ...(9.8)$$

The value of n generally lies between 2.5 and 5 at the ground surface. The stress concentration factor (n) increases with time of consolidation and decreases along the length of the stone column. Higher n value at ground surface may result if load is applied to the composite ground through a rigid foundation as compared to the flexible foundation.

(*iii*) The stress concentration factor, n, may be predicted using elastic theory as a function of the modular ratio of the stone and the clay assuming equal vertical displacements. However, as the modular ratio can vary within wide limits, it should be selected from section 9.7.1.1 (*ii*).

9.7.1.2 Replacement ratio

For purpose of settlement and stability analysis, the composite ground representing an infinitely wide loaded area may be modelled as a unit cell comprising the stone column and the surrounding tributary soil. To quantify the amount of soil replaced by the stone, the term replacement ratio, a_s, is used. Replacement ratio (a_s) is given by:

$$a_s = \frac{A_s}{A} = \frac{A_s}{A_s + A_g} \qquad \text{...(9.9)}$$

A_s = area of the stone column,

A_g = area of ground surrounding the column, and

A = total area within the unit cell.

The area of replacement ratio may also be expressed as follows:

$$a_s = 0.907\ (D/S)^2 \qquad \text{...(9.10)}$$

where the constant 0.907 is a function of the pattern used which, in this case, is the commonly employed equilateral triangular pattern.

9.7.1.3 Settlement of the untreated ground

It should be computed as per IS 8009. Settlements of the treated ground may be estimated using the reduced stress method based on the stress concentration factor *n* and the replacement ratio, a_s. Stress concentration factor should be suitably arrived as per section 9.7.1.1. Area replacement ratio may be obtained from the column diameter and the spacing of stone columns (section 9.7.1.2). Following this, the settlement of the treated ground and reduction factor *b* can be worked out as follows:

(*a*) The applied stress, σ is shared between the columns and the surrounding soft ground in proportion to the relative stiffness of the 2 materials, the cross-sectional area of the columns (A_s) and their spacing.

(*b*) Sharing of the applied stress between the column and the tributary soil is expressed in terms of the stress concentration ratio, *n*, given below:

$$n = \frac{\sigma_s}{\sigma_g} \qquad \text{...(9.11)}$$

where,

σ_s = vertical stress in compacted columns, and

σ_g = vertical stress in surrounding ground.

(*c*) If A_g is the plan area of the soil for columns, then:

$$(A_s + A_g)\,\sigma = A_s\,\sigma_s + A_g\,\sigma_g$$

(*d*) The replacement of soil with stone is expressed in terms of the replacement ratio, a_s:

$$a_s = \frac{A_s}{A_s + A_g} \qquad \text{...(9.12)}$$

(*e*) The sharing of applied load between the soil and stone column is determined from the following formulae:

$$\sigma_g = \frac{\sigma}{1 + (n-1)\,a_s} = \mu_1 \sigma \qquad \text{...(9.13)}$$

$$\sigma_g = \frac{n\,\sigma}{1 + (n-1)\,a_s} = \mu_s \sigma \qquad \text{...(9.14)}$$

(*f*) Consolidation settlement of the composite (treated) soil S_t is given by:

$$S_t = m_v \sigma_g H = m_v \mu_g \sigma H \qquad ...(9.15)$$

where,

m_v = modulus of volume decrease of soil, and

H = thickness of treated soil.

Note: This is based on the assumption that the settlement of the strata underlying the column tip is added to the settlement of the reinforced ground.

(*g*) Consolidation settlement S of the unreinforced ground is computed from the one dimensional consolidation theory as given below:

$$S = m_v \sigma H \qquad ...(9.16)$$

(*h*) Settlement reduction ratio β is defined as:

$$\beta = \frac{\text{Settlement of treated soil, } S_t}{\text{Settlement of untreated soil, S}}$$

$$\beta = \frac{S_t}{S} = \frac{1}{1+(n-1)\,a_s} \qquad ...(9.17)$$

9.8 SIMPLIFIED DESIGN OF STONE COLUMN

The length of stone column should be decided as per the site condition and the load to be transferred by super structure. The length of stone column may also be decided on the basis of Standard Penetration tests and/or Dynamic Cone Penetration test result. However, the length of stone column (L_c) should be greater than:

$$L_c \geq \frac{P - A_0(9\,C_{pt})}{\pi\, d.c.} \qquad ...(9.18)$$

where,

P = total expected load on stone column

A_0 = cross-sectional area of stone column

d = average diameter of stone column

c, C_{pt} = side and point cohesion of soil respectively

Note: Though the ground also takes some load, but for the conservative design it is assumed that the foundation loads are carried only by total no. of stone columns.

9.8.1 Load capacity of granular pile/stone column

The load capacity of granular pile or stone column may be calculated by the following formula:

$$q_a = \frac{K_p\,(4c + \sigma'_r)}{FOS} \qquad ...(9.19)$$

where,

K_p = passive pressure coefficient = $\tan^2(45 + \phi/2)$

ϕ = angle of internal friction of stone ballast in drained condition

c = drained cohesion

σ'_r = effective radial stress as measured by a pressure meter (may be taken as $2c$ if pressure meter data is not given)

FOS = factor of safety = 2 to 2.5

On the basis of theory proposed by Gibson & Anderson, the load supported by the stone column (granular pile) is given as:

$$Q_d = K_p (8C_u + O_e + O_v) \, . \, A_p \quad \text{...(9.20)}$$

where, K_p = passive pressure coefficient = $\tan^2 (45 + \phi/2)$.

(Here, for computation of K_p, ϕ is normally taken as angle of internal friction of material used. For stone ballast, ϕ is 44°, but for conservative design ϕ is taken as 2/3 of 44° i.e. 29.5°.)

C_u = undrained shear strength = $2 \times O_e$

O_e = effective average stress at critical pile depth

= $d_c \times \gamma_b$ (γ_b = bulk density)

d_c = critical pile depth (may be taken as 4 to 5 times of pile diameter)

O_v = effective normal pressure (kg/cm²) = $k_B q$

$$k_B = 1 - \left\{ \frac{1}{1 + \left[\frac{a}{z}\right]} \right\}^{3/2}$$

a = diameter of circular load (if circular foundation is there, otherwise the shorter dimension of footing size)

z = critical depth of pile (d_c) = 4 to 5 times of pile diameter

q = load intensity = $\dfrac{\text{Total load of super structure}}{\text{Bearing area}}$

A_p = cross-sectional area of pile stem

$$Q_{safe} = \frac{Q_d}{FOS} \quad \text{...(9.21)}$$

9.8.2 Design Example (Mittal, 2012)

A granular pile foundation has to be designed for a 23.8 m diameter molasses storage steel tank proposed to be laid on ground resting on silty clayey strata. The tank should satisfy both shear and settlement criteria as applicable to liquid storage tanks. The dynamic cone test data is as shown in the Fig. 9.9. The other data are:

(*i*) Submerged unit weight = 0.95 t/m³ and Liquid limit of soil = 49%

(*ii*) Undrained shear strength (c) = 2.7 t/m² (determined from unconfined compressive strength test)

(*iii*) Ground water table (GWT) at 0.5 m depth below natural ground level (NGL).

SOLUTION: From the dynamic cone test results (Fig. 9.11), the N-values indicate up to 10 m depth the soft to medium compact strata and thereafter it tends to become stiffer. This chart can give fairly good idea of deciding depth of stone column (granular pile).

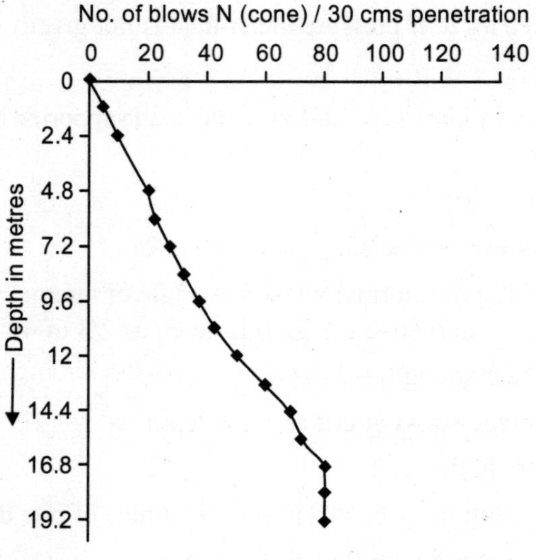

Fig. 9.11: Dynamic cone resistance curve

(i) Computation of safe bearing capacity of soil

(a) Net ultimate bearing capacity from shear criteria (IS: 6403-1981, reaffirmed 1997)

The net ultimate bearing capacity for foundations on fairly saturated cohesive soil is calculated by using following relationship (IS: 6403-1981)

$$q_d = CN_cS_cd_ci_c \quad ...(9.22)$$

where,

q_d = net ultimate bearing capacity (t/m²)

c = cohesion of the soil (2.7 t/m²)

N_c = 5.14 (IS: 6403-1981)

S_c = shape factor = 1.3 (for circular foundations)

d_c = depth factor = 1 (assumed)

i_c = inclination factor = 1 (assumed)

Thus, $q_d = 2.7 \times 5.14 \times 1.3 \times 1 \times 1 = 18.04$ t/m²

Applying a factor of safety as 2.5

$$Q_{safe} = \frac{18.04}{2.5} = 7.21 \text{ t/m}^2$$

(ii) Computation of settlement

Firstly, the influence zone below foundation (equivalent to 1.5 times the foundation width) is divided into different layers (Fig. 9.12). Lesser the thickness of layer, more accurate will be the computation of settlement. Then settlement (S) in each layer is computed. The load dispersion may be taken as 2 : 1 i.e., 2 vertical to 1 horizontal. This becomes nearly 26° which has been taken for computations of settlement. For settlement computations in this example the influence zone below foundation has been taken as B i.e., diameter of the tank = 23.8 m ≈ 25 m.

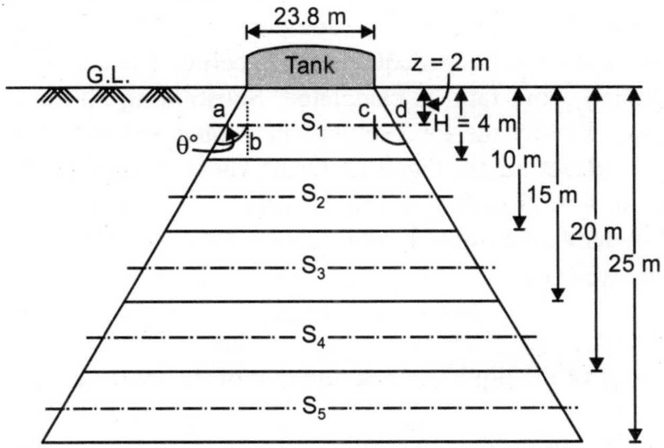

Fig. 9.12: Influence zone divided Into different layers

(b) The settlement in each layer is given by (IS: 8009 Part 1, 1976, reaffirmed 1993)

$$S = \frac{C_c \times H}{1 + e_0} \log_{10} \frac{p_0 \times \Delta P}{p_0} \qquad ...(9.23)$$

where, C_c = compression index = 0.009 (w_L – 10) ...(9.23 A)

w_L being the liquid limit of the soil in percentage (= 49%)

H = thickness of the compressible layer (read from Fig. 9.12)

e_0 = initial void ratio (computed from *e* log *p* curve, *Ref.* Soil Testing for Engineers by Mittal & Shukla)

p_0 = initial stress at centre of the layer = γZ

Z = depth of the centre of the layer below natural ground

= submerged density of the strata (0.95 t/m³)

γ = pressure increment (due to total liquid & structural load in the tank) at centre of each layer

ΔP = Total load including that of the structure (Q) multiplied by the original area & divided by the area of spread at the centre of layer.

$$= \frac{\frac{\pi}{4} B^2}{\frac{\pi}{4} (B + H \tan \theta)^2} \times \frac{Q}{\frac{\pi}{4} \times B^2} \qquad ...(9.24)$$

B being width of footing (≈ diameter of tank = 23.8 m),

The total settlement (S) = $S_1 + S_2 + S_3 + S_4 + S_5$

where, S_1, S_2, S_3, S_4, S_5 are settlements in different layers of influence zone.

Note: θ = 26° (before piles) and θ = 44° (after piles).

The soil samples are collected at various depths in the field and they are tested in the laboratory to determine their Atterberg limits. From the various values of liquid limits at various depths, from equation 9.23 A, the C_C is calculated from *e* log *p* curves (computed from consolidation tests). C_c can also be computed using liquid limit test values as per equation 9.25 below. The value of C_c can also be computed from *e* log *p* curve. Similarly, void ratio is calculated. Substituting all the values, the settlement is calculated at various layers. Finally the total settlement is obtained by adding the various values of settlement at different layers. Referring the Fig. 9.12 the thickness of compressible layers within influence zone work out to 4 m, 6 m, 5 m, 5 m & 5 m respectively. From the laboratory test conducted at the sample collected from 2 m depth, the liquid limit i.e., $w_L = 49\%$, $e_0 = 0.95$.

Therefore, $C_c = 0.009\ (49 - 10)$ ...(9.25)

$= 0.35$

(*i*) Here Q = total load of the liquid in tank and that of the structure = 5336 t

(*ii*) q = intensity of loading at the bottom of tank $= \dfrac{Q}{A} = \dfrac{Q}{\frac{\pi}{4} \times D^2} = \dfrac{5336\text{ t}}{\frac{\pi}{4} \times 23.8^2} = 11.994 = 12\text{ t/m}^2$

***(iii)* Settlement computations**

Settlement in first layer

$H = 4\text{ m}$

$p_0 = \gamma z_1$ (where z is depth at centre of each layer below foundation)

$= 0.95 \times 2 = 1.90\text{ t/m}^2$

Here, $z_1 = \dfrac{H}{2} = \dfrac{4}{2} = 2\text{ m}$

$$\Delta p = \frac{B^2}{(B + H\tan\theta)^2} \times q \quad ...(9.26)$$

$$\Delta p = \left(\frac{D}{D_2}\right)^2 \times q \quad ...(9.27)$$

where, D = diameter of tank

D_1 = diameter of centre of the layer

Here $D_1 = ab + bc + cd$ (*Ref.* Fig. 9.12)

$= z_1 \tan\theta + 23.8 + z_1 \tan\theta$

$= 2\tan 26° + 23.8 + 2\tan 26°$

$= 23.8 + 4\tan 26° = 25.75\text{ m}$

Taking q as 12 t/m². ...(9.28)

Settlement in first layer

$$\Delta p_1 = \left(\frac{23.8}{25.75}\right)^2 \times 12 = 10.25 \text{ t/m}^2$$

$$\Delta p_0 = \gamma z_1 = 0.95 \times 2 = 1.90 \text{ t/m}^2$$

$$S_1 = \frac{0.35 \times 4}{1 + 0.95} \log \frac{1.90 + 10.25}{1.90} = 0.579 \text{ m}$$

Settlement in second layer

Here, $H = 6 \text{ m}, \Delta p_2 = \left(\frac{D}{D_2}\right)^2 \times 12, z_2 = 4 + 3 = 7 \text{ m}$

$$\Delta p_0 = 0.95 \times 7 = 6.65 \text{ t/m}^2$$

$$D_2 = (23.8 + 14 \tan 26°) = 23.8 + 6.83 = 30.63 \text{ m}$$

$$\Delta p_2 = \frac{23.8^2}{30.63} \times 12 = 7.25 \text{ t/m}^2$$

$$S_2 = \frac{0.35 \times 6}{0.95} \log \frac{6.65 + 7.25}{6.65} = 0.345 \text{ m}$$

Settlement in third layer

Here, $H = 5 \text{ m}, z_3 = 4 + 6 + 2.5 = 12.5 \text{ m}$

$$\Delta p_0 = 0.95 \times 12.5 = 11.875 \text{ t/m}^2$$

$$D_3 = (23.8 + 2 \times 12.5 \tan 26) = (23.8 + 12.19) = 35.99 \text{ m}$$

$$\Delta p_3 = \left(\frac{D}{D_3}\right)^2 \times 12$$

$$\Delta p_3 = \frac{23.8^2}{35.99} \times 12 = 5.25 \text{ t/m}^2$$

$$S_3 = \frac{0.35 \times 5}{1.95} \log \frac{11.875 + 5.25}{11.875} = 0.143 \text{ m}$$

Settlement in fourth layer

Here, $H = 5 \text{ m}, z_4 = 4 + 6 + 5 + 2.5 = 17.5 \text{ m}$

$$\Delta p_0 = 0.95 \times 17.5 = 16.625 \text{ t/m}^2$$

$$\Delta p_4 = \left(\frac{D}{D_4}\right)^2 \times 12$$

$$D_4 = (23.8 + 2 \times 17.5 \tan 26) = 23.8 + 17.07 = 40.87 \text{ t/m}^2$$

$$\Delta p_4 = \frac{(23.8)^2}{(40.8)} \times 12 \text{ t/m}^2$$

$$S_4 = \frac{0.35 \times 5}{1.95} \times \log \frac{16.625 + 4.07}{16.625} = 0.0853 \text{ m}$$

Settlemment in fifth layer

Here, $H = 5\text{ m}, z_5 = 4 + 6 + 5 + 5 + 2.5 = 22.5\text{ m}$

$$\Delta p_0 = 0.95 \times 22.5 = 21.375\text{ t/m}^2$$

$$\Delta p_5 = \left(\frac{D}{D_5}\right)^2 \times 12$$

$$D_5 = (23.8 + 2 \times 22.5 \tan 26) = (23.8 + 21.95) = 45.75\text{ m}$$

$$\Delta p_5 = \left(\frac{23.8}{45.75}\right)^2 \times 12 = 3.25\text{ t/m}^2$$

$$S_5 = \frac{0.35 \times 5}{1.95} \log_{10} \frac{21.375 + 3.25}{21.375} = 0.0552\text{ m}$$

Total settlement, $S = S_1 + S_2 + S_3 + S_4 + S_5 = 0.579 + 0.345 + 0.143 + 0.0853 + 0.0552$

$= 120.75\text{ mm} \approx 121\text{ mm}$

This settlement is on higher side hence to reduce this, the granular piles have to be provided. It is evident from the above results that settlement in the upper 10.5 m depth is around 0.924 m. Thus more than 75 per cent of the settlement is expected within top 10.5 m depth of strata. The allowable soil pressure is 7.2 t/m² against the design load of about 12 t/m². Also the total settlement is on higher side. Hence it is essential to share the load on piles up to 10.5 m depth.

Note: This also gives a fairly good assessment of deciding the pile length. Since up to 10.5 m depth, the virgin ground settlement is of the order of 0.9 m, hence pile length should be more than 10.5 m depth so that the influence zone may be made strong and intact.

After strengthening the ground by stone column:

$$\gamma_{sub} = 1.25\text{ t/m}^3 \text{ and } \theta° = 44°$$

Settlement in first layer (*Ref.* Fig. 9.13)

Here, $H = 4\text{ m}, z = 2\text{ m}$

$$\Delta p_0 = 1.25 \times 2 = 2.5\text{ t/m}^2$$

$$\Delta p_1 = \left(\frac{D}{D_1}\right)^2 \times 12 \; D_1 = (23.8 + 4 \tan 44) = (23.8 + 3.86) = 27.66\text{ m}$$

$$\Delta p_1 = \left(\frac{23.8}{27.66}\right)^2 \times 12 = 8.88\text{ t/m}^2$$

$$S_1 = \frac{0.35 \times 4}{1.95} \log \frac{2.5 + 8.88}{2 \times 1.25} = 0.473\text{ m}$$

Settlement in second layer

Here, $z = 4 + 3 = 7\text{ m}$

$$\Delta p_0 = 1.25 \times 7 = 8.75\text{ t/m}^2$$

$$\Delta p_2 = \left(\frac{D}{D_2}\right)^2 \times 12, D_2 = (23.8 + 14 \tan 44) = 37.32\text{ m}$$

$$\Delta p_2 = \left(\frac{23.8}{37.32}\right)^2 \times 12 = 4.88\text{ t/m}^2$$

$$S_2 = \frac{0.35 \times 6}{0.95} \log \frac{8.75 + 4.88}{8.75} = 0.207 \text{ m}$$

Settlement in third layer

Here, $\quad H = 5 \text{ m}, z = 4 + 6 + 2.5 = 12.5 \text{ m}$

$$\Delta p_0 = 1.25 \times 12.5 = 15.625 \text{ t/m}^2$$

$$\Delta p_3 = \left(\frac{D}{D_3}\right)^2 \times 12, D_3 = (23.8 + 2 \times 12.5 \tan 44) = 47.94 \text{ m}$$

$$\Delta p_3 = \left(\frac{23.8}{47.94}\right)^2 \times 12 = 2.96 \text{ t/m}^2$$

$$S_3 = \frac{0.35 \times 5}{1.95} \log \frac{15.625 + 2.96}{15.625} = 0.0676 \text{ m}$$

Settlement in fourth layer

Here, $\quad H = 5, z = 4 + 6 + 5 + 2.5 = 17.5 \text{ m}$

$$\Delta p_0 = 1.25 \times 17.5 = 21.875 \text{ t/m}^2$$

$$D_4 = (23.8 + 2 \times 17.5 \times \tan 44) = 57.6 \text{ m}$$

$$\Delta p_4 = \left(\frac{23.8}{57.6}\right)^2 \times 12 = 2.049 \text{ t/m}^2$$

$$S_4 = \frac{0.35 \times 5}{1.95} \log \frac{21.875 + 2.049}{21.875} = 0.0349 \text{ m}$$

Settlement in fifth layer

Here, $\quad H = 5, z = 4 + 6 + 5 + 5 + 2.5 = 22.5 \text{ t/m}^2$

$$\Delta p_0 = 1.25 \times 22.5 = 28.125 \text{ t/m}^2$$

$$D_5 = (23.8 + 2 \times 22.5 \times \tan 44) = 67.26 \text{ m}$$

$$\Delta p_5 = \left(\frac{D}{D_5}\right)^2 \times 12$$

$$\Delta p_5 = \left(\frac{23.8}{67.26}\right)^2 \times 12 = 1.50 \text{ t/m}^2$$

$$S_5 = \left(\frac{0.35 \times 5}{1 + 0.95}\right) \log \frac{28.15 + 1.50}{28.125} = 0.0203 \text{ m}$$

$$S = S_1 + S_2 + S_3 + S_4 + S_5 = 0.473 + 0.207 + 0.0676 + 0.0349 + 0.0203$$

$$= 0.8028 \text{ m} = 80.28 \text{ mm} \approx 80 \text{ mm}$$

Settlement in natural strata of the influence zone = 121 mm.

Settlement after strengthening the ground using stone column = 80 mm. Thus it is noticed that the foundation on settlement after introduction of stone column reduces to about 65%.

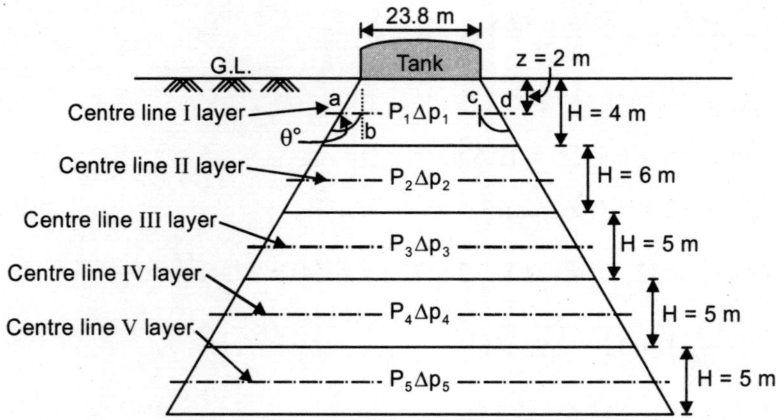

Fig. 9.13: Settlement diagram before piles construction

Note: (θ = 26°). Before using stone column and θ = 44° after strengthening of ground by stone column.

(iv) Design and layout of piles

It is assumed that 80% of the load shall be shared by the piles and balance by the soil.

The critical depth of granular pile diameter after construction. If the average pile diameter is taken as 0.5 m (say), this depth is taken as 0.5 × 4 = 2.0 m.

$$\text{Now, } O_e = d \times \gamma_{sub}$$

$$d = 2.0 \text{ m}$$

$$\gamma_{sub} = 0.95 \text{ t/m}^3$$

$$O_e = 2 \times 0.95 = 1.9$$

$$Cu = 2 \times O_e = 2 \times 1.9 = 3.8$$

$$O_v = K_B \cdot q$$

$$O_v = \left[1 - \left\{\frac{1}{1 + \left(\frac{a}{z}\right)^2}\right\}\right]^{3/2} \cdot q \qquad \ldots(9.29)$$

[Here, z = critical depth (d_c = 2 m), a = tank dia, q = loading intensity]

$$O_v = \left[1 - \left\{\frac{1}{1 + \left(\frac{23.8}{2.0}\right)^2}\right\}\right]^{3/2} \times 12$$

$$= 0.999 \times 12 = 11.95 \approx 12$$

Passive pressure coefficient $K_p = \tan^2 (45 + \phi/2)$

Taking ϕ' as $\frac{2}{3}$ of 44° = 29.3° i.e., 29.5° (say)

$$K_p = \tan^2\left(45 + \frac{29.5}{2}\right)$$
$$= 2.95$$

Substituting the values in equation,

$$Q_d = K_p\,(8Cu + O_e + O_v).\ A_p \qquad \ldots(9.30)$$

$$Q_d = 2.95\,(8 \times 3.8 + 1.4 + 12) \times \frac{\pi}{4} \times (0.5)^2 = 25.37\ \text{t}$$

Applying the factor of safety 2.0

$$Q_{safe} = \frac{25.37}{2.0} = 12.68\ \text{t}$$

Load shared by them piles = 0.80 × 5336 =4269 t

Load shared by the soil = 0.20 × 5336 = 1067 t

$$\text{No. of piles} = \frac{4269}{12.68} = 336.7,\ \text{say } 340$$

These piles shall be constructed in a zig-zag pattern at the corners of isosceles triangles. The layout is shown in Fig. 9.14. The spacing of the pile is decided by hit and trial method. One pile is made centrally beneath the tank.

$$\text{The intensity of load to be shared by soil} = \frac{1067}{\frac{\pi}{4}(23.8)^2} = 2.39 \approx 2.4\ \text{t/m}^2$$

This is less than 7.21 t/m². Hence safe.

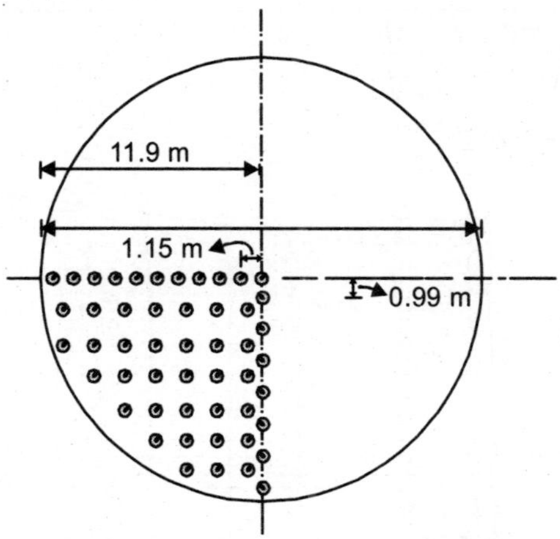

Fig. 9.14: Layout of piles below tank

Note: Only one quarter has been shown with piles. In other part also the layout shall be same.

This is the complete design of the granular pile foundation for tanks. Similarly, the foundation may be designed for other structures also like transmission towers, multistoreyed buildings, etc.

A RCC skirt wall can also be provided circumscribing all the piles. This wall reduces differential settlement and provides a positive confinement to a greater extent. The design of this skirt wall is beyond the scope of this book. This wall is designed like a cantilever. Besides this wall, a ring beam is also provided as shown in Fig. 9.15. The whole tank rests on a soil padding made of local material adequately compacted. The Fig. 9.15 illustrates the cross-section of the skirted granular pile foundations.

Settlement after piles

Considering the depth of stone column as 12 m and load distribution as 44° the total settlement calculated as above comes out to 80 mm only. Due to R.C.C. skirt wall (if provided), the author's experience is that it further reduces to half say 40 mm only. Besides the R.C.C. skirt, a ring beam may also be provided as shown in Fig 9.15, which reduces the differential settlement also.

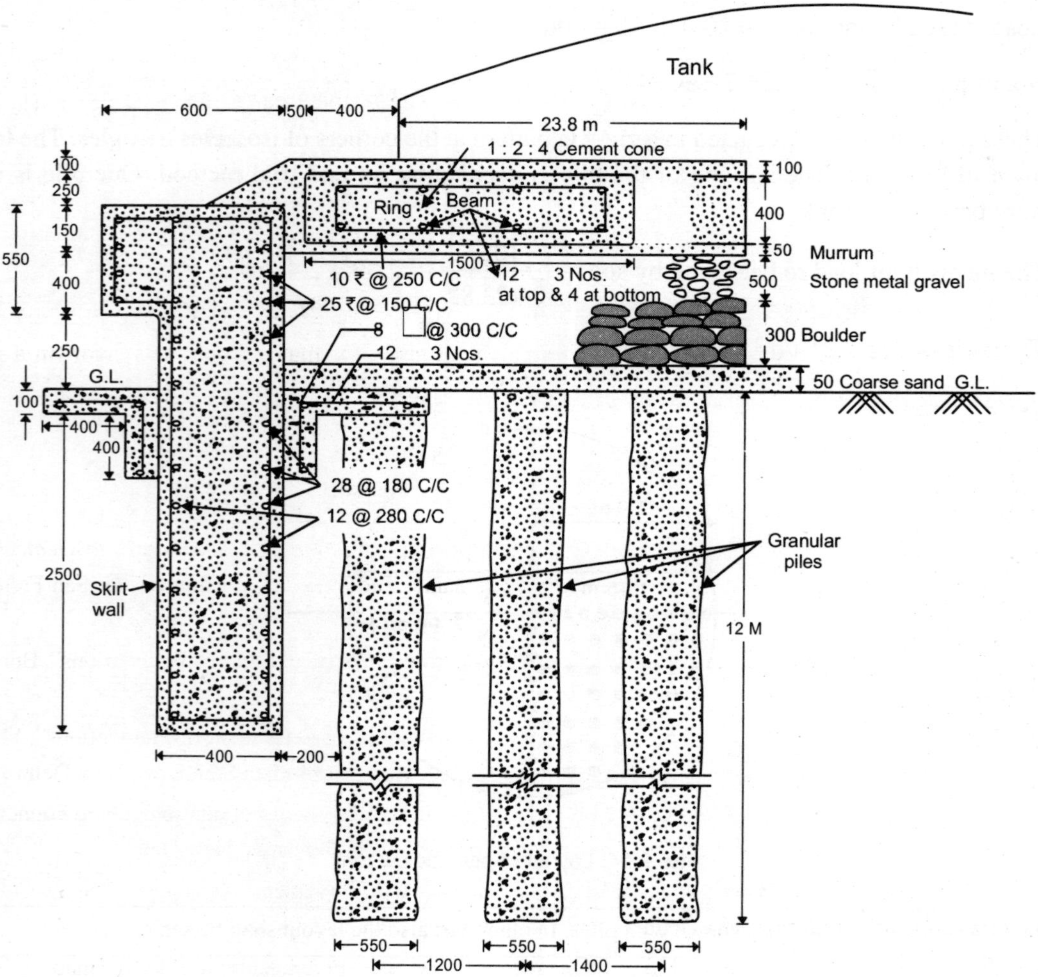

Fig. 9.15: Schematic of granular pile (Stone column), skirt wall and raft

Note: All dimensions are in mm unless specified.

9.9 APPROXIMATE METHOD FOR DETERMINING SPACING OF STONE COLUMN

Author has designed stone columns for many projects in Haldia (West Bengal) refinery, sugar factories etc. for various applications. Based on his experience, the approximate method of determining stone column spacing is explained below (readers should note that it has no technical basis. Rather it gives only a good start for determining spacing of columns to accommodate required no. of columns in available space).

In this example, the tank area = $\frac{\pi}{4} \times (23.8)^2 \approx 446 \text{ m}^2$

The number of stone columns = 340

Hence approximate area occupied by each column = $\frac{446}{340} = 1.31 \text{ m}^2$

Half of this area = $\frac{1.31}{2} = 0.655 \text{ m}^2$

The columns should be placed at equilateral triangle pattern, the area of which is $\frac{\sqrt{3}}{2} y^2$ (y, being the length of each arm).

Take $\frac{\sqrt{3}}{2} y^2 = 0.655$

Hence, $y = 1.23$ m

Thus approximate spacing between two consecutive columns may be 1.23 m.

Users or practitioners may adopt this figure as a beginning for making trial and error on a graph paper to decide placement of stone columns in field.

References

1. Datye, K.R., and Nagaraju, S.S., (1985). "Ground Improvement", "*Indian Contribution to Geotechnical Engineering*", XI International Conference on Soil Mechanics and Foundation Engineering, San Francisco, Aug. 1985, Indian Geotechnical Society, New Delhi, 1985, pp. 121–125.
2. IS: 6403: 1981, "Code of Practice for Determination of Bearing Capacity of Shallow Foundations", Bureau of Indian Standards, New Delhi.
3. IS: 8009 (Part 1) <197> (1976). "Code of Practice for Calculations of Settlement of Foundations", Shallow Foundations Subjected to Symmetrical Static Vertical Loads", Bureau of Indian Standards, New Delhi.
4. IS: 8009 (Part 2), (1980). "Code of Practice for Calculations of Settlement of Foundations. Deep Foundations, Subjected to Symmetrical Static Vertical Loading", Bureau of Indian Standards, New Delhi.
5. IS: 15284 (Part 1), (2003). "Design and Construction for Ground Improvement—Guidelines" Stone Columns, Bureau of Indian Standards, New Delhi.
6. Madhav, M.R. (2012). "Soft Ground Improvement with PVDs", Keynote Lecture in IIT Guwahati.
7. Mittal, Satyendra (2012). "Pile Foundations Design and Construction" CBS Publishers & Distributors.

8. Monograph on Particular Approach to Analysis of Stone Columns with and other Geosynthetics Encasing, CBIP Manual, Chanakyapuri, New Delhi, 1993.
9. Nayak, N.V. "Foundation Design Manual", Third Edition, Dhanpat Rai and Sons, New Delhi, 1985.
10. Ranjan, G. (1989). "Ground Treatment with Granular Piles and its Response under Load", Eleventh IGS Annual Lecture, Indian Geotechnical Journal, Vol. 19, No. 1, January.
11. Rao, B.G. (1983). "Behaviour of Skirted Granular Pile Foundation", Ph.D. Dissertation, University of Roorkee.
12. Vesic, A.S. "Expansion of Cavities in Infinite Soil Mass Journal of Soil Mechanisms Foundations, Div, ASCE, Vol. 98, No. SM3, pp 265–290.

Improvement of Black Cotton Soil and Use of Admixtures

CHAPTER 10

10.1 INTRODUCTION

Black cotton soils constitute a large part of world. In India alone, many states like Madhya Pradesh, Maharashtra, Andhra Pradesh etc. have most of the areas occupied with black cotton soil. Black cotton soil deposits are found under conditions wherein the slope of the terrain is less than 3% and the rainfall is in the range of 300–900 mm. The parent material is normally derived from weathering of basalt. However, there are black cotton soil deposits derived from other types of rocks and also from very old sedimentary deposits. Black cotton soils pose a lot of difficulty to foundation engineers due to its frequent swelling and shrinking nature. When in contact with water, black cotton soil swells, while on drying it shrinks. The liquid limit of these soils range from 52.9–92% while the plasticity index vary between 15.9–53.9% indicating a wide range of plasticity. The shrinkage limit range from 7–18%. Thus soils undergo considerably large volume changes with change in moisture content. This may lead to development of wide cracks, leading to desiccation. The major constituents of Black Cotton soil are silica, alumina and iron. Most of the soils contain silica to the extent of 40–45% and alumina about 13.6–22%.

Katti, et al. (2000) conducted study on Black Cotton (BC) soils collected from various parts of the country. The study as aimed for following:

1. Determination of the properties such as consistency, density, swelling pressure, consolidation, shear, permeability and thixotropic properties.
2. Effect of inorganic chemicals on the various properties of these soils and the feasibility of field applications.
3. Studies concerning engineering behaviour of Black Cotton soils treated with lime, cement and a combination of both.

The properties of Black Cotton soils collected from two places were as follows (Table 10.1).

Table 10.1: Properties of black cotton soils collected for study

Physical property	*Amravati soil*	*Vijayawada soil*
Liquid limit, LL (%)	81	92
Plastic limit, PL (%)	47	38
Shrinkage limit, SL (%)	10	9.8
G	2.1	2.2
Classification	Clay	Clay
Proctor density (kN/m^3)	13.2	13.9
OMC (%)	33	30.4
pH	7.4	8.8
Organic matter (%)	0.6	0.4
Carbonate content (%)	0.2	0.4
SiO_2 (%)	53.2	57
Al_2O_3 (%)	15.7	17.5
(SO_3) (%)	1.2	1.3

10.2 EFFECT OF INORGANIC CHEMICALS ON VARIOUS PROPERTIES OF BLACK COTTON SOIL (after Katti, et al. 2000)

It was realised that detrimental property of Black Cotton soil should be altered for foundation purpose. The change in engineering properties was studied by addition of chemicals. The chemicals added were KOH, NaOH, Ca $(OH)_2$, $CaCl_2$ etc. The results are shown vide (Figs. 10.1 and 10.2). The percentage of chemical added was based on over dry weight of soil. Addition of KOH ranging from 1.5–7% made the soil non-plastic. KOH beyond the addition of 0.75–3% reduced the Liquid Limit (LL) and thus Plasticity index (PI). Beyond 1% addition, Shrinkage limit (SL) values significantly increased, indicating that the tendency towards volume change had been considerably reduced. SL values went as high as 40 from initial values of 8 or 10 for other soils. By addition of chlorides, a decrease in LL values and consequently a decrease in PI value was observed. Chloride solution enhances surface tension and provides better lubricant.

Tests were done with cement also. Cement has an effect similar to that of Ca $(OH)_2$ but to a lesser degree. This might be due to the lesser amount of free lime available from cement. Even with 10% cement, the soils did not become non plastic. SL values however considerably increased with the addition of cement.

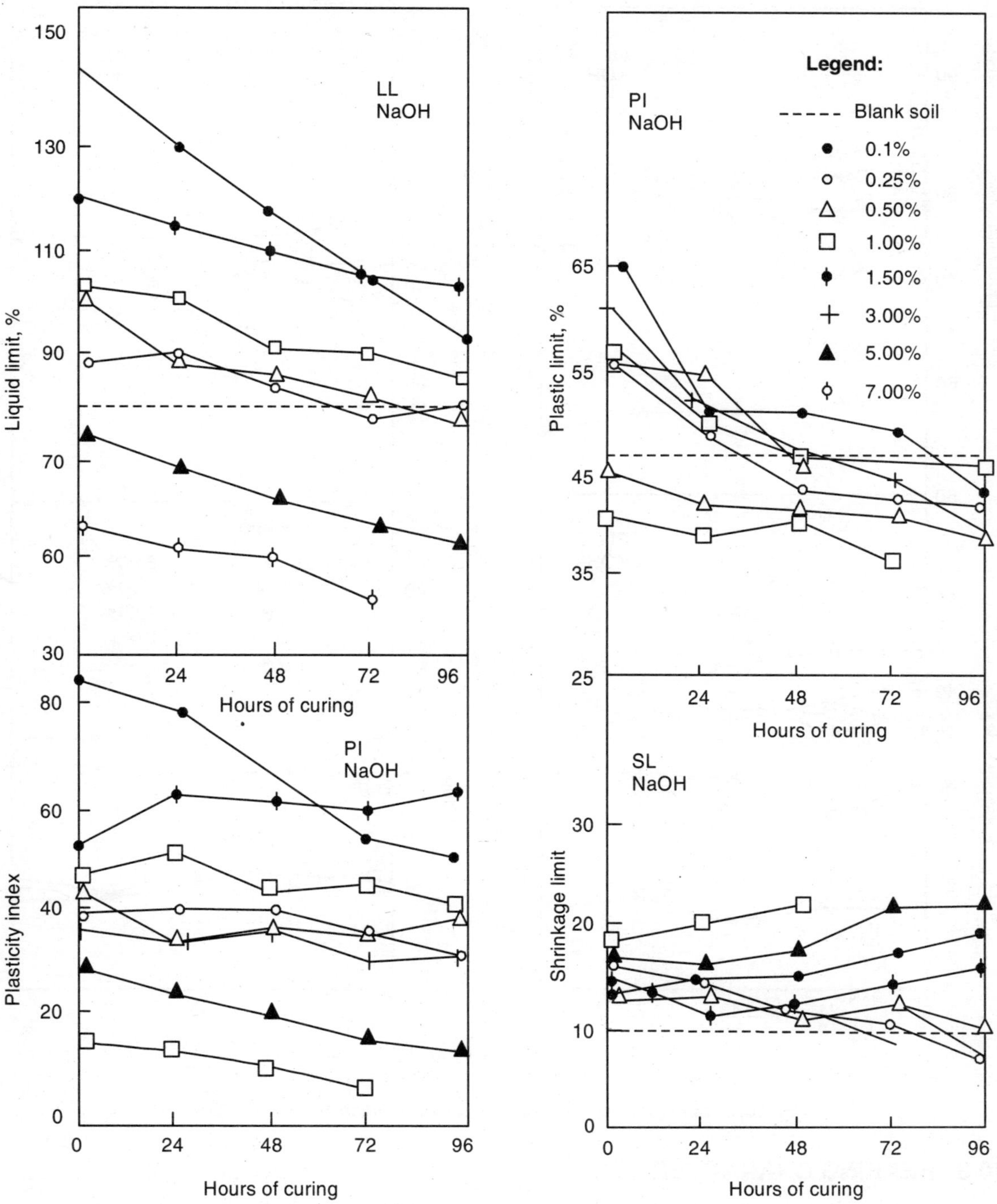

Fig. 10.1: Variation in consistency properties with the addition of NaOH

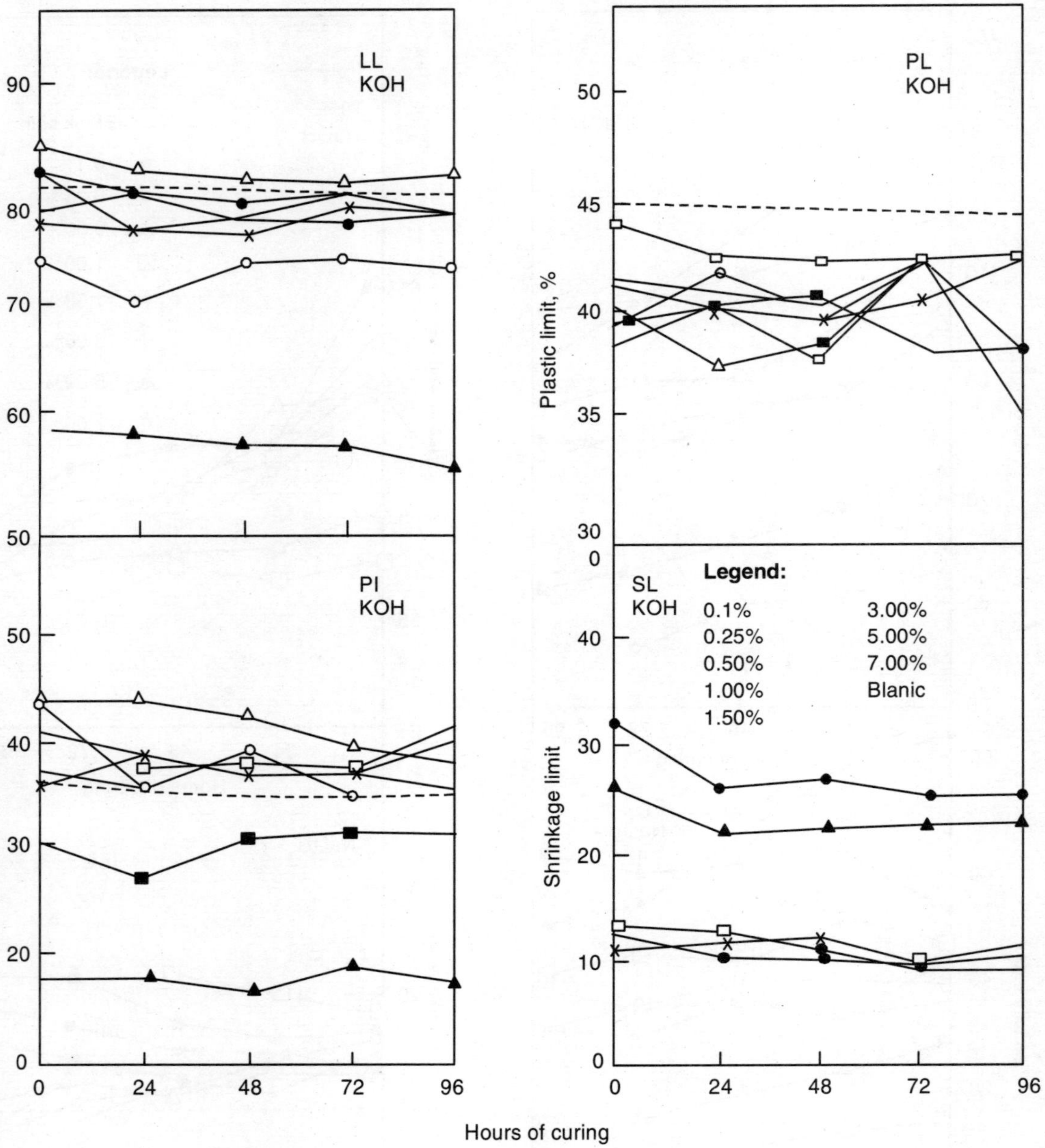

Fig. 10.2: Variation in consistency properties with the addition of KOH

10.3 BEARING CHARACTERISTICS

To understand the change in bearing capacity, CBR tests at standard proctor density with surcharge on samples soaked for 4 days were conducted on Amravati soil. The results are as follows (Table 10.2).

Table 10.2: Effect of different chemicals on BC soil								
Name of chemical/admixture added	Percentage addition (%)							
	0.1	0.1	0.25	0.5	0.75	1.0	3.0	7.0
KOH	3.84	3.72	4.66	5.19	4.22	3.9	16.46	26.18
NaOH	3.84	4.50	3.99	3.87	3.86	4.28	21.44	23.59
$Ca(OH)_2$	3.84	4.54	3.84	4.86	5.90	8.38	47.20	83.18
Cement	3.84	4.23	4.13	4.59	4.67	5.64	11.39	42.33
Na_2Co_3	3.84	3.78	3.33	3.27	4.07	4.11	5.16	6.31

The addition of 2–3% NaOH brought down the permeability values in soil than that for blank soil. Beyond this percentage, the permeability values continued to increase to about 10%, the increment being very rapid beyond 5%.

Decrease in permeability at lower percentage may be due to dispersive action of NaOH. KOH also showed the similar trend of that of NaOH.

- Ca $(OH)_2$ addition showed significant increase in permeability (k).
- $CaCl_2$ increased k at all percentages.
- Na_2Co_3 being highly dispersive agent decreased the values of k even at low percentage.

10.4 UNCONFINED COMPRESSIVE STRENGTH STUDIES

10.4.1 Lime treated black cotton soil

The maximum compressive strength of black cotton soil increased by addition of lime. It was observed that 28 days strength was much more than 7 days strength. The 7 days strength was less than 20.7 kg/cm^2. While that of 28 days was more than 20.7 kg/cm^2. The increase in strength was due to long-term pozzolanic reaction taking place in soil.

It was further observed that for a given amount of lime, the strengths decreased with an increment in clay content for soils containing almost the same silica content. The presence of large amount of silica seemed to increase the 7 days strength as well as the percentage gain in strength over time. The optimum lime content for various soils studied fell in the range of 4–10% and appeared to shift to higher percentages with an increase in curing time: A 2–4% addition of lime produced little gain in strength over time.

10.4.2 Cement treated soil

In general soil cement mixed showed an increase in strength with each increment in cement with untreated BC soil. But with addition of 15% cement, the 7 days strength did not reach desired strength of 20.7 kg/cm^2 however with 28 days curing period, this strength was obtained. But in lime this amount was of higher order. Up to 10% additive limit, the lime treated soils yielded better results than cement treated soils.

10.4.3 Lime-cement treated soils

These tests proved that addition of lime and cement in combination produced distinctly high strengths. Several combinations in all the soils were tried, comprising of less than 15% total admixture gave 7 days strength as high as 34.5–41.3 kg/cm^2. In general, however a 6–12% of total admixture proved sufficient to give more than 20.7 kg/cm^2 after 7 days curing, the amount of lime content in combination varied between 2–4% and cement 4–10%. The cement content of 4–10% is necessary to obtain strength of 20.7 kg/cm^2 for silty and sandy soils. It has been found in studies that addition of lime changes texture of soils, giving rise to a reduction in surface area and making the mixes behave like silt or sandy soils.

10.5 SHEAR STRENGTH CHARACTERISTICS WITH ADMIXTURES

10.5.1 Lime

In untreated state, Amravati soil had $c = 0.48$ kg/cm^2 and Φ as 4°. However, with addition of lime the c and Φ values changed as below (Table 10.3).

Table 10.3: Qualitative change in Amravati soil with addition of lime

Increase lime (%)	*c (kg/cm^2)*	Φ
1	1.38	10°
4	3.9	25
6	4.5	36
8	3.6	28

10.5.2 Cement

With addition of cement, c and Φ showed a tendency to increase. The changes in values were as below (Table 10.4).

Table 10.4: Qualitative change in Amravati soil with addition of cement

Cement (%) added	*c (kg/cm^2)*	Φ
1.0	0.62	6
4.0	0.62	25
6.0	0.76	25
8.0	3.4	35
15.0	6.9	28

It is seen that lower percentage of admixture of lime (up to about 6–8%) gave relatively higher value of shear and Φ than did cement.

It is also noted that cohesion value continued to increase beyond 8% addition of cement but Φ reduced, indicating that beyond a certain amount of additive, hydration products of cement cannot effect much change in roughness and interlocking.

10.5.3 Lime + Cement

The combination of lime and cement in general resulted in comparatively large improvement in strength characteristics. An optimum lime cement, varying between 6–8% gave (for a given amount of cement) the maximum cohesion and internal friction. Value of *c* was observed more than 6.9 kg/cm^2 for mixes containing more than 20% total admixture while the value of Φ was of the order to 20°–30°. Thus, as the data shows since lime brings about aggregation, the surface area greatly reduce, creating a greater opportunity for cement to cover aggregation and to develop strength more effectively.

10.6 SWELLING PROPERTIES OF EXPANSIVE SOILS

The swelling potential of expansive of soils is as shown below (Mittal and Shukla, 2012).

Table 10.5: Relationship between Atterberg limits and swelling potential

PI	*SL*	*LL*	*Swelling potential classification*
< 18	< 15	< 39	Low (DFS = < 20%)
15–28	10–16	39–50	Medium (DFS = 20–35%)
25–41	7–12	50–63	High (DFS = 35–55%)
> 35	> 11	> 63	Very high (DFS = > 50%)

DFS: Differential free swell

Typical swell classification of soil based on swell test results is explained below (Table 10.6).

Table 10.6: Swell potential and Swell classification

Swell potential (%)	*Swell classification*
< 0.5	Low
0.5–1.5	Marginal
> 15	High

Anderson and Lade (1981) suggested expansion index test to be conducted in a ring of 4″ diameter and 1″ height. At a degree of saturation of 50%, a surcharge load of 1 lb/inch^2 was applied and then sample is saturated and allowed to stand until the rate of swelling reaches a certain value or until 24 hrs, whichever is longer. The amount of swell is expressed in terms of expansion index which is defined as:

$$\text{EI} = 1000\, h.\text{F} \quad ...(10.1)$$

where, EI = expansion index

h = expansion of soil (inch)

F = % of sample by weight that passes through 4.75 mm sieve.

The values of EI and corresponding potential expansion are defined as below (Table 10.7).

Table 10.7: Relationship between expansion index and potential expansion (Anderson and Lade, 1981)

EI	*Potential expansion*
0–20	Very low
21–50	Low
51–90	Medium
91–130	High
> 130	Very high

10.7 PREVENTIVE MEASURES FOR BLACK COTTON SOILS

10.7.1 Building foundations

Buildings especially those that are light weight are prone to damage from expansive soils. The magnitude of heaving and shrinking generally varies across the building thus causing problems similar to those associated with excessive settlements. These include cracks, inoperative doors and windows etc. Common preventive measures include:

(*i*) Extending foundation to greater depths, thus bypassing the zone of greatest moisture change and supporting the building on more stable soil.

(*ii*) Adding extra reinforcement steel to foundation and slabs. In some cases, pre-stressed or post-tensioned slabs are used.

(*iii*) Being careful to provide (and also maintain) good surface drainage around building.

(*iv*) Avoiding placement of irrigated landscaping close to building.

(*v*) Pre-moistening the soil, prior to construction thus causing it to expand before building is erected.

10.7.2 Highway pavements

Highway pavements are also prone to damage from expansive soils. Common preventive measures include:

(*i*) Providing extensive surface and subsurface drainage to keep water away from subgrade soil.

(*ii*) Providing a non expansive sub-base material.

(*iii*) Treating the sub-grade soils with lime or some other material to reduce their expansive properties.

(*iv*) Providing steel reinforcement if budget permits.

10.8 IMPROVEMENT OF HIGH COMPRESSIBLE CLAY WITH ADMIXTURES (After Mittal et al., 2011)

Mittal et al., (2011) studied to study the behaviour of ground with black cotton after treatment using deep mixing technique with lime or cement as admixtures. The steps included in their study were:

1. Preparation of soil samples (passing through 425 microns IS sieve size) mixed with lime at varying percentage (from 2%–6% by weight of soil) keeping soil-lime-water ratio constant.

2. Preparation of soil samples mixed with varying percentage of lime at water content equivalent to liquid limit.
3. Preparation of soil-cement mixtures by adding 5, 7.5, 10, 20, 30, 40, 50, 60, 70 and 80% of cement (Grade 43) in dry condition and 61% of water equivalent to liquid limit of soil.
4. Pouring the soil-admixture paste into specially fabricated perspex cylindrical moulds (Fig. 10.3) and tampted to expel air from mix and tightly capped.

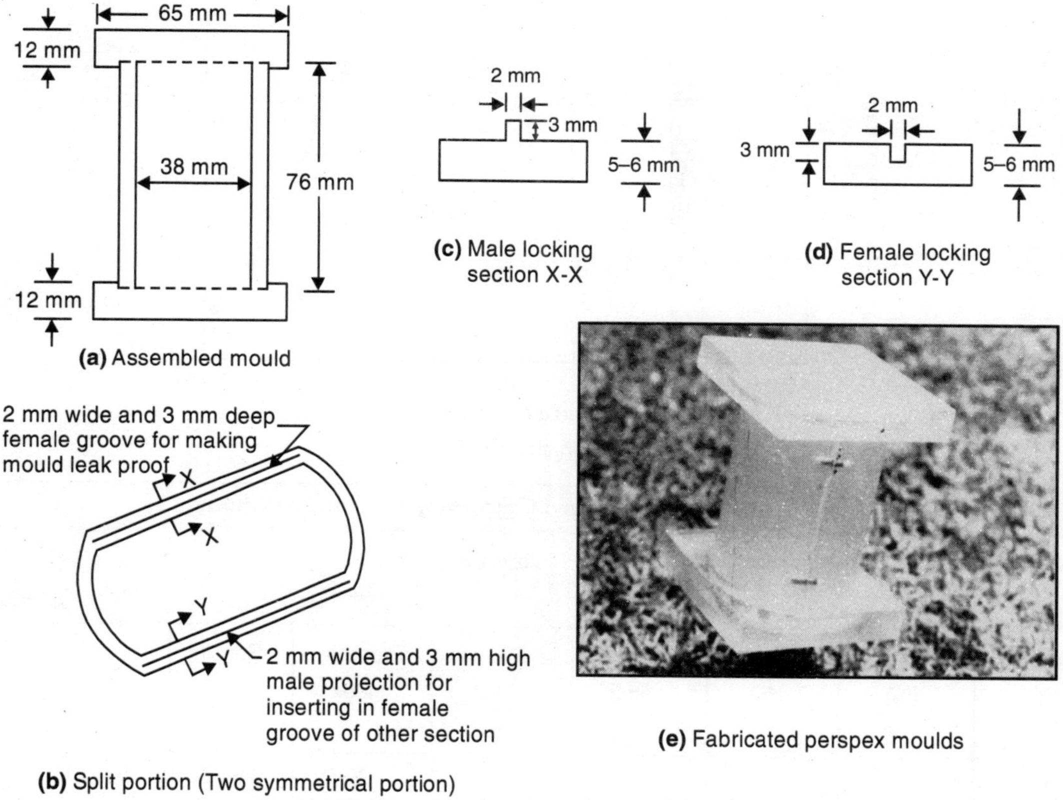

(a) Assembled mould

(b) Split portion (Two symmetrical portion)

(c) Male locking section X-X

(d) Female locking section Y-Y

(e) Fabricated perspex moulds

Fig. 10.3: Design of perspex mould

5. Keeping samples thus prepared in humidity chamber for various setting (rest) periods (7, 14, 21 and 28 days) to study the improvement in strength of soil-admixture mix.
6. Determination of strength properties of casted soil samples by conducting unconfined compression test at different rest periods after curing in humidity chamber.
7. Study of behaviour of soil in respect of its unconfined compressive strength and recommendations for deep mixing technique in fields.

10.8.1 Laboratory investigations

The soil collected from Navi Bagh, Bhopal (India) has been selected for study as this soil had reasonably high liquid limit and compressibility. The soil properties are shown in Table 10.8 and the grain size distribution curve is shown in Fig. 10.4.

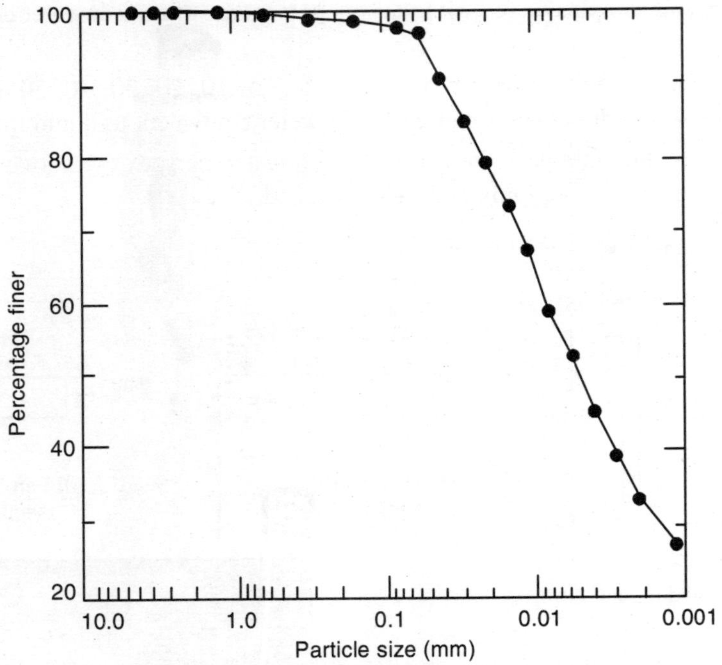

Fig. 10.4: Grain size distribution curve

Table 10.8: Properties of Navi Bagh, Bhopal soil used in study

Grain size distribution			***Atterberg's limits***				***Soil classification***
Sand (%)	Silt (%)	Clay (%)	Liquid limit (%)	Plastic limit (%)	Plasticity index	Specific gravity	
02	65	33	61	28	33	2.70	CH

10.8.2 Preparation of test samples

The samples were prepared by mixing lime and cement admixtures at different percentages by weight and were kept in special leak proof perspex mould of 38 mm diameter and 76 mm height (Fig. 10.3). The admixtures used were (*a*) The hydrated lime in quantity–2%, 4% and 6% by weight of soil (Katti & Katti), (*b*) Cement (Grade 43) in quantity–5, 7.5, 10, 20, 30, 40, 50, 60, 70 and 80% by weight of soil. At liquid limit, soil started gaining its shear strength. Therefore, as a datum, liquid limit (61% of water) had been selected to compare strength improvement behaviour of composite soil.

10.8.3 Casting of soil samples

One kg of soil was dried at 105 ± 5°C temperature in oven. Varying percentages of lime or cement were thoroughly mixed with soil in dry condition. Distill water equivalent to liquid limit, was gently added and homogeneous paste was made by thorough mixing in a glass bowl. The composite soil water paste was put in perspex moulds (Fig. 10.3). For every percentage mix, a total of 48 specimens were casted for lime mix and 30 specimens were made for cement mix. The casted samples were gently tamped on table to ensure that no void had been left in the casted samples. The casted samples were put in humidity chamber and covered to maintain constant humidity and temperature, later to be tested at different rest periods *i.e.,* 7, 14, 21 and 28 days. The stabilised soil samples are shown in Fig. 10.5.

Fig. 10.5: View of stabilised soil samples

10.8.4 Tests conducted

The unconfined compression strength tests had been carried out on all samples prepared to determine shear strength of lime treated soil. The tests had been carried out at low strain rate and each test was continued till the failure of test sample occurred.

10.9 RESULTS OF STUDY ON NAVI BAGH BC SOIL

10.9.1 Clay-lime mix

The data obtained from Unconfined Compression Strength Tests were analysed after making area correction. The stress-strain curves were plotted for each test. Total 4 plots have been prepared taking average value from soil specimens at varying percentage of lime and rest periods. The peak stress in each curve gives the unconfined compression strength (q_u) value. These values are presented vide Table 10.9 to Table 10.12. The strength improvement (as observed) *versus* percentage of lime for different rest periods (curing time) i.e., 7, 14, 21 and 28 days are illustrated vide Figs. 10.6–10.9.

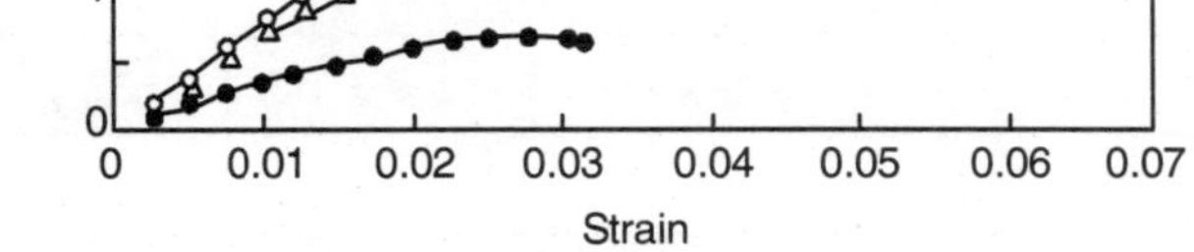

Fig. 10.6: Stress-strain plot for 7 days

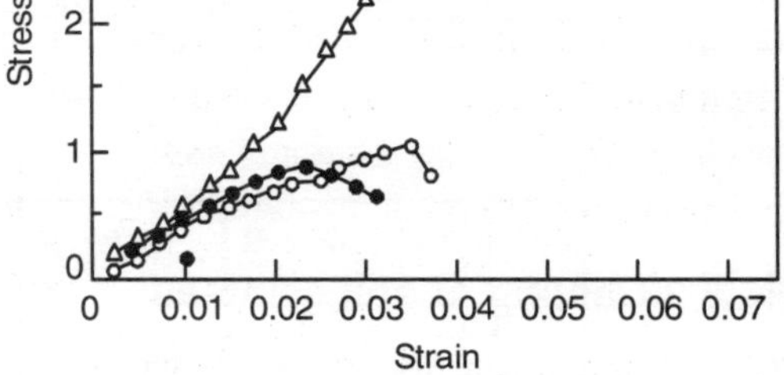

Fig. 10.7: Stress-strain plot for 14 days

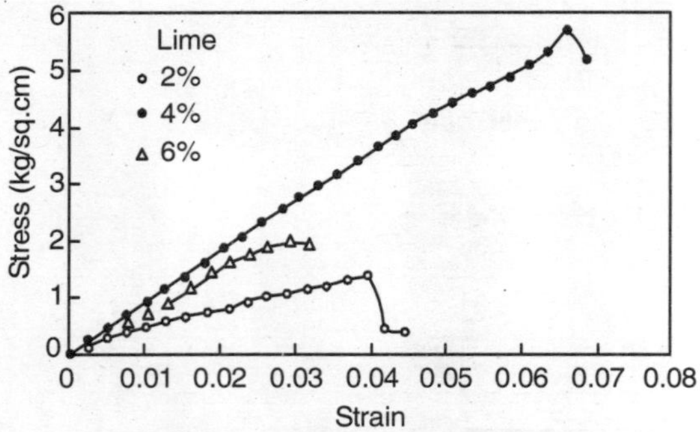

Fig. 10.8: Stress-strain plot for 21 days

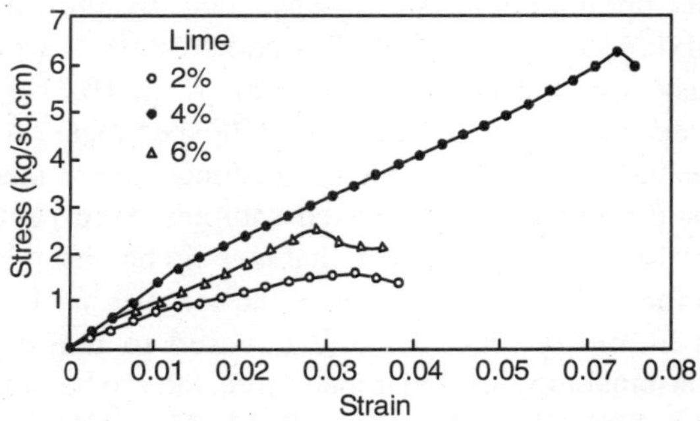

Fig. 10.9: Stress-strain plot for 28 days

The 7 days test data show that the improvement starts from 2% mixing of lime in 7 days curing. At 2% mixing of lime, strengths was 0.62 kg/cm^2 which by making of 4% lime, increased to 3.34 kg/cm^2 in 7 days curing (Fig. 10.8). The 14 days test data (Fig. 10.9) show the improvement starts from 2% mixing of lime, which shows unconfined compressive strength as 0.95 kg/cm^2 in 14 days. By mixing of 4% lime, the strength had been increased to 4.78 kg/cm^2. The 21 days test data show the improvement starts from 2% mixing of lime, which shows unconfined compressive strength as 1.38 kg/cm^2 in 21 days. By mixing of 4% lime the strength has been increased to 5.57 kg/cm^2. But by mixing of 6% lime the strength was observed as 1.95 kg/cm^2. The 28 days test data show that the improvement in unconfined compressive strength starts from 2% mixing of lime, which shows strength as 1.53 kg/cm^2. By mixing of 4% lime the strength had been increased to 6.27 kg/cm^2. But by mixing of 6% lime the strength was limited to 2.50 kg/cm^2. The unconfined compressive strength of the specimen at 6% mixing of lime was less than that at 4% of mixing of lime. This might be due to leaching of lime water in soil-lime mix.

Table 10.9: Unconfined compressive strength, clay with lime—7 days curing

Sample No.	*Constituents (by weight of soil)*			*Unconfined compressive strength of stabilised clay (q_u kg/cm^2)*			
	Soil %	*Lime %*	*Water %*	*Sample 1 (S1)*	*Sample 2 (S2)*	*Sample 3 (S3)*	S_{avg}
1–3	100	2	61	0.66	0.59	0.60	0.62
4–6	100	4	61	3.69	2.81	3.51	3.34
7–9	100	6	61	1.10	1.25	1.13	1.16

Table 10.10: Unconfined compressive strength, clay with lime—14 days curing

Sample No.	*Constituents (by weight of soil)*			*Unconfined compressive strength of stabilised clay (q_u kg/cm^2)*			
	Soil %	*Lime %*	*Water %*	*Sample 1 (S1)*	*Sample 2 (S2)*	*Sample 3 (S3)*	S_{avg}
10–12	100	2	61	0.90	1.01	0.94	0.95
13–15	100	4	61	4.64	5.20	4.50	4.78
16–18	100	6	61	1.81	1.92	1.81	1.85

Table 10.11: Unconfined compressive strength, clay with lime—21 days curing

Sample No.	*Constituents (by weight of soil)*			*Unconfined compressive strength of stabilised clay (q_u Kg/cm^2)*			
	Soil %	*Lime %*	*Water %*	*Sample 1 (S1)*	*Sample 2 (S2)*	*Sample 3 (S3)*	*S_{avg}*
19–21	100	2	61	1.42	1.33	1.45	1.40
22–24	100	4	61	5.26	5.81	5.69	5.58
25–27	100	6	61	1.96	1.93	1.97	1.95

Table 10.12: Unconfined compressive strength, clay with lime—28 days curing

Sample No.	*Constituents (by weight of soil)*			*Unconfined compressive strength of stabilised clay (q_u Kg/cm^2)*			
	Soil %	*Lime %*	*Water %*	*Sample 1 (S1)*	*Sample 2 (S2)*	*Sample 3 (S3)*	*S_{avg}*
28–30	100	2	61	1.58	1.64	1.44	1.55
31–33	100	4	61	6.12	6.30	6.36	6.26
34–36	100	6	61	2.45	2.45	2.50	2.46

10.9.2 Clay-cement mix

The unconfined compression tests on soil cement mixture prepared with varying percentage of cement at various rest periods were also carried out by authors (Mittal, et al., 2011). The stress-strain relationship plots with varying percentage of cement at 7, 14, 21 and 28 days rest periods are shown vide Figs. 10.10–10.13 showing the strength behaviour. The gain in strength with increasing cement content and the rest period indicate from ductile to brittle behaviour of clay-cement samples shown in Figs. 10.14 and 10.15.

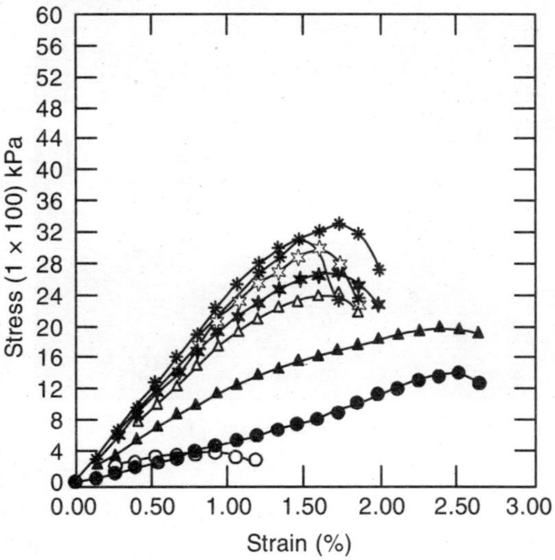

Fig. 10.10: Stress *vs.* Strain (Rest period 7 days)

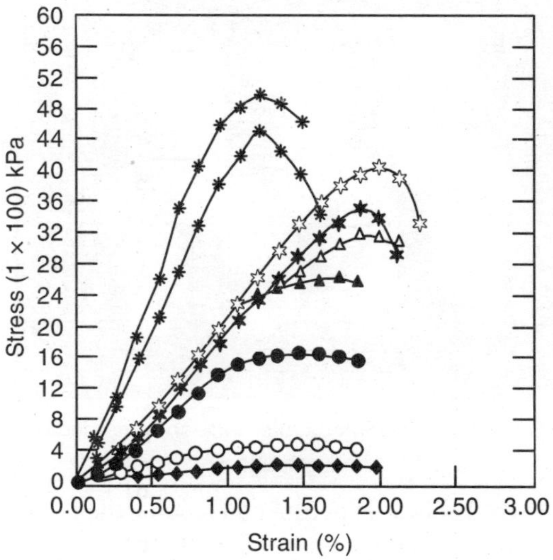

Fig. 10.11: Stress *vs.* Strain (Rest period 14 days)

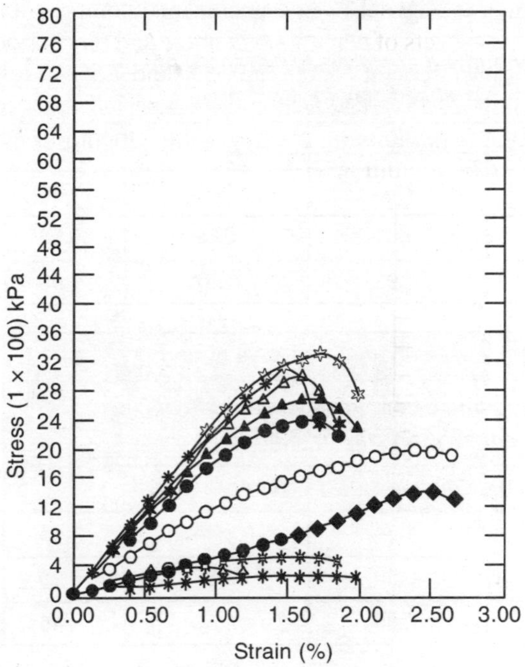

Fig. 10.12: Stress *vs.* Strain (Rest period 21 days)

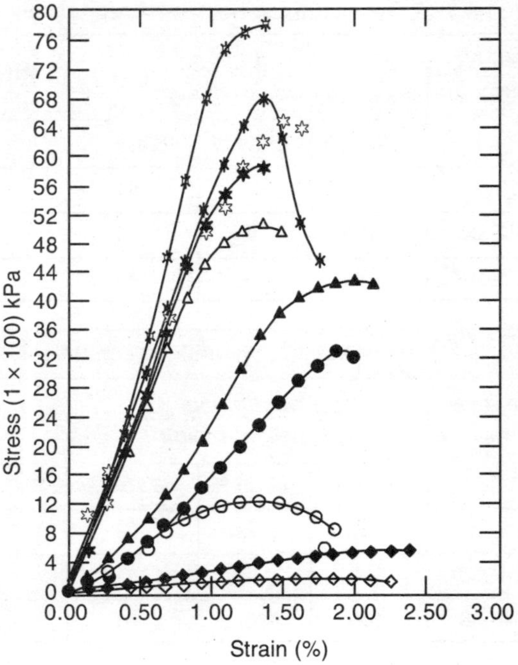

Fig. 10.13: Stress *vs.* Strain (Rest period 28 days)

◆	7.5% Cement	○	10% Cement	●	20% Cement
▲	30% Cement	△	40% Cement	✱	50% Cement
☆	60% Cement	✱	70% Cement	☆	80% Cement

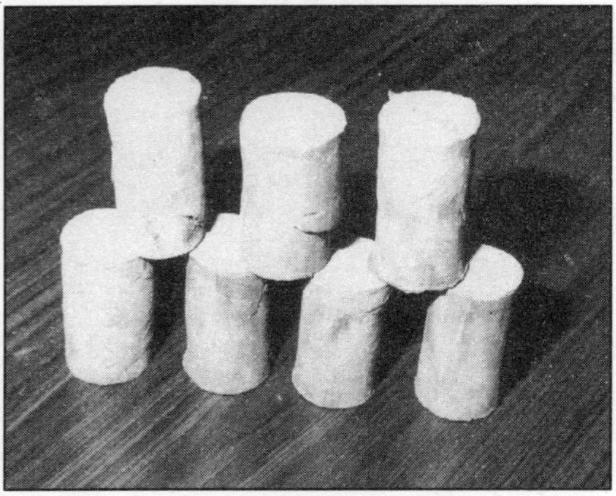

Fig. 10.14: Specimens after test, showing ductile failure pattern

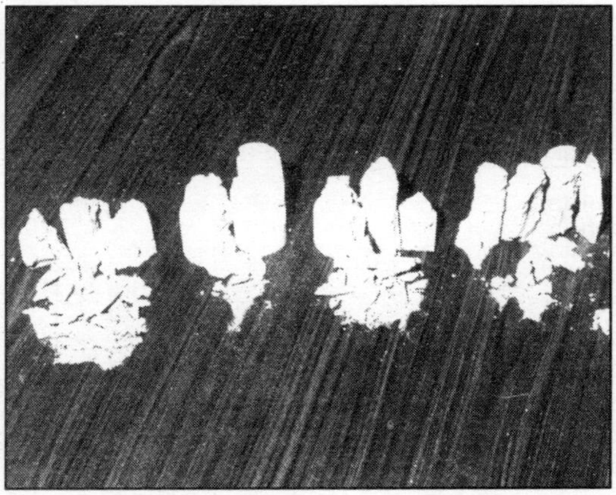

Fig. 10.15: Specimens after test, showing brittle failure pattern

On the basis of experimental results, the relationship plots between unconfined compression strength (q_u) and water-cement ratio (Fig. 10.16) show the decreasing trend in strength with the increase of water cement ratio at 7, 14, 21 and 28 days rest period which is in agreement with Abraham's Law applicable in cement concrete mixture. Similarly, the relationship plots between unconfined compression strength and percentage increase of cement in total mix are shown in Fig. 10.17 and unconfined compression strength *vs.* rest period are shown in Fig. 10.18 showing the effects of cement percentage and rest period on the improvement in strength. The gain in strength at lower cement content i.e., 5% and 7.5% at rest period, 7 and 14 days did not produce induced cementation. However, at higher rest period induced cementation was observed. It was also mentioned that the soil without admixing with cement did not indicate any strength even up to 28 days rest period in similar condition.

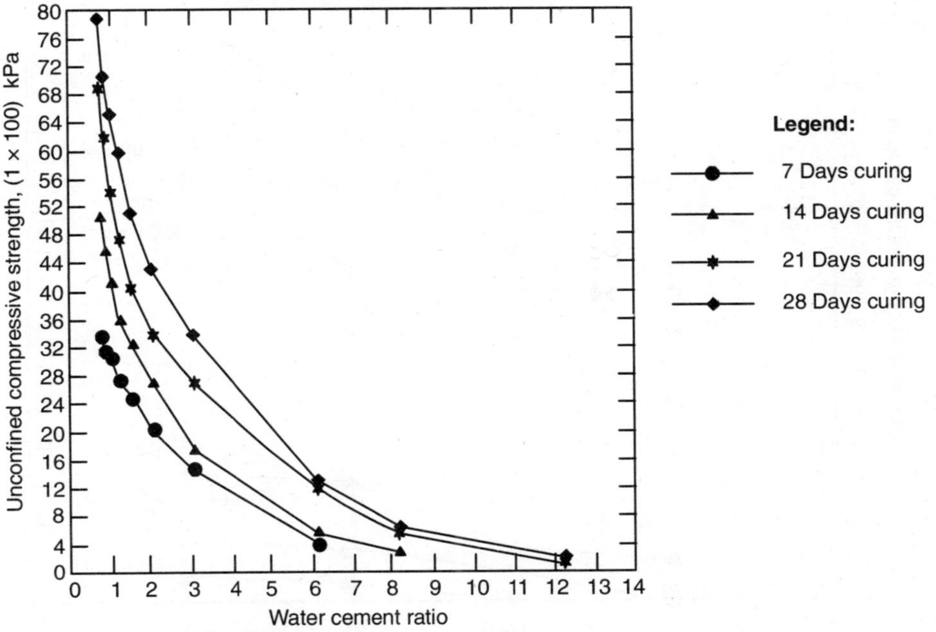

Fig. 10.16: Unconfined compressive strength (q_c) *vs.* Water cement ratio

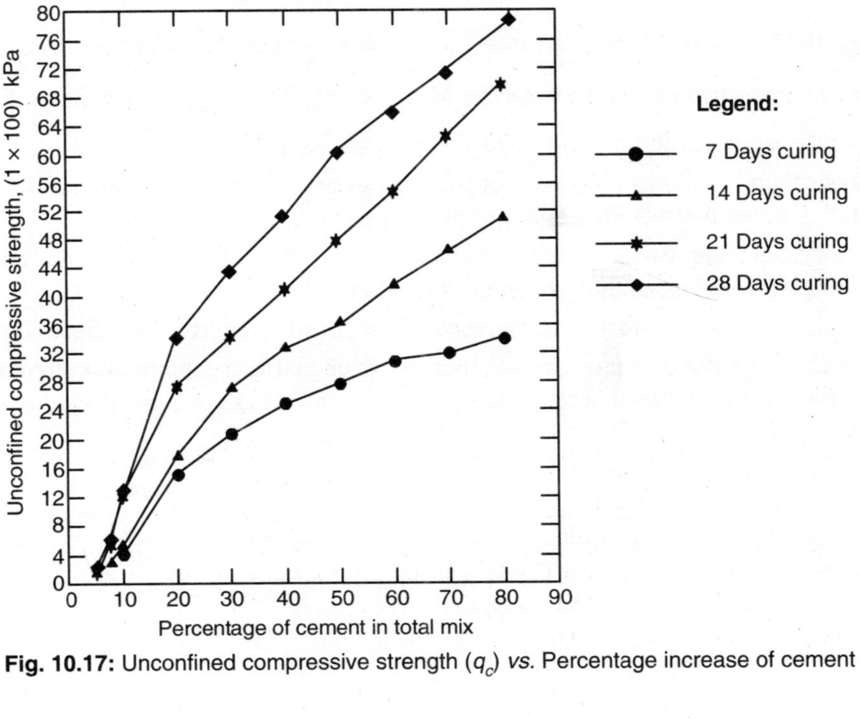

Fig. 10.17: Unconfined compressive strength (q_c) *vs.* Percentage increase of cement

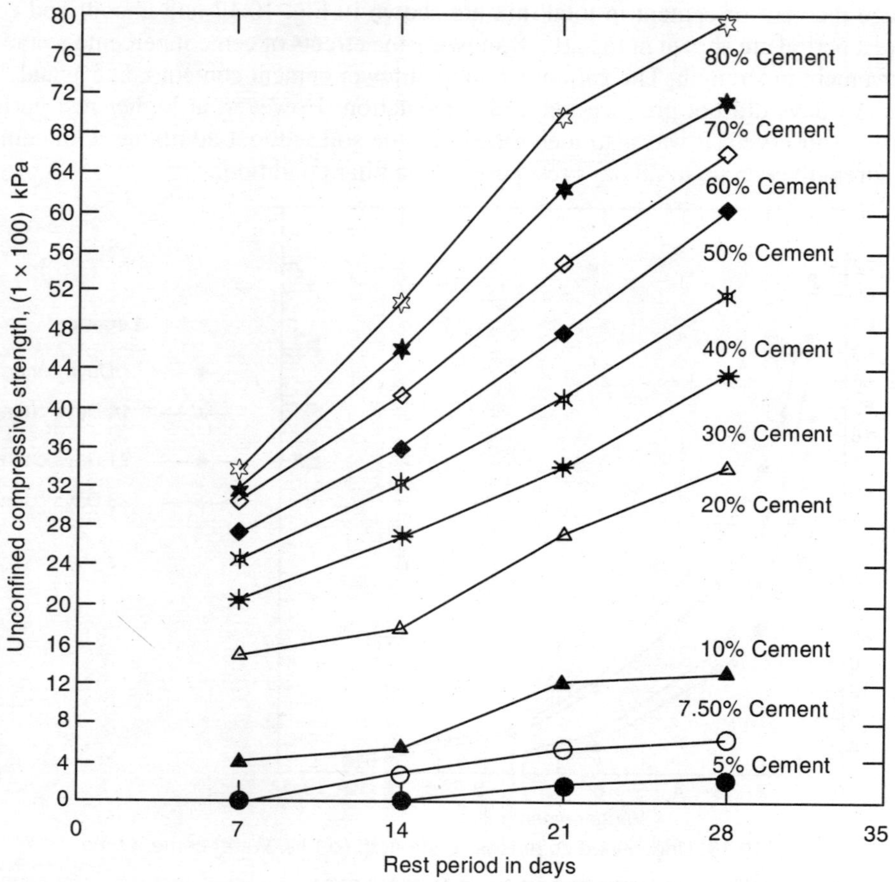

Fig. 10.18: Relationship between unconfined compressive strength *vs.* rest period in days

10.9.3 Conclusions drawn from study by Mittal et al., (2011) on Navi Bagh BC soil

On the basis of the study carried out following conclusions were drawn:

1. The physical and chemical properties of soil such as grain size, water content, Atterberg's limits, type of clay mineral affect the properties of soil treated with admixtures.
2. The unconfined compression strength of black cotton soil treated with lime and cement increases with curing time similar to the behaviour of concrete.
3. The increase of water content in soil adversely affects the strength of soil treated with lime.
4. By increase of cement content there was increase in unconfined compressive strength and change of soil mass from its behaviour from ductile to brittle as in case of cement sand mortar.
5. The unconfined compressive strength of black cotton soil could be increased considerably by addition of admixtures like cement and lime.

Mittal et al. (2006) also conducted an extensive research work on black cotton soil (collected from Navi bagh, Bhopal) to study its shrinkage and swelling behaviour with use of biodegradable jute reinforcement. The soil classification was CH with liquid limit as 61%. The jute used was woven and non-woven, with its properties as defined below:

Table 10.13: Properties of woven JGT used in study

Sl. No.	Properties	Values
1	Weight (g/m^2) at 20% moisture retention	760
2	Threads (Warp × Weft)	102 × 39
3	Thickness (mm)	4
4	Width (mm)	760
5	Strength kN/m (Warp × Weft)	20 × 20
6	Elongation at break (%) (Warp × Weft)	10 × 10
7	Pore size (O_{90}) micron	300
8	Water permeability at 10 cm water head cm/sec	0.03
9	Puncture resistance N/cm^2	380
10	Permittivity under 10 cm water head	50 l-s/m^2/sec

Table 10.14: Properties of non-woven JGT used in study

Sl. No.	Properties	Values
1	Weight (g/m^2) at 20% moisture retention	500
2	Thickness (mm)	4
3	Width (mm)	1500
4	Strength kN/m (Warp × Weft)	4 × 5
5	Elongation at break (%) (Warp × Weft)	20 × 25
6	Pore size (O_{90}) micron	500
7	Coefficient of water permittivity (m/s)	3.4×10^{-3}
8	Durability (yrs) Max.	1

The tests on synthetic reinforcement were also conducted in the laboratory. The synthetic reinforcement was non-woven geotextile, supplied by Ocean Non Woven Pvt. Ltd., New Delhi. The properties of non-woven geotextile are summarised in Table 10.15.

Table 10.15: Properties of non-woven geotextile			
Sl. No.	**Properties**	**Testing methods**	**Values**
1	Weight (N/m^2)	ASTMD-3776	4.30
2	Apparent opening size (U.S. standard sieve)	ASTMD-4751	112
3	Permeability of water (m/s)	ASTMD-4491	0.0037
4	Grab tensile strength (N) (Machine direction)	ASTMD-4632	1440
5	Grab tensile strength (N) (Transverse direction)	ASTMD-4632	1460
6	Grab tensile elongation (%) (Machine direction)	ASTMD-4632	76
7	Grab tensile elongation (%) (Transverse direction)	ASTMD-4632	72
8	Trapezoidal tear strength (N) (Machine direction)	ASTMD-4633	560
9	Trapezoidal tear strength (N) (Transverse direction)	ASTMD-4633	570
10	Puncture resistance (N)	ASTMD-4633	760
11	Mullens burst strength (kN/m^2)	ASTMD-3786	4600
12	Seam strength (N)	ASTMD-1683	1390

Some tests were also conducted by using sand as reinforcing material. The properties of sand used in the study are given in Table 10.16.

Table 10.16: Properties of sand used in study		
Sl.No.	**Properties**	**Values**
1	Coefficient of uniformity (C_u)	6.48
2	Coefficient of curvature (C_c)	0.89
3	Moisture content (%)	1.2
4	Classification	SP

The grain size analysis curve of sand is shown in Fig.10.19.

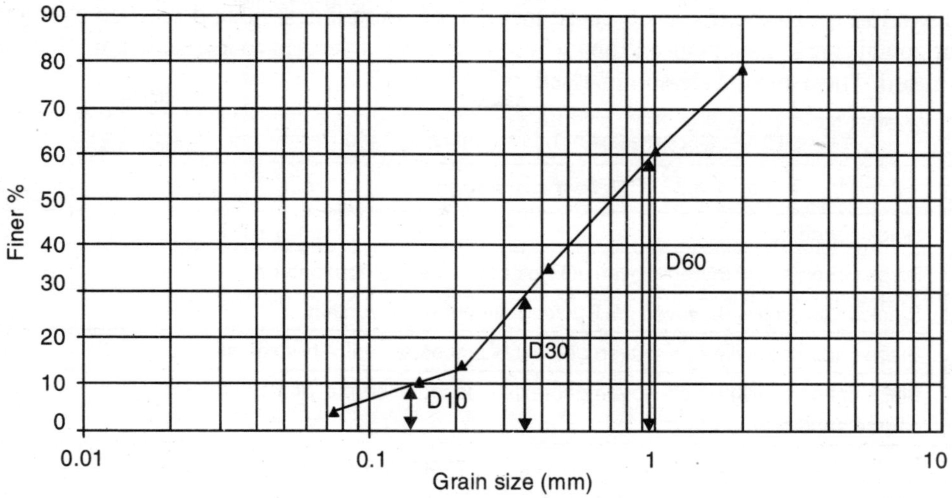

Fig. 10.19: Grain size analysis of sand

10.10 TESTS CONDUCTED

In IIT Roorkee, the tests were conducted under following two conditions:

(*i*) Consolidation test at constant rate of loading

(*ii*) C-U triaxial test at axial deformation control

The result of the tests conducted in these conditions are given as follows:

10.11 RESULTS OF STUDY DONE (Mittal et al., 2006)

10.11.1 Settlement behaviour of reinforced black cotton soil

The compression index of fine grained soil is related to its index properties, especially the liquid limit. For remoulded soils, $C_c = 0.007$ (LL-10). By this expression, the compression index was determined as 0.357. Consolidation tests were conducted for different test conditions on GDS consolidation testing system to evaluate the compression index and the same was found as 0.34 for virgin black cotton soil. Results in Table 10.17 show compression index values for various test conditions. It is seen that compression index value with woven JGT was less than that with non-woven JGT and compression index value C_c, of non-woven JGT was less than that of virgin soil as also shown in Fig. 10.20. It was found that compression index value decreases by about 8% in case of non-woven JGT and 16.5% in case of woven JGT when compared with virgin soil. Fig. 10.21 shows that compression index value of non-woven geotextile was less than that with woven JGT. It is seen that compression index value decreased by about 25% when compared with that of virgin soil. When non-woven geotextile is compared with non-woven JGT, it is seen that compression index value with non-woven geotextile is less than that of non-woven JGT as shown in Fig. 10.22. It had been seen by authors that compression index value with woven JGT was less than that with non-woven JGT when 10 mm diameter vertical coarse sand drain was provided at centre of consolidation sample as shown in Fig. 10.23. When black cotton soil was reinforced with woven JGT or non-woven geotextile and a vertical coarse sand was also provided (Fig. 10.24), the compression index value was less for non-woven geotextile than that for woven JGT. When virgin soil was reinforced with 4 mm horizontal coarse sand drain and in other test with 10 mm vertical coarse sand drain, it was found that in case of vertical coarse sand drain, compression index values was less as shown in Fig. 10.25. It was observed that compression index value decreased by about 20% in case of horizontal coarse sand drain and about 36.76% in case of vertical coarse sand drain when compared with virgin soil. These results are summarised in Table 10.17.

Table 10.17: Compression index values C_c for different test conditions

Sl. No.	Test conditions	C_c values
1	Black cotton soil	0.34
2	Black cotton soil with non-woven JGT placed at centre horizontally	0.312
3	Black cotton soil with woven JGT placed at centre horizontally	0.284
4	Black cotton soil with non-woven geotextile placed at centre horizontally	0.255
5	Black cotton soil with non-woven geotextile placed at centre with 10 mm vertical coarse sand drain	0.212
6	Black cotton soil with non-woven JGT placed at centre with 10 mm vertical sand drain	0.219
7	Black cotton soil with woven JGT placed at centre with 10 mm vertical sand drain	0.216
8	Black cotton soil with 4 mm coarse sand drain placed at centre horizontally	0.272
9	Black cotton soil with 10 mm vertical sand drain at centre	0.215

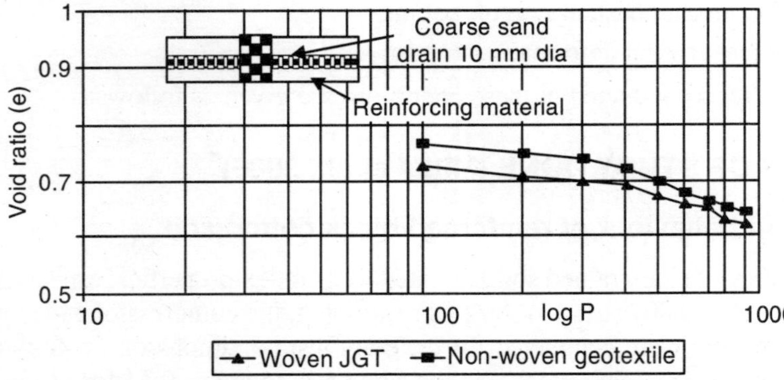

Fig. 10.20: Void ratio (e) *vs.* log P curve with non-woven JGT and woven geotextile with 10 mm vertical coarse sand drain

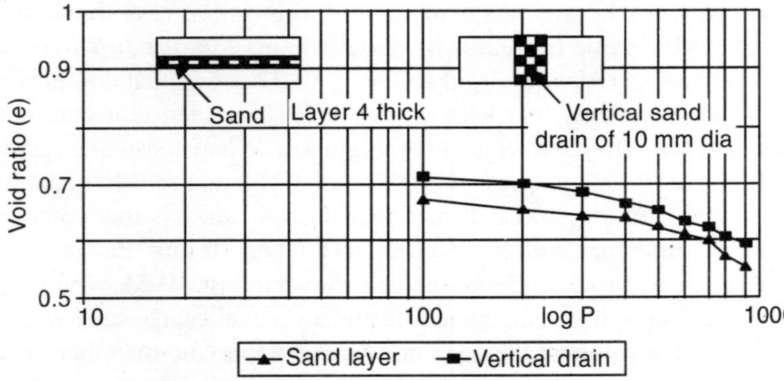

Fig. 10.21: Comparison of (e) *vs.* log P curve with horizontal and vertical coarse sand drain

10.11.2 Shear strength behaviour of reinforced black cotton soil

For determining the shear strength parameters (c' and Φ') of reinforced black cotton soil, C-U triaxial tests were conducted on GDS Triaxial Testing System. Results are summarised in Table 10.18. It was found that angle of internal friction, Φ' increased in all the cases as shown in Figs. 10.26–10.30. In case of non-woven geotextile, cohesion intercept also increased. It was observed that shear strength increases by about 13% in case of black cotton soil plus non-woven geotextile when compared with virgin soil at 100 kPa load. When shear strength behaviour of virgin soil was compared with woven JGT, it is seen that it increases by about 9% at 100 kPa load. In case of 10 mm vertical coarse sand drain, shear strength increases marginally by about 1.5% at 100 kPa load as compared with virgin soil but in case of 4 mm horizontal coarse sand drain it increased by less than 1% as compared with virgin soil.

Table 10.18: Shear strength parameters for different test conditions

Sl. No.	*Test conditions*	*c′ (kPa)*	*ϕ′ (Degree)*
1	Black cotton soil only	102	13
2	Black cotton soil with woven JGT placed at centre horizontally	92.5	16
3	Black cotton soil with non-woven geotextile placed at centre horizontally	106	19
4	Black cotton soil with 4 mm coarse sand at centre horizontally	70	22
5	Black cotton soil with 10 mm diameter vertical coarse sand drain	82	24

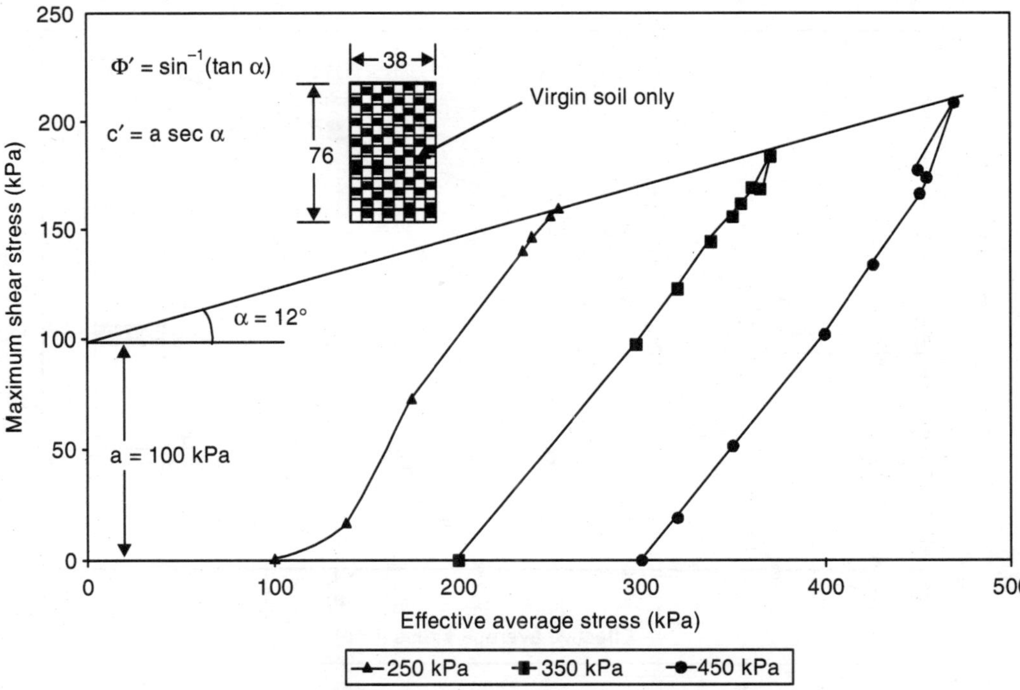

Fig. 10.22: Maximum shear stress *vs.* effective stress for black cotton soil

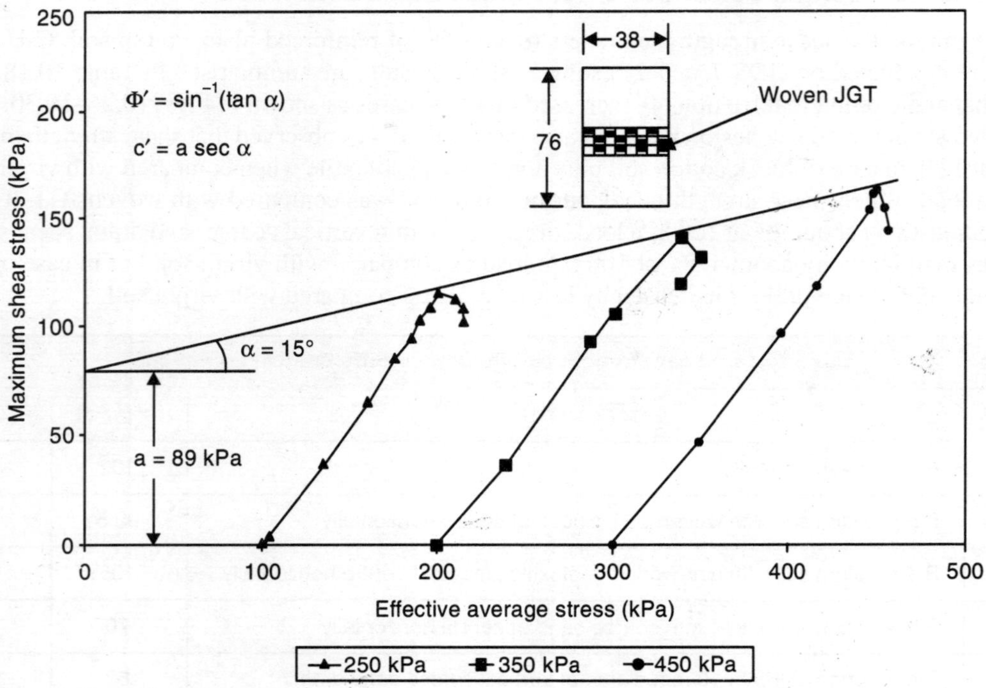

Fig. 10.23: Maximum shear stress *vs.* effective stress for woven JGT placed at centre

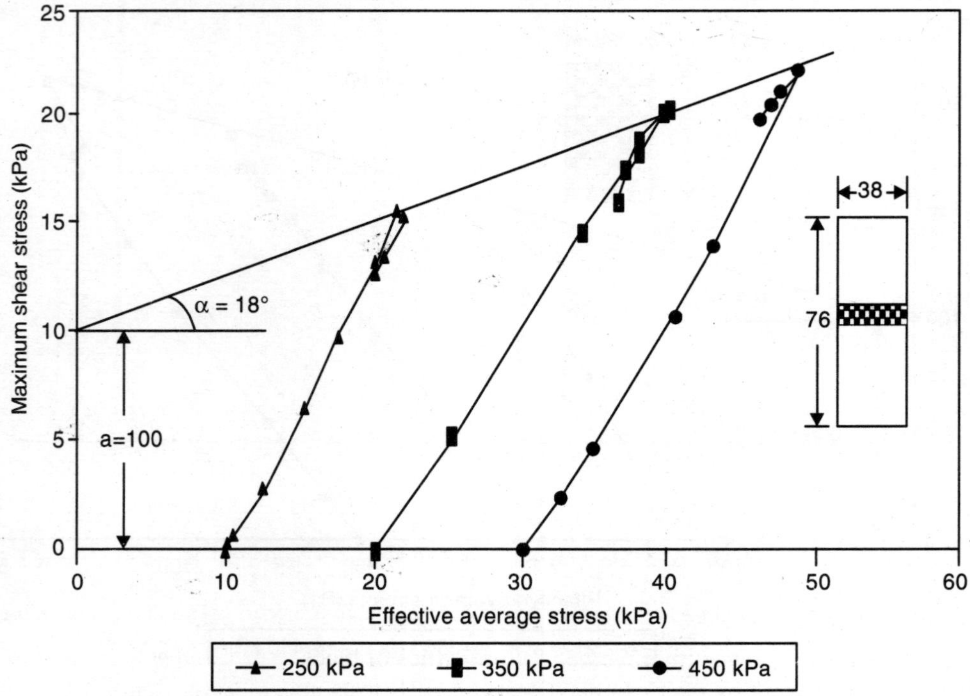

Fig. 10.24: Maximum shear stress *vs.* effective stress non-woven geotextile placed at centre

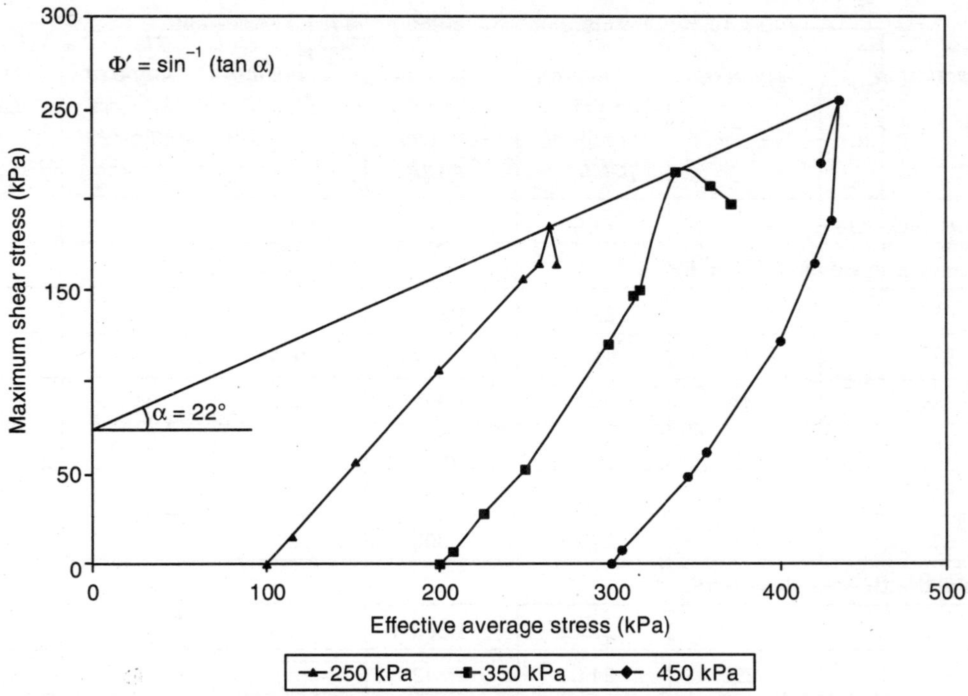

Fig. 10.25: Maximum shear stress *vs.* effective stress with 10 mm vertical sand drain placed at centre

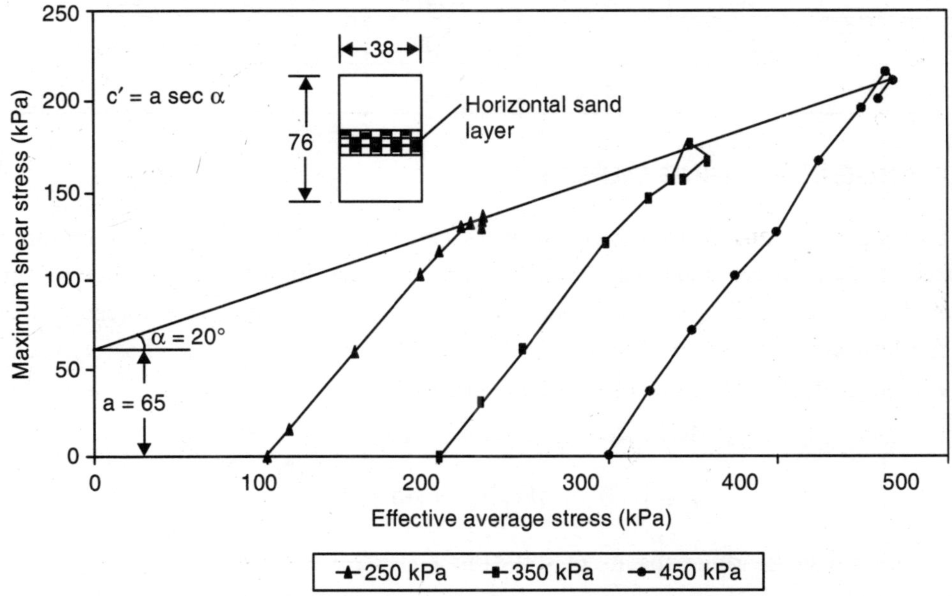

Fig. 10.26: Maximum shear stress *vs.* effective stress curve with 4 mm horizontal sand drain placed at centre

Gosavi et al. (2004) showed improvement of properties of black cotton soil sub-grade through fibres of synthetic reinforcement by their extensive tests on CBR tests under soaked condition. The results of the same are as below (Table 10.19).

Table 10.19: Summary of the results of the tests performed

Description	*Aspect ratio*	*Optimum moisture content (OMC), %*	*Maximum dry density (MDD)*	*Cohesion (c), t/m²*	*Angle of internal friction (Φ), degree*	*California bearing ratio (CBR), %*
Soil without any fibre	-	21.04	1.440	4.1	14.0	4.90
Soil mixed with woven fabrics' fibres						
1%	25	22.40	1.371	4.6	12.8	7.30
2%	25	23.73	1.369	5.2	12.5	7.65
3%	25	46	1.387	4.8	12.5	7.50
1%	50	2.55	1.382	5.2	13.0	7.85
2%	50	24.03	1.367	5.8	12.0	8.20
3%	50	21.73	1.390	5.3	12.5	8.05
Soil mixed with fibre glass fibres						
1%	250	22.15	1.350	4.6	12.8	7.00
2%	250	24.03	1.342	5.4	13.0	7.25
3%	250	21.73	1.360	5.1	13.5	7.10
1%	500	22.64	1.357	4.9	12.5	7.15
2%	500	24.78	1.330	5.6	12.0	7.60
3%	500	20.32	1.380	5.4	13.0	7.45

10.12 CHANGE IN BEARING CAPACITY

Knowing the values of c and Φ from direct shear test, the bearing capacity of a strip footing of 1.0 m width located at 1.0 m below ground surface was determined. The footing was analysed for three cases resting on (*i*) black cotton soil ($c = 4.1$ t/m^2 and $\Phi = 14°$) (*ii*) black cotton soil reinforced with 2% fibres of woven fabrics (Aspect Ratio* = 500) ($c = 5.6$ t/m^2 and $\Phi = 12°$). The factor of safety used was 2.0.

(Aspect Ratio = Ratio of Length to Diameter of Fibre)

The safe bearing capacity (SBC) was obtained using equation

$$q_u = \left[c'N_{c'} + \gamma D_f N_{q'} + 0.5\,\gamma B N_{\gamma'}\right]\frac{1}{F} \qquad ...(10.2)$$

where q_u is the ultimate bearing capacity of soil, (t/m^2); $c' = 2/3$ c, c = cohesion, (t/m^2); D_f, depth of footing, (*m*); B, width of footing, (*m*); γ, maximum dry density, 18 t/m^3; $N_{c'}$, $N_{q'}$, N_{γ} bearing capacity factors corresponding to local shear failure condition (Fig 10.27); and F, the factor of safety taken as 2.0.

The results are summarised in Table 10.20.

Table 10.20 indicates that, due to addition of 2% woven fabrics with aspect ratio 50, the SBC of black cotton soil increases by 33.58%. Similarly, increase in SBC of black cotton with addition of 2% fibre glass was 29.67% for aspect ratio 500.

10.12.1 Thickness of base course

The basic parameters affecting the reinforcement function of geotextiles are (*i*) soil strength, (*ii*) elastic characteristics of the geotextile, (*iii*) contact pressure below the wheels. For the cost benefit analysis of the reinforcement function it is necessary to evaluate the pavement thickness reduction that are achieved by use of geotextiles.

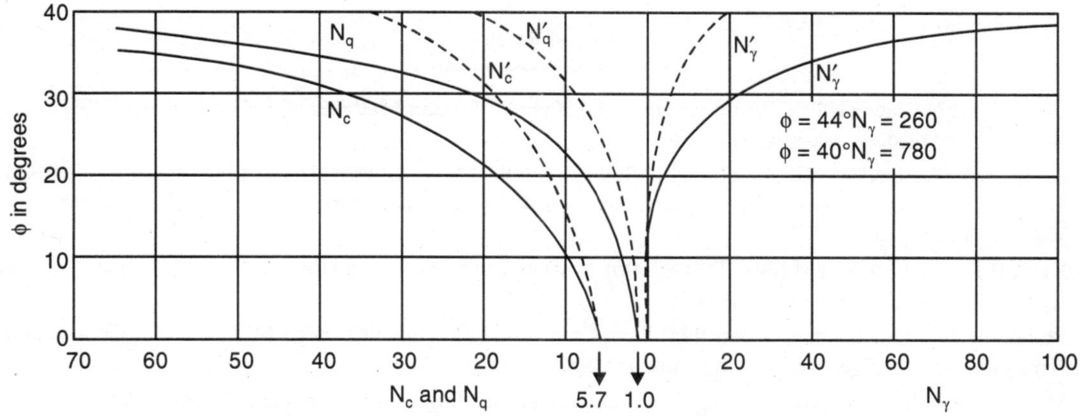

Fig. 10.27: Bearing capacity factors (General and Local shear failure)

Table 10.20: Effect of reinforcement on safe bearing capacity (SBC) of black cotton soil

Description	***Bearing capacity***
Black cotton (BC soil without reinforcement)	13.49
Black cotton soil with 2% woven fabrics reinforcement t/m² (AR = 50)	18.02
Black cotton soil with 2% fibre glass reinforcement t/m³ (AR = 500)	17.49

Design calculations of reinforcement function was done by authors using the Giroud and Noirays (1994) pavement design method for unpaved roads. This method is applicable to unpaved roads subjected to light to medium traffic (1–10,000 passes over the lifetime of the road). Equation (10.3) is given by Mandal and Mhaiskar (1994) for calculating the thickness of base course.

$$c = \frac{P}{\left\{2\pi\left[\sqrt{\left(\frac{P}{P_c}\right)} + 2h_0 \tan\alpha\right]\left[\sqrt{\left(\frac{P}{2P_c} + 2h_0 \tan\alpha\right)}\right]\right\}} \qquad \text{...(10.3)}$$

where h_0 is the thickness of base course; α, angle of load dispersion taken as 26°; P, axle load i.e., 8 t; $P_{c,}$ tyre inflation pressure, 62 t/m²; and *c*, the cohesion, t/m².

Putting the value of cohesion, c in equation (10.3) for the three different cases considered earlier (black cotton soil without reinforcement, with 2% woven fabrics and with 2% fibre glass) and values of α, P and P_c the values obtained were:

$h_0 = 0.5$ m (for black cotton soil without reinforcement, $c = 4.1$ t/m^2); $h_0 = 0.38$ m (for black cotton soil with 2% woven fabrics AR = 50, $c = 5.8$ t/m^2) and $h_0 = 0.39$ m (for black cotton soil with 2% fibre glass AR = 500, $c = 5.6$ t/m^2).

These results indicate significant change in pavement thickness while using the woven fabric/fibre glass as a reinforcing material when mixed randomly with black cotton soil for construction of unpaved roads.

References

1. Giroud and Noirays (1994). "An Overview of Pavement Design Methods with Geosythetics', *Geosynthetic World,* Wiley Eastern Limited, Mumbai.
2. Gosavi, M., Patil, K.A., Mittal, Satyendra, Saran, Swami (2004). "Improvement of Properties of Black Cotton Soil Subgrade through Synthetic Reinforcement", *Journal of Institution of Engineers*, Civil Engineering Division, Vol. 88, pp. 257–261.
3. Katti, R.K., Katti, A.R. and Katti, D.R. (2000). "Influence of Gravity on Granular Soil Mechanics", Oxford & IBH Publishing Co. (P) Ltd., New Delhi.
4. Mandal, J.N. and Mhaiskar, S.Y. (1994). "An Overview of Pavement Design Methods with Geosynthetics", *Geosynthetic World,* Wiley Eastern Limited, Mumbai.
5. Mittal, Satyendra and Shukla, J.P. (2012). "Soil Testing for Engineers", Khanna Publishers, Delhi.
6. Mittal, S., Sharma, A.K. and Rao, Manmadha, N. (2011). "Improvement of High Compressible Clay with Admixtures", Unpublished paper.
7. Mittal, S., Sharma, A.K. and Kumar, Binod (2006). "Behaviour of Black Cotton Soil with Bio-degradable Reinforcement", Indian Highways, Vol. 34, No. 2, pp. 5-12.
8. Rao, Manmadha, N. (2006). "Behaviour of High Compressible Clay with Admixtures" M. Tech. Dissertation, IIT Roorkee.

Construction Dewatering

CHAPTER 11

11.1 INTRODUCTION

Under many field situations during construction operation it may be necessary to eliminate seepage pressure to increase the shearing resistance or to reduce the danger of frost damage. These could be done by reducing the neutral stresses effectively by adopting any one of the drainage methods.

Dewatering systems and drains are the two methods which are used to improve the ground condition before, during and after construction.

Dewatering systems essentially consists of lowering the water table to required elevation and to establish below this level a system of collectors located in wells, galleries or ditches. The collected water at collectors are pumped. Continuous pumping from excavations or from natural ground is a costly affair and the continued flow from the surrounding ground may endanger the stability of adjacent structures. Further heavy inflow may lead to erosion or collapse of the sides in an open excavation. In certain situations there may be instability of the base of excavation due to upward seepage.

On the other hand, drains are provided to control the flow, in some cases for lowering the water table, in others for reducing pore pressure and seepage forces.

In order to ensure in selecting a proper drainage method, it is important to obtain all the necessary information before commencing work, and this aspect should not be neglected at the site investigation stage.

The yield of water from the ground is also function of permeability of the ground. The approximate values of the permeability of different ground is given in the following Table 3.1. The readers are however advised to refer any book of soil mechanics to understand permeability properties of ground.

<table>
<tr><th colspan="13">Table 11.1: Permeability and drainage characteristics of soils coefficients of permeability k in cm per sec</th></tr>
<tr><td></td><td>10^2</td><td>10^1</td><td>1.0</td><td>10^{-1}</td><td>10^{-2}</td><td>10^{-3}</td><td>10^{-4}</td><td>10^{-5}</td><td>10^{-6}</td><td>10^{-7}</td><td>10^{-8}</td><td>10^{-9}</td></tr>
<tr><td>Drainage</td><td colspan="6">Good</td><td colspan="2">Poor</td><td colspan="4">Practically impervious</td></tr>
<tr><td rowspan="2">Soil types</td><td colspan="3">Clean gravel</td><td colspan="3">Clean sands, clean sand and gravel mixtures</td><td colspan="4">Very fine sands, organic sand inorganic silts, mixtures of sand silt and clay, glacial till, stratified clay deposits, etc.</td><td colspan="2">"Impervious" soils, e.g., homogeneous clays below zone of weathering</td></tr>
<tr><td colspan="3"></td><td colspan="7">"Impervious" soils modified by effects of vegetation and weathering</td><td colspan="2"></td></tr>
<tr><td rowspan="2">Dirtect deter-mination of k</td><td colspan="7">Direct testing of soil in its original position pumping tests. Reliable if properly conducted considerable experience required.</td><td colspan="5"></td></tr>
<tr><td colspan="6">Constant-head permeameter. Little experience required.</td><td colspan="6"></td></tr>
<tr><td rowspan="2">Indirect deter-mination of k</td><td colspan="2"></td><td colspan="4">Falling-head permeameter. Reliable. Little experience required.</td><td colspan="3">Falling-head permeameter. Unreliable. Much experience required.</td><td colspan="3">Falling-head per-meameter. Fairly reli-able. Considerable experience necessary.</td></tr>
<tr><td colspan="6">Computation from grain-size distribution. Applicable only to clean cohesionless sands and gravels.</td><td colspan="4"></td><td colspan="2">Computation based on results of consolidation tests. Reliable. Considerable experience required.</td></tr>
</table>

11.2 DEWATERING SYSTEMS

Ground water can be controlled by adopting one or more types of dewatering systems or drains suiting to the size and depth of excavation, geological conditions and characteristics of soil. In earlier days, the control of ground water in construction work was accompanied by one or a combination of following methods:

(*i*) Constructing ditches and dykes to intercept seepage from the slopes combined with a controlled rate of excavation.

(*ii*) Sheeting combined with pumping from sumps.

(*iii*) Pumping from deep sheeted sumps dug outside of the working area.

Various types of drainage systems which may be considered most suitable based on grain size distribution of soils is presented in Fig. 11.1 (IS: 9759–1981).

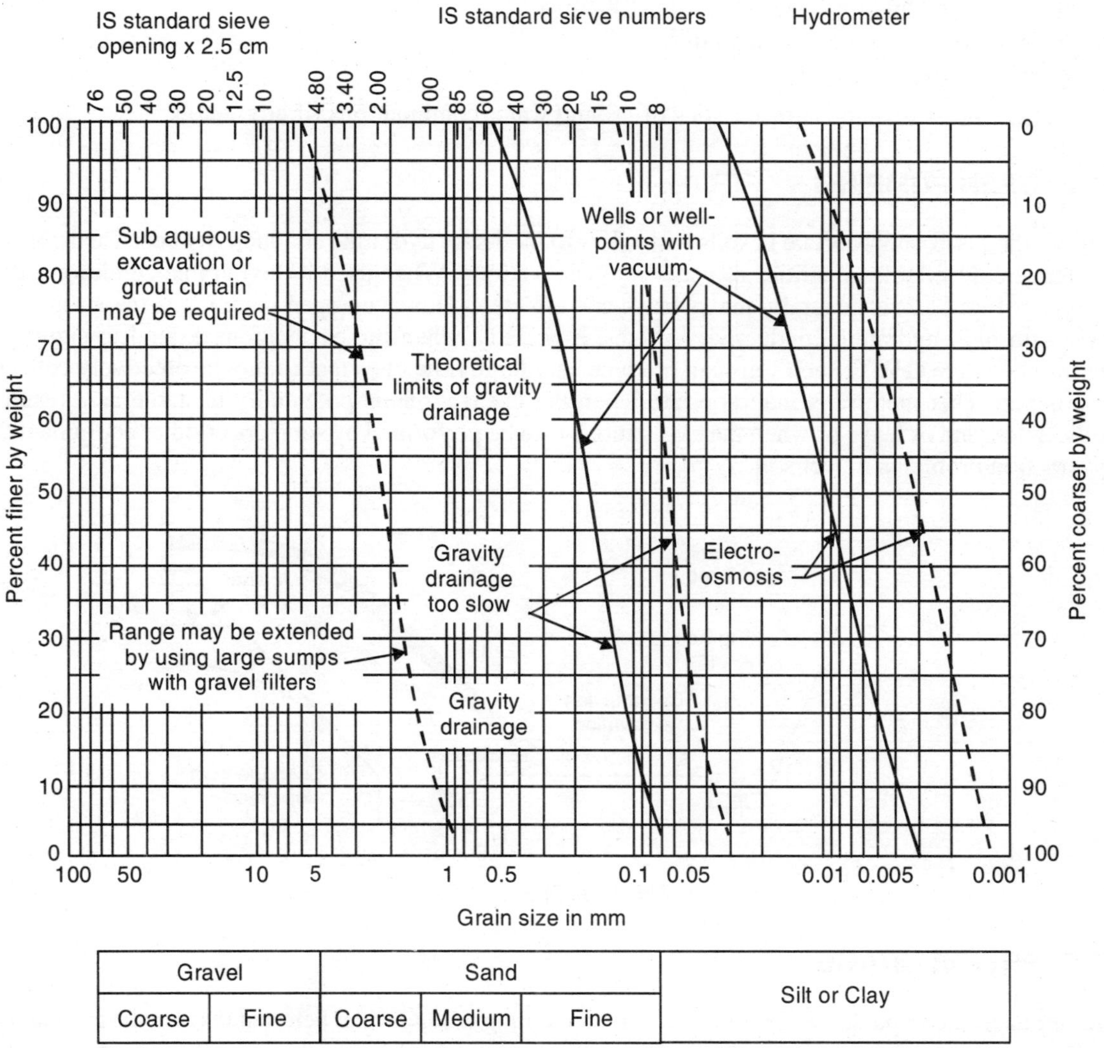

Fig. 11.1: Dewatering systems applicable to different soils (adopted from IS: 9759–1981)

11.3 CATEGORIES OF CONSTRUCTION DEWATERING

Construction dewatering is done at the time of construction. There are mainly four basic categories of construction dewatering methods e.g.,

1. Open pumping
2. Pre-drainage
3. Cut offs
4. Exclusion

The selection of these methods depends on the following:

1. The soil conditions especially the hydrualic conductivity.
2. The size and depth of the construction excavations, especially depth below ground water table.
3. The planned method of excavations and ground support.

4. The type and proximity of nearby structures.
5. The type of structure being built.
6. The planned schedule.
7. The pressure and characteristics of ground water continuations (if any).

11.4 OPEN PUMPING

This method is not appropriate in soils with a low to moderate hydraulic conductivity because these soils discharge controllable quantities of water into excavations. When used in favourable conditions, open pumping (Fig. 11.2) is generally inexpensive and effective. However, open pumping is inappropriate in soils with high hydraulic conductivity values, especially, when the excavations extend to significant depths below the G.W.T. Such soils can produce very large flow rates that cannot be efectively collected or pumped. This method is inappropriate when the excavations is potentially unstable as a result of boils, quicksand or leave or when the excavations is to be performed by scrapers or other equipment that has traction problems in wet soil.

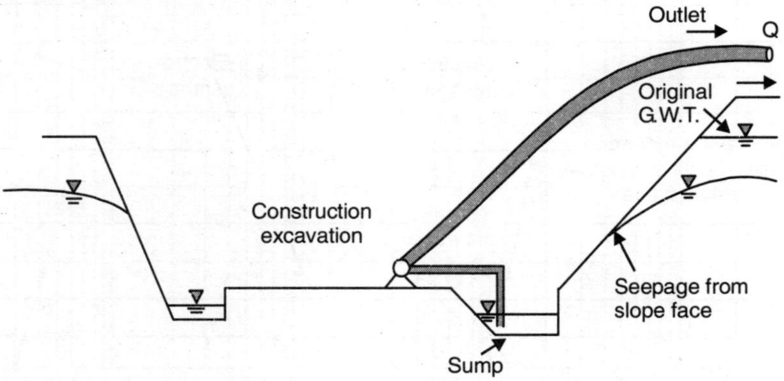

Fig. 11.2: Open pumping

11.5 PRE-DRAINAGE

Pre-drainage encompasses several methods of intercepting ground water before it reaches the excavations. Unlike open pumping, these methods allows the contractor to lower the ground water table before excavations begins, which often solves equipment mobility problems. They are also suitable for a wider range of soil conditions. Most pre-drainage methods use well located at a short distance away from the periphery of excavations (Fig. 11.3).

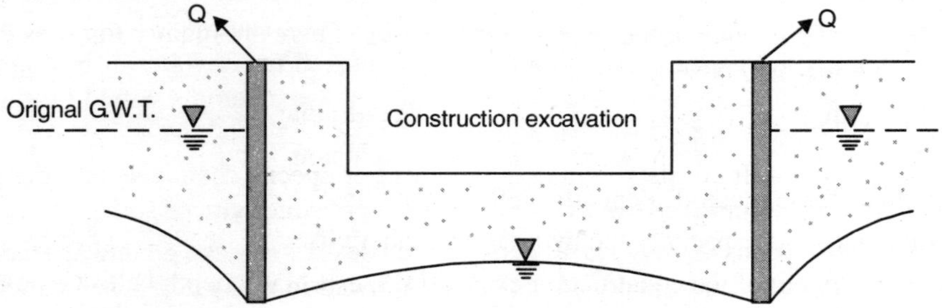

Fig. 11.3: Pre-drainage

Several methods are available to extract water from well including:

1. Pumpled well systems—With submersible pump in each well. This can accommodate large flow rates. This is expensive.
2. Well point systems—Inexpensive, but limited to depths up to 5 to 6 m.
3. Ejector system—Uses a nozzle and a venturi in each well to lift water. These systems do not have the depth limitations of well points, and often are less expensive than pumped well systems. However, they are inefficient particularly when flow rate is high.

11.6 CUT OFFS

Cut offs (also known as seepage barriers) are meant to block the ground water before it reaches the excavations. Although stoppage of complete ground water flow is impossible, cut offs can be very effective and they reduce the water flow to a small fraction. The small amount of ground water table if left behind can be removed by pumping. There are many methods and techniques to cut off ground water (Fig. 11.4). These are:

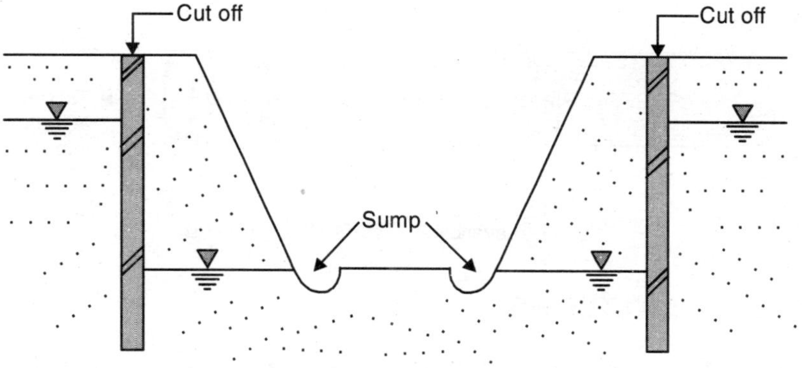

Fig. 11.4: Cut offs

- Steel sheet piles
- Diaphragm walls
- Slurry trenches—Same as diaphragm wall, except that they are filled with low permeability soils instead of concrete. They provide ground water control but do not provide structural support.
- Secant drilled shaft walls—These are drilled shaft foundations (large diameter borings filled with concrete) placed on close centres, thus creating a continuous underground wall.
- Tremie seal—These are concrete barriers placed at the bottom of excavations to block water and resist hydrostatic pressure. They are typically used in combination with sheet piles or diaphragm walls and are often placed under water using a tremie. A tremie is a long tube used to place concrete. They are used when the concrete must be placed under water or through a bentonite slurry, because simply dropping the concrete through these materials would produce excessive segregation and result in a very poor product.
- Permeation grouting—It consists of injecting coment or special chemicals into the ground. This fills have voids and significantly reduces the hydraulic conductivity of soil.
- Ground freezing—It consists of inserting refrigeration devices into the ground and freezing ground water. This is an expensive method, and is being followed in many developed nations.

11.7 EXCLUSION

This method uses compressed air to prevent ground water from entering construction excavations. This method has been used in tunnel and foundation construction because these excavations can be sealed and pressurised. The air pressure is maintained at a level approximately equal to the pore water pressure, thus preventing or at least greatly reducing seepage into the excavation. Workers, equipment and excavated soil pass through a decompression chamber located between the excavation and the atmosphere (Fig. 11.5).

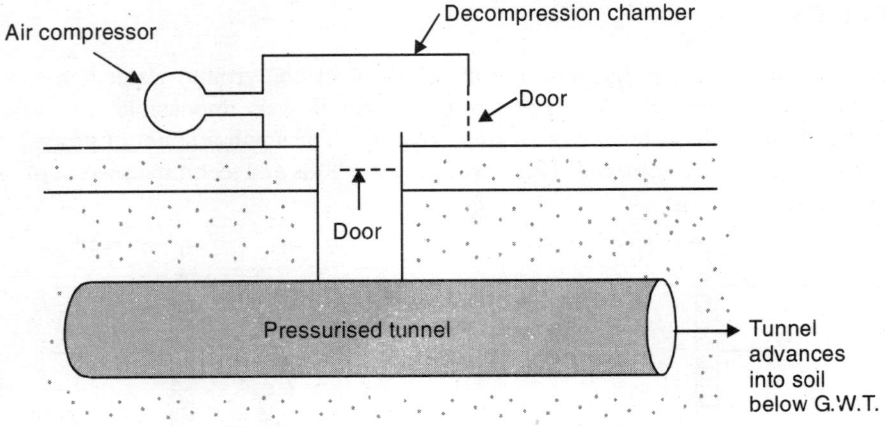

Fig. 11.5: Exclusion method of dewatering

11.8 PERMANENT DEWATERING

Permanent dewatering measures are necessary when ground water must be controlled throughout the useful life of a civil engineering project. Some of the construction dewatering methods also may be used for permanent dewatering. However, economics and reliability concerns generally prevent the use of extensive long-term pumping. The specific methods of permanent controlling ground water depend on the application. Following techniques are used in specific applications.

(a) Underground structures

Buildings with basements, tunnels, utility vaults, sewage pumping stations and other structures need to be protected from flooding, especially when they extend below the ground water table. These structures have thick concrete wall that are reasonably watertight. The outside areas of these walls are coated with a waterproof material and drain pipes are sometimes installed around the perimeter. However, some water may still enter the structure. Therefore, it is recommended to direct such water to a sump and pump it out from there.

(b) Retaining walls

If ground water is allowed to build up behind retaining walls, the resulting hydrostatic pressures may be much greater than the lateral pressure due to soil. Sometimes it becomes necessary to design for these pressure, but this can significantly increase the cost of wall. The other option is to install drain pipes or weep holes to drain ground water, thus maintain ground water table at suitable low elevation.

11.9 PAVEMENTS

Pavements for highways, parking, airports and other facilities are very sensitive to the accumulation of water in the underlying soils and many pavement failure may be attributed to poor drainage of sub-grade soils. Water inevitably passes through joints, cracks and other openings in pavements and in the case of certain open graded asphalt it may pass through pavement itself. Therefore, this water need be drained to a safe location carefully.

11.10 EARTH SLOPES

Permanent drainage measures are often installed in earth slopes to control the ground water level and thus enhance stability.

11.11 EARTH DAM

Earth dams include extensive drainage facilities to control the ground water flow through the dam. This enhances its stability.

11.12 DIFFERENT DEWATERING SYSTEMS

11.12.1 Open sumps and ditches

The essential features of method are a sump below ground level of excavation at one or more corners or sides. In order to prevent standing water on the floor of excavation, a small grip on ditch is cut around the bottom of excavation, falling towards sump. In case of large excavations which have to remain open over long periods, it is advisable to design these drainage ditches with much more care. This is most widely used method of ground water lowering. The greatest depth to which water table can be lowered by this method is 8 m below sump.

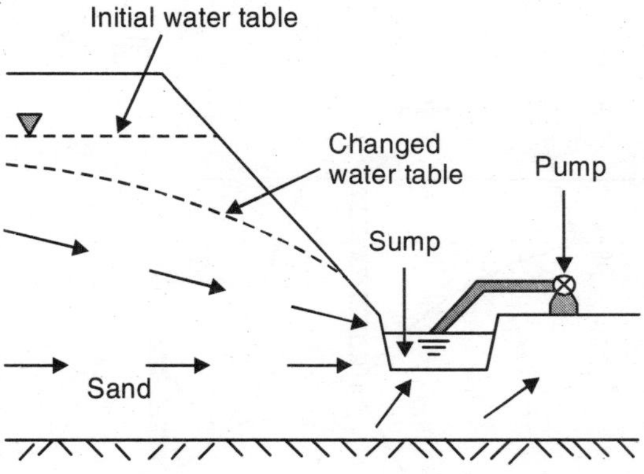

Fig. 11.6: Open sump

The method has the disadvantage that ground water, flows towards excavation with a high head or a steep slope and hence there is a risk of collapse of sides (Fig. 11.6).

11.12.2 Well point system

Well point system is a very good and effective technique. In this method, a boring is done in which a perforated pipe is lowered. This pipe may be wrapped around with geosynthetic filter cloth. On this pipe, a pump and motor are mounted to suck water from the ground. The water thus extracted from ground is disposed of through a header pipe. The selection of place for installation of suction pipe is chosen judiciously.

The multistage well point system is again a very effective method of extraction of ground water for construction of underground structures, laying of RCC raft etc. In this method, the suction pipes are installed at different places covering the entire area to be treated. These pipes are attached to a common header pipe (delivery pipe). The delivery pipe should be fixed in such a manner that it does not disturb neighbour's property. If well point system is installed near a river/canal, the delivery pipes' outlet can be focused towards downstream flow of river/canal (Figs. 11.7 and 11.8).

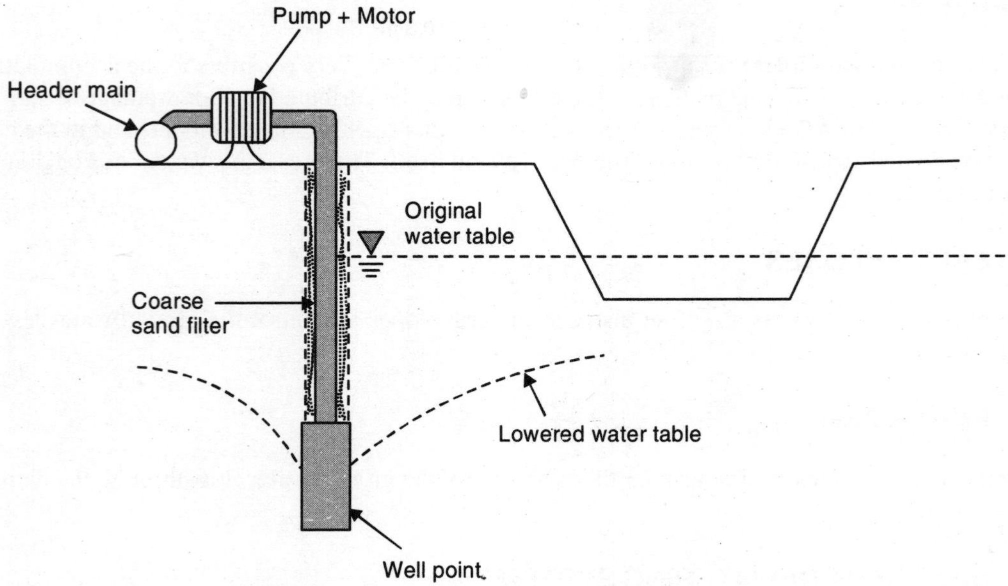

Fig. 11.7: Single stage well point

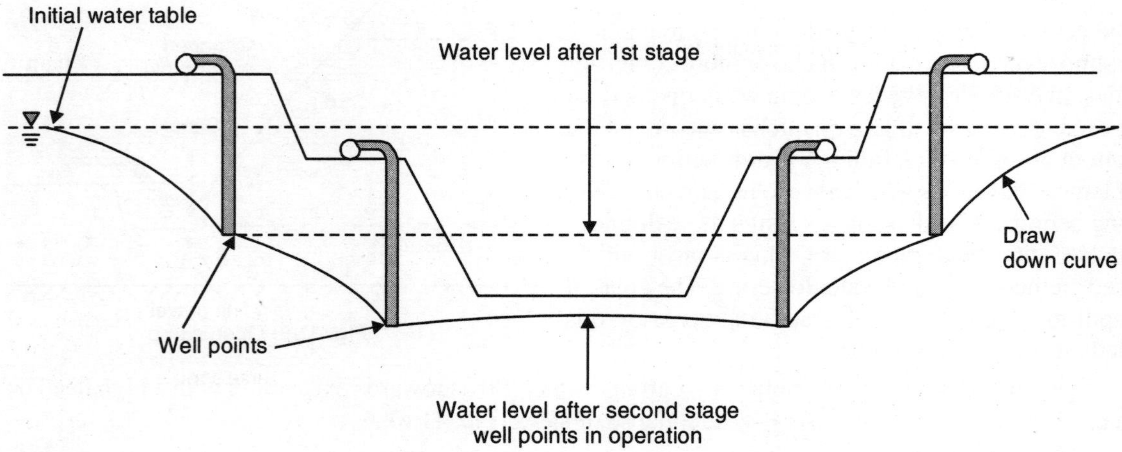

Fig. 11.8: Multi stage well point

The other methods of dewatering are:

1. Vacuum dewatering system (useful for fine grained soils with permeability in the range of 0.1–10^{-2} mm/sec).
2. Dewatering by electroosmosis (also useful for fine grained soils)

   Well point spacing can also be determined from a flow net (IS: 9759–1981). An approximate spacing of well point required to produce a given ground water lowering in various soils can also be estimated from the monographs shown in Figs. 11.9 and 11.10. These monographs are based on empirical data and are used for fairly good judgement of water quantity. Therefore, designers may use their experience and engineering assessment.

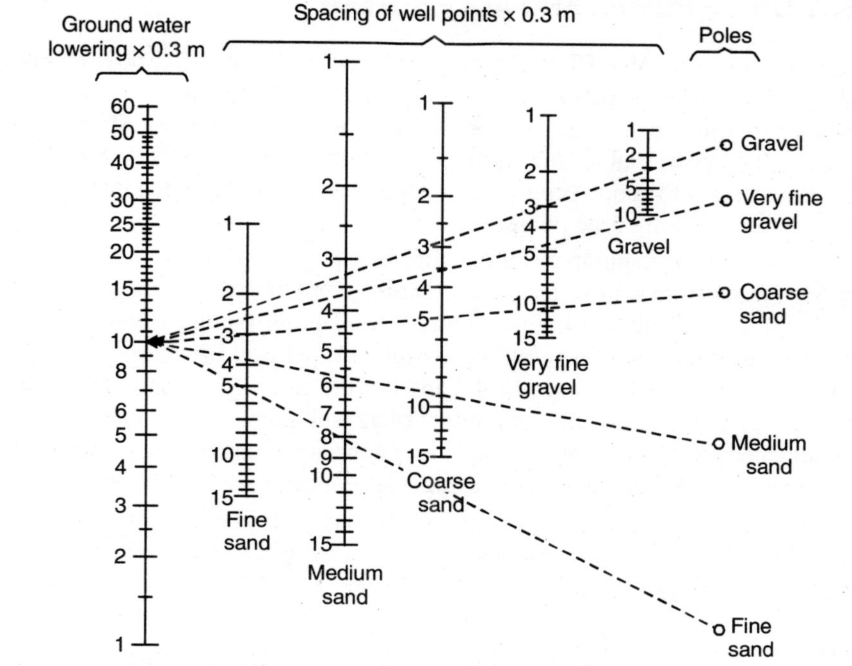

Fig. 11.9: Well point spacing for stratified clean sands and gravels (IS: 9759–1981)

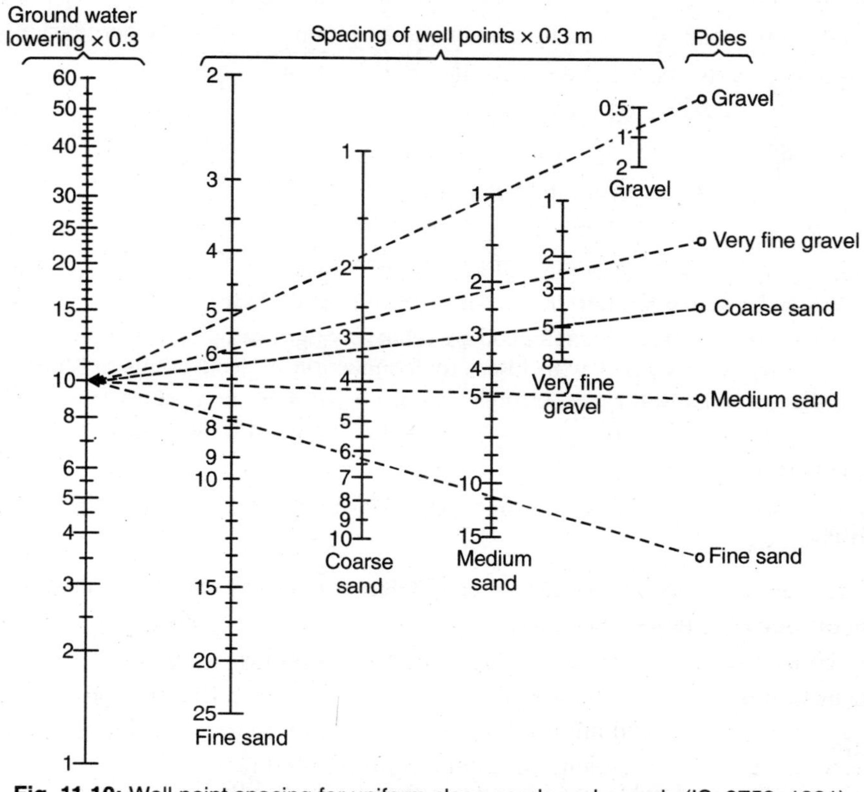

Fig. 11.10: Well point spacing for uniform clean sands and gravels (IS: 9759–1981)

11.13 SELECTION OF PUMPS AND ACCESORIES

Pumps and power unit must be of sufficient capacity to extract the required flow of the wells or well points and conduit it away from the dewatered area. The selection of a pump and power unit depends on various factors such as required discharge, suction lift plus positive head including hydraulic head losses, air-handling capacity, power available, fuel economy and durability of units. While selecting a pump and motor unit, it is necessary to adopt a conservative approach as the estimated flow is not exact and the efficiency may be reduced in long time pumping.

Centrifugal pumps are used to pump water in collector pipes connected to wells or well point system and removing surface water from sumps. The selected pumps should have sufficient air handling capacity, and they should be able to produce a high-vacuum. A well point pump consisting of a self-priming centrifugal pump with attached vacuum pump is generally adequate and develops 6–7.5 m of vacuum. If the depth of water table lowering is large (> 4.5 m) but the rate of pumpage for each well point is relatively small (< 10–15 gpm) installation of a single stage well point system at the top of excavation or water table, with attached jet-educator pump may prove to be advantageous than a multistage well point system. Water table up to 30 m can be lowered using a jet-educator well point system.

In selecting the power unit to be used for driving the pumps, consideration should be given to the initial cost of the unit, and the cost of operation, including maintenance and fuel.

The capacity of a pump required for a dewatering system is decided by the horse power as obtained from the formula.

$$\text{Horse power} = \frac{\text{Total discharge in gpm} \times \text{Total dynamic load}}{3960 \times \text{Efficiency of the pump and engine}} \qquad ...(11.1)$$

where, total dynamic load = operating vacuum at the pump in take + discharge friction losses.

Pump's capacity can also be estimated by following equation:

$$P(\text{kW}) = 9.81 \times Q \times H \times \eta \qquad ...(11.2)$$

where, P is power in kW

H = hydraulic head (m)

Q = Discharge (m^3/sec)

η = efficiency of system (generally 0.8–0.9)

Power is kW can be converted into horse power by standard relationships.

Headers pipes are of 15–30 cm diameter and are relatively light weight steel or plastic pipes. Headers for well point systems are provided with inlets for connection of well points at short intervals. The connection of collector lines to the header pipes should be through non-return valves. Discharge line may be of steel, aluminium or plastic pipes which should conduct the flow with relatively small head loss.

11.14 DRAINS

The drain consists of three components viz. filter, conduit or collector and disposal system.

Drains are of following types:

1. Open drains
2. Closed drains
3. Horizontal drains
4. Drainage after construction

1. Open drains

These are open drains through which seepage water may be diverted.

2. Closed drains

When seepage erosion or piping is troublesome or where a permanent drain is desired, perforated pipe can be laid at a required depth in ditches and the ditch is backfilled with a suitable filter material. As far as possible pipes should be laid in straight lines. Opening should be provided for every 30–50 m to flush out the pipe occasionally. Also manholes should be provided at changes in direction and at intervals of 100–150 m along straight sections (Fig. 11.11).

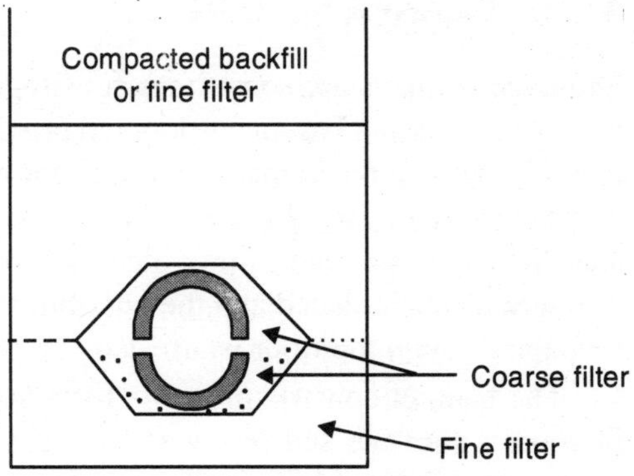

Fig. 11.11: Closed drains

3. Horizontal drains

If field situations warrant to avoid open cut work or non-availability of adequate submergence, the ground water can be lowered by means of Ranny drainage systems. This system consists of reinforced concrete shaft or wells from which a number of horizontal perforated pipes are fixed. These pipes may be extended to a required length in any direction. Water collected in the well is pumped out by means of a turbine pump.

4. Drainage after construction

Preventing ground water from seeping into an structure or through a finished structure may be necessary in order to obtain proper use of structure or to protect it from drainage. In some situations, the dewatering system utilised during construction may be further used to protect the structure.

11.15 FOUNDATION DRAINS

Where ground water is present in the vicinity of a structure, provision should be made to quickly carry away the water for building. In the worst condition, ground water effect may be allowed only on exterior side of building. When depth below water table is not much, it is feasible to control the water by foundation drains as shown. However, these drains should not be placed lower than the bottom of footing. These drains consists of perforated pipes or pipes with open joints so that the ground water can enter into pipe (Fig. 11.12).

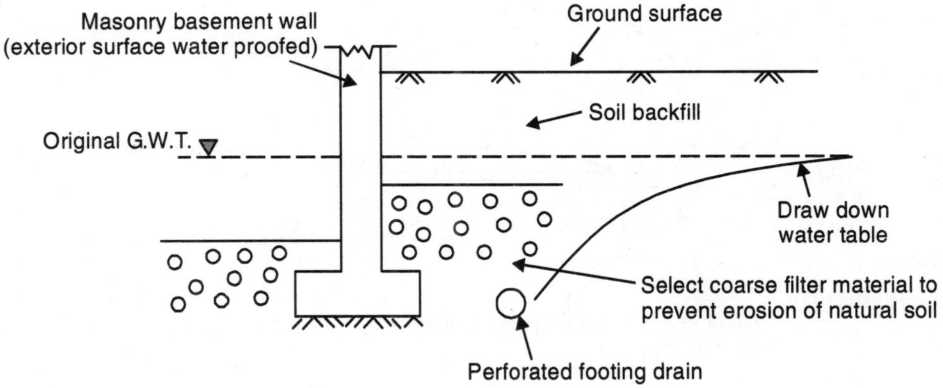

Fig. 11.12: Foundation drain

11.16 BLANKET DRAIN

Drainage blankets are sometimes constructed beneath drains and basement floors to provide a highly permeable drainage path for removal of ground water acting against the bottom of slab. If an escape path is provided the uplift pressure can be reduced and the possibility of seepage through the floor is arrested.

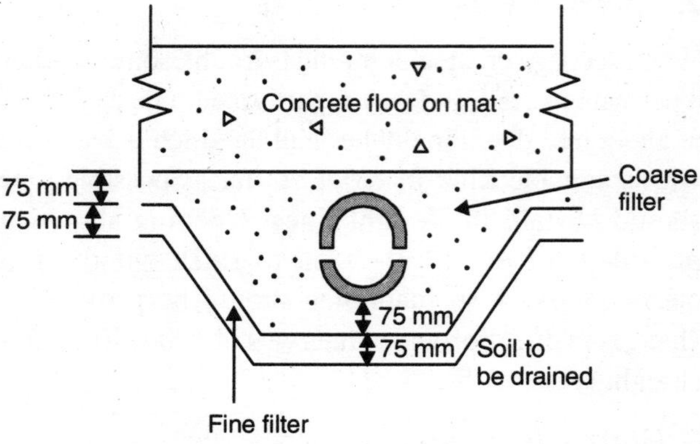

Fig. 11.13: Blanket drain

The blanket consists of a fine filter layer in contact with the soil followed by a coarse filter cum collector layer and the latter is in contact with the underside of a masonry dam or basement floor. The blanket is connected by conduits to a sump where the collector water is pumped out or to drainage pipe where disposal is by gravity. Such blankets are also provided beneath pavements to prevent capillary flow upward (Fig. 11.13).

11.17 INTERCEPTOR DRAINS

In paved highways and airfield runways, trench drains are provided parallel to shoulder. Such drains are termed as interceptor drains which are provided to lower the ground water table to a level beneath the pavement and to permit easy lateral drainage for water finding its way into the coarse base material.

The purpose of such drainage facility is to keep the base and sub-grade soils dry so as to maintain adequate strength and stability.

The drains also provide a means for disposal for surface and near surface wear and also shall help to intercept underground flow trying to enter the pavement from side areas. Similarly open drainage ditches located adjacent to shoulder area may help to intercept surface and near surface water flowing towards the roadway area from the sides also prevent development of excess pore water pressure (Figs. 11.14 and 11.15).

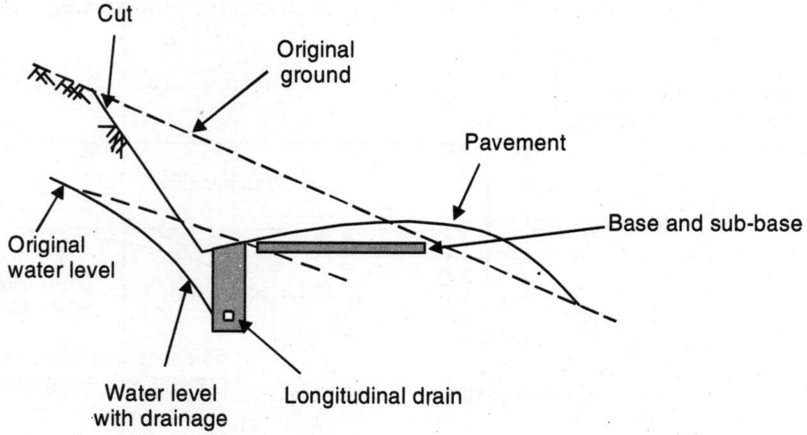

Fig. 11.14: Interceptor drain for highway constructed in a side

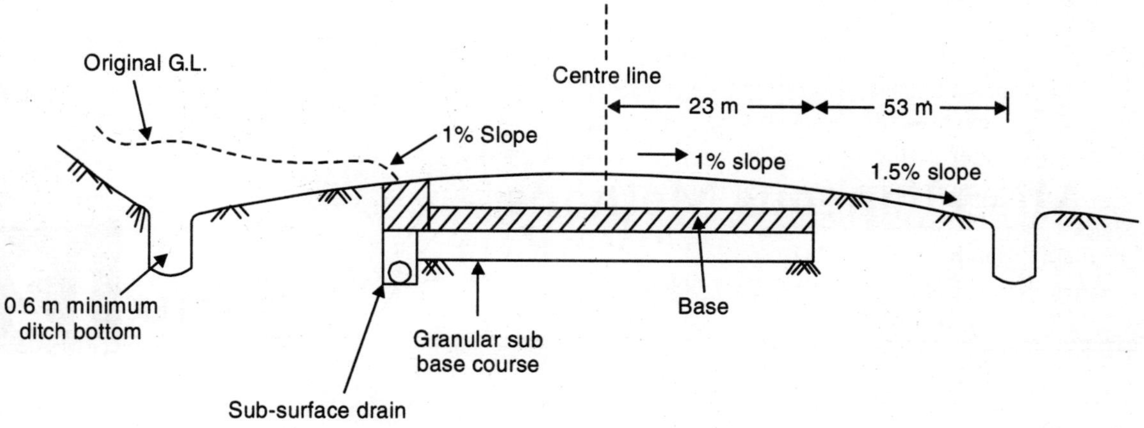

Fig. 11.15: Interceptor drains on airport runways

References

1. IS 9759: 1981 Guidelines for De-watering during Construction Reaffirmed 2008.
2. Mittal, S. and Shukla, J.P. (2012). "Soil Testing for Engineers", Khanna Publishers, Delhi.
3. Raj, Purushothama, P. (2009). "Ground Improvement Technique", University Science Press, New Delhi.

Miscellaneous Methods

CHAPTER 12

12.1 LANDFILLS

12.1.1 Geosynthetic applications in landfills

Until 1980, clay liners were used as lining material which had hydraulic conductivity of 10^{-7} cm/sec or less. Subsequently it was realised globally that landfills should have double liners and systems for leachate collection and removal. Now geosynthetic products are also being used as lining material. When geosynthetic product is considered for use in design and construction of landfill liners, some of the following properties should be computed by suitable laboratory tests.*

'Mass per unit area, apparent opening size of apertures, thickness, U-V resistivity, permittivity, transmissivity, puncture resistance, resistance to abrasion, compressibility, tensile strength and elongation properties, chemical resistance, etc'.

12.1.2 Geomembranes

Geomembranes provide a good solution to landfill liner system. Geomembranes are practically impermeable liquid barriers. These are primarily made from continuous polymeric sheets. The type of polymeric material used for geomembranes may be thermoplastic or thermoset. The thermoplastic polymers include PVC, polyethylene, chlorinated polyethylene and polyamide while thermoset polymers include ethylene vinyl acetate and polychloroprene.

Fig. 12.1 (*a*) shows the cross-section of a single clay liner system for a landfill. It consists of a compacted clay liner over the native foundation soil (the thickness of compacted clay liner varies

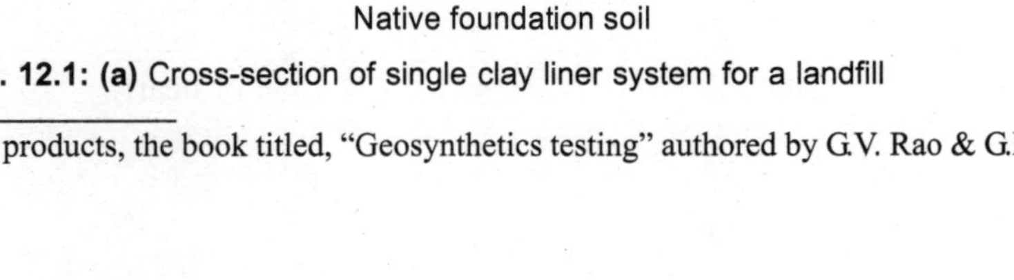

Fig. 12.1: (a) Cross-section of single clay liner system for a landfill

*For testing of geosynthetic products, the book titled, "Geosynthetics testing" authored by G.V. Rao & G.K. Pothal may be referred.

between 0.9 and 1.8 m). The maximum required hydraulic conductivity (k) is normally to the range of 10^{-7} cm/sec. Over clay liner, a layer of gravel with perforated pipes for leachate collection and removal thereof are placed.

Over gravel layer, a layer of filter soil is kept (the filter is used to protect the holes in the perforated pipes against the movement of fine soil particles). In most cases, filter is medium coarse to fine sandy soil. But this system is now a days modified with use of geomembrane, which is shown in Fig. 12.1 (*b*).

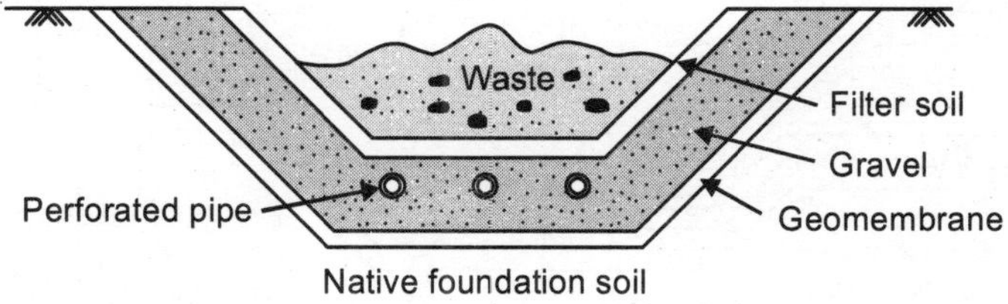

Fig. 12.1: (b) Cross-section of single geomembrane liner system for a landfill

The old system [Fig. 12.1 (*a*)] does not have leak-detection capability. Hence geomembrane is now a days used as a liner material for landfill site [Fig. 12.1 (*b*)]. The geomembrane is laid over native foundation soil. Over the geomembrane, a layer of gravel with perforated pipes is placed, same as in Fig. 12.1 (*a*). The modern liner system for landfills is with double liners (Fig. 12.2).

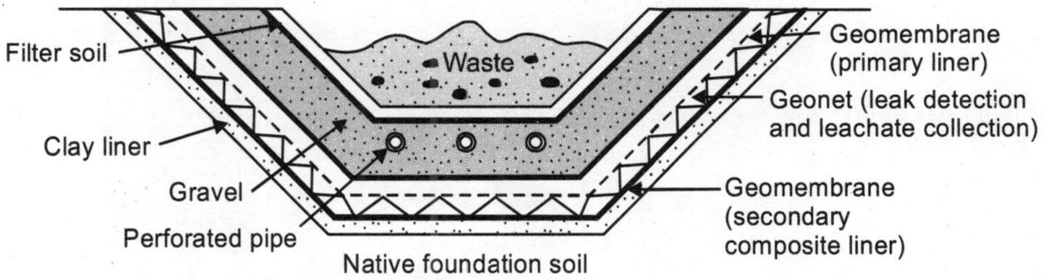

Fig. 12.2: Cross-section of double-liner system (note the secondary composite liner)

The two liners are an upper primary liner and a lower secondary liner. Above the top liner is a primary leachate collection and removal system. For design of liner system, the compacted clay layer should be at least 1 m with $k \leq 10^{-7}$ cm/sec. Figs. 12.2 and 12.3 show the c/s of double liner system.

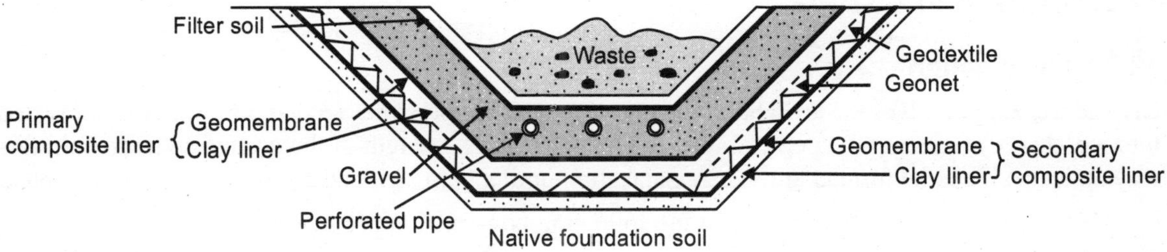

Fig. 12.3: Cross-section of double-liner system (note the primary and secondary liner)

Author designed a rainwater harvesting pond for storage of nearly 34 lacs litres of water by use of geomembrane and geosynthetics. The soil at the site was highly permeable, hence the pond with conventional technique would have probably not served the purpose. Hence geomembrane was used to provide a totally impermeable surface on the side walls and bottom of tank. Provision for air vent was also made to avoid rupture of geomembrane if at any stage tank is empty and there is accumulation of pore water in the soil. The cross-section of the same is shown in Fig. 12.4 and its photo is shown in Fig. 12.5.

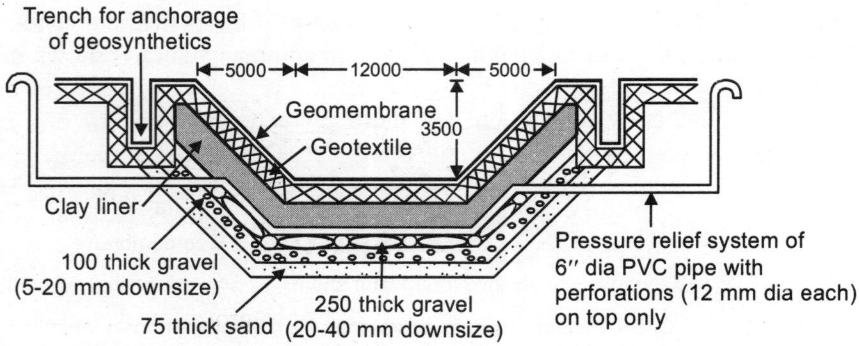

Fig. 12.4: Cross-section of RWH pond; also showing pressure relief system

Fig. 12.5: Completed view of RWH pond

12.2 GROUND FREEZING

12.2.1 Introduction

Ground freezing involves the artificial lowering of ground temperatures so that pore water is converted into ice, thereby reducing the permeability and increasing the strength of the ground so treated. In this way an ice wall can be formed around an excavation until a permanent structure has been constructed.

This technology finds application in the following construction projects:

- Underpinning
- Tunnel roof freezing
- Freezing of cross-cuts in tunnel tubes
- Clearing out of tunnel fall-ins

- Forcing of framework constructions into railway embankments
- Foundation skirtings
- Removal of intact soil samples
- Rehabilitation measures

Frozen walls are frequently used in shaft sinking and to a lesser extent in tunnel construction. The design of a frozen wall is basically a problem in heat transfer and, in the absence of moving ground water, transfer in the ground is by conduction. The thermal conductivity and the heat capacity of the ground governs the rate at which freezing proceeds. Fortunately the values of thermal conductivity for all types of frozen soil fall within quite narrow limits. This is one of the reasons why artificial ground freezing is a versatile technique, being able to deal effectively with a great variety of soil types (Jones, 1981). Artificial freezing can be employed as an exclusion technique to stop the flow of groundwater into excavations, thus producing dry working conditions. In this context, frozen walls can be used in conjunction with bored pile walls or slurry trenches. The frozen ground provides a clear area that is free of bracing and earth supports within which excavation can occur. Ground freezing is influenced by the geological and hydrogeological conditions at the site, the soil types present, and their thermal properties and water content. To be able to use freezing technology, a complex understanding of the thermal, mechanical and hydraulic properties of the frozen soil, the equipment used to freeze the ground, and the assessment of the construction process, is required. Thermal conductivity varies vertically with each change in water content and lithology. It also changes horizontally from the frozen to the unfrozen ground. Obviously the position of the water table and its fluctuation are important, as are the temperature and velocity of groundwater flow. Although complete saturation of the soil is desirable, freezing may be carried out safely when the degree of saturation is as low as 10%. Freezing of pore water in cohesive soils never reaches completion; and attempts at ground freezing may be doomed to failure if there is significant groundwater flow.

The mechanical properties of frozen ground are much more dependent on the time taken in freezing and the ground temperature than on the nature of the soils involved. Consequently, artificial ground freezing is less sensitive to geological conditions than other methods of ground treatment and may be used in any moist soil, irrespective of its structure, grain size or porosity. Weak, running or bouldery ground, and ground in which the pore-water pressures are high, lend themselves to freezing. Soil profiles, too heterogeneous to grout predictably, are also amenable to freezing. However, in heterogeneous soils the frozen zone tends to be irregular in shape. For a certain refrigerant temperature the frozen ground is usually thinner in silts, clays and organic soils than in sand gravels. As these commonly are the weaker strata, design may be dictated by the former three soil types. When clay soils are frozen, vertical cracks may form in the soil between the frozen columns and the soil may bulge near the zone being frozen. The fibrous structure of organic soils tends to be altered permanently by freezing.

Ground freezing consists of inserting refrigeration devices into the ground and freezing the groundwater. This method has been used successfully, and is very effective. It also is very expensive.

12.2.2 Methods of ground freezing

Artificial freezing of ground is usually carried out in two stages, referred to as the active and passive stages of freezing. Active freezing involves freezing the ground to form the ice wall, while passive freezing is that required to maintain the established thickness of the ice wall against thawing. The refrigeration plant has to operate at a much higher capacity during the active than the passive stage of freezing. The primary plant with pumped loop secondary coolant is the system most frequently used to freeze ground. Ammonia or Freon refrigeration plant provides the primary source of refrigeration. The system is worked by either electricity or diesel motor and the condensers may be air- or water-cooled. These plants have a wide range of capacities. Two stages are required to produce temperatures below 25°C. Several different types of coolant have been used with this system; however, the most common type of refrigerant is calcium chloride brine. Other brines have been made from chlorides of sodium, magnesium or lithium.

The growth of ice is more rapid at the base of the freeze probes. This can be overcome by circulating the coolant more quickly or by installing a ring line about 0.3 m below the ground surface immediately inside the perimeter of the freeze pipes, thereby forming a beam of ice. A warmer brine is discharged from the head of each probe into a collection main, whence it is piped back to the refrigeration plant. There it is pumped through a chiller and is delivered, re-cooled, via a distribution main to the inner tubes of the freeze probes. Such a system, with two compressors, can ultimately lower the brine temperature to – 40° C if required. All piping above the surface in a freezing system must be insulated to minimise heat loss. Insulation is also necessary in large shallow excavations. A temporary cover is desirable for large excavations which may be open for months. However, in a deep excavation of small diameter such as a shaft, the air is cooled appreciably and, as a result, protection of the exposed surface of the frozen soil may be unnecessary. Expendable refrigerants are very attractive for individual projects which only last a few days or for projects where the cost of delay is high (for instance, for emergency situations where rapid formation of a frozen wall is required or where particularly low temperatures are needed to deal with heavy groundwater flow). Although no refrigeration plant is required, the cost of refrigerant is high. In expendable refrigerant systems the refrigerant is lost to the atmosphere after it.

12.2.3 Freezing soil

It is adopted in cold regions to bind the soil grains in a mass of frozen pore water. Such modification is useful if, for example, we need to make a deep excavation. During construction such treatment will stabilise the sides of the excavation. Bore holes are drilled at regular intervals and cooling tubes or freeze pipes are installed into the soil mass through which liquid nitrogen or brine solution is circulated. The soil around each borehole freezes as shown in Fig.12.6. Overlapping frozen columns result in a

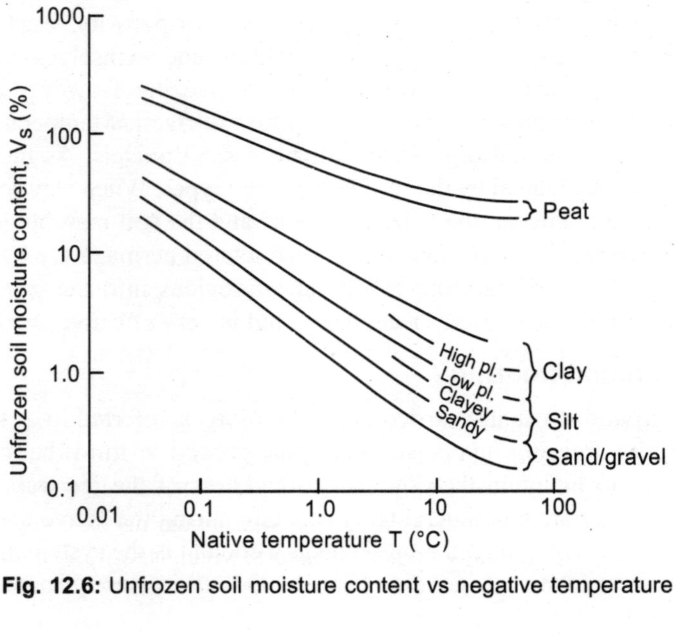

Fig. 12.6: Unfrozen soil moisture content vs negative temperature

stabilised soil mass. Excavation is carried out in the stabilised soil. The frozen soil mass acts as a block which keeps the soil behind the vertical cuts stable and also prevents water from flowing into the excavated area. Once the permanent structure has been constructed in the excavation including the arrangement for retaining the adjacent soil in position, ground freezing is terminated.

Ground freezing layout for an excavation has absorbed energy in the ground and vaporised. Liquid nitrogen (LN_2) provides the fastest and thermally most efficient means of freezing the ground [Fig. 12.7 (*a*) and (*b*)]. It evaporates at a temperature of 195°C at atmospheric pressure and in doing so draws 162.4 kJ/l from its surroundings. About the same amount of heat is needed to raise the cold gas to ambient temperature. In fact such systems can sometimes freeze the ground in a matter of hours. The use of liquid nitrogen also gives rise to fewer ice lenses, to less heaving and to greater strength than freezing with brine. The use of liquid carbon dioxide is thermally less efficient and is normally more difficult to control. A tanker, from which the liquid nitrogen is injected into the freeze pipes, may be used on small or emergency projects, otherwise one or more storage tanks are needed on site. Liquid nitrogen is introduced directly into the circuit, by utilising the pressures developed in the storage tanks, through a distribution pipe carrying the liquid nitrogen to the freeze probes. These are grouped and connected in series. There is a valve downstream from the terminal freezing pipe through which the nitrogen gas is vented to the atmosphere. The ground freezing with liquid nitrogen pressure and consumption of liquid nitrogen are regulated at the storage tank in accordance with the desired rate of heat extraction to form the frozen wall. It is possible to combine brine and liquid nitrogen freezing.

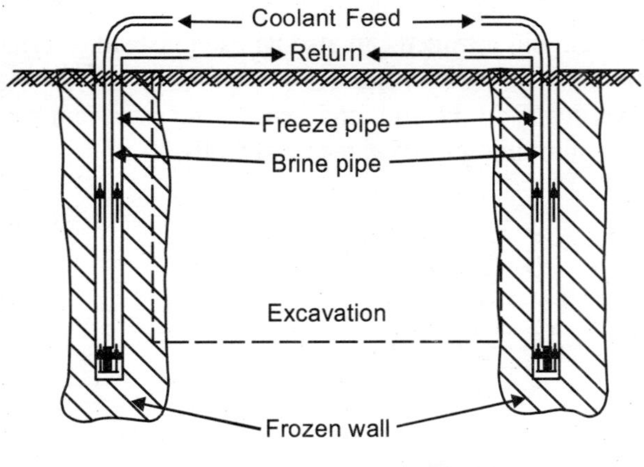

Fig. 12.7: (a) Freezing with liquid nitrogen or brine

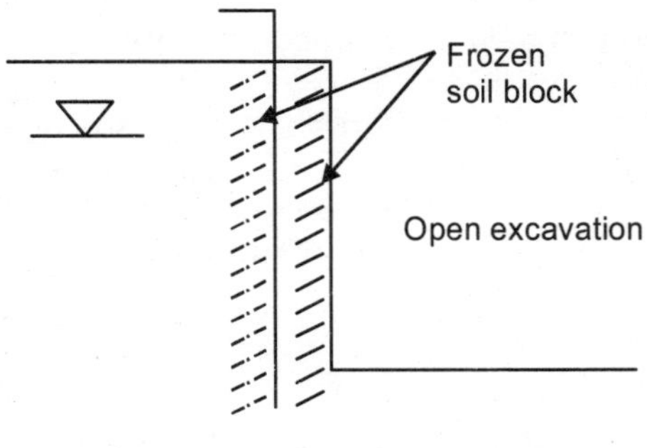

Fig. 12.7: (b) Vertical cut stabilised by freezing

In this way the freezing capacity can be raised to meet peak demands, or liquid nitrogen can be used as a standby in the event of failure. Another possibility involves freezing the ground with liquid nitrogen and then keeping it frozen with cooled brine. Liquid nitrogen can be very quickly re-applied if unforeseen difficulties arise. Most ground-freezing contractors have evolved their own method of calculating the thickness of a frozen wall. When the thickness of the frozen wall has been decided, the number of freeze holes and their spacing together with the refrigeration capacity is ascertained. The most important factors governing cost are the size and spacing of freeze pipes. Estimation of the time required to freeze a radius R by using freeze pipes with radius r_0 can be obtained from Fig. 12.8. Relative spacings, R, in excess of 15 normally are not used. The energy requirement needed to freeze soil in kcal/m^2 is approximately 2200–2800 times the percentage natural moisture content. A frozen ground-support system is a massive relatively rigid structure, which is formed generally before excavation commences.

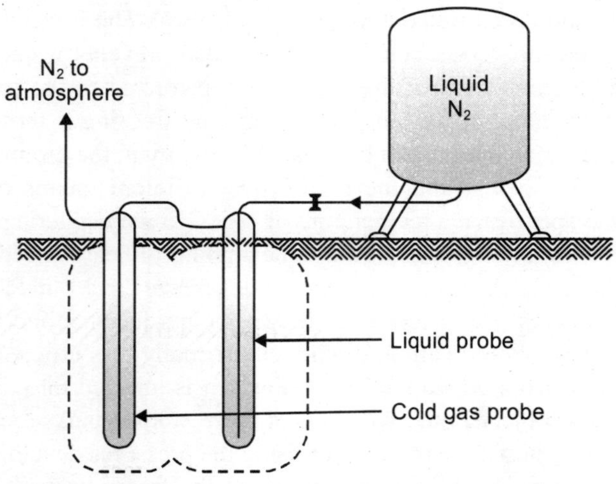

Fig. 12.8: Control process of liquid nitrogen freezing

Due to its expense, this is a rarely used technique. By inserting brine pipes into the surrounding saturated soil, a completely unstable area can be frozen solid, and excavation can make cutting tools brittle and break easily. It was used by a Brazilian architect to stop a 24-storey skyscraper weighing 25,000 t from toppling over in San Paulo in 1941. The building was later jacked back the vertical after strengthening the foundations with new *in situ* piles. Liquid nitrogen can also be used instead of brine in the pipes. This results in quicker freezing, but is more difficult to handle.

Typically, a row of freeze pipes were placed vertically in the soil, and heat energy was removed through them, in a process remarkably analogous to pumping groundwater from wells.

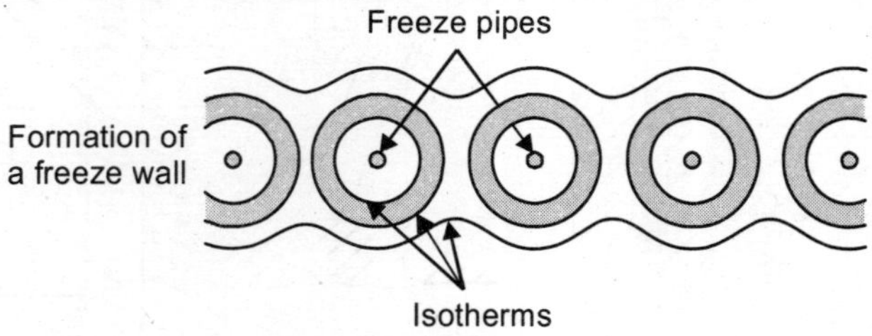

Fig. 12.9: Formation of frozen wall

When the earth temperature reaches 32° F (0°C), water in the soil pores turns to ice. Then further cooling proceeds. With granular soils, the groundwater in the pores freezes readily, and a saturated sand, for example, achieves excellent strength at only a few degrees below the freezing point. With clays, however, the ground water is molecularly bonded at least in part to the soil particles.

12.2.4 Freezing applications

12.2.4.1 The freezing method is remarkably versatile, and with ingenuity it can be adapted to a great many project conditions.

12.2.4.2 The penetration of a freeze does not vary greatly with permeability, so it is much more effective as a cut off than grout.

12.2.4.3 Freezing can perform the dual function of water cut off and earth support, eliminating sheeting and bracing.

Fig. 12.10 shows a circular excavation supported by a freeze wall.

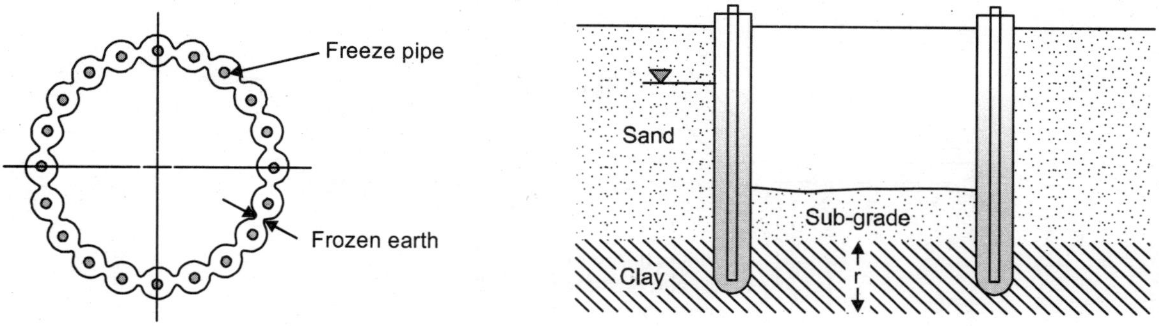

(a) Plan **(b)** Section

Fig. 12.10: Circular excavation supported by freeze wall

Fig. 12.11 shows an excavation supported by gravity retainning wall of frozen earth. A combination of vertical and inclined freeze pipes is typical to achieve the shape illustrated.

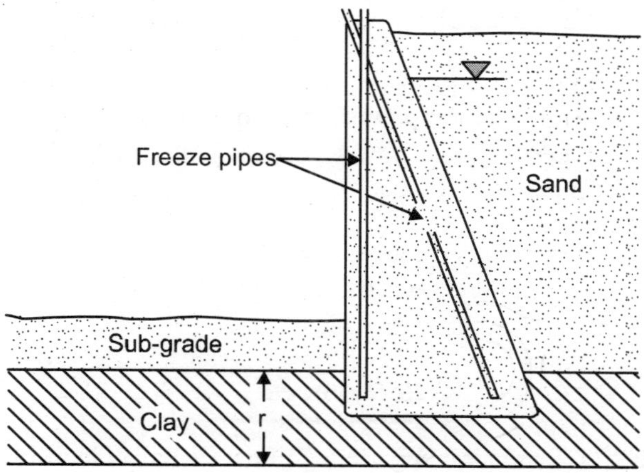

Fig. 12.11: Excavation supported by gravity retainning wall of frozen earth

> **Note:** In both cases the freeze ball toes into an impermeable clay layer below the proposed sub grade.

12.3 STRENGTHENING OF EXISTING FOUNDATIONS

The preservation of old buildings often entails the strengthening or even complete renewal of existing foundations. In many cases, it is necessary to underpin and build a new foundation under the existing sections. Jacked-in sectional piles are used in some instances. A method which has been used successfully on some heavily loaded foundations which have shown signs of failure is the one shown in Fig. 12.12. This is made possible by modern post-tensioning methods. These techniques utilise specially designed jacks to tension and wedge the wires or cables passed through holes drilled in the old walls and anchored in pile head anchor blocks. The wall is thus sandwiched under pressure and the load transferred to the new piles formed on each side in pairs.

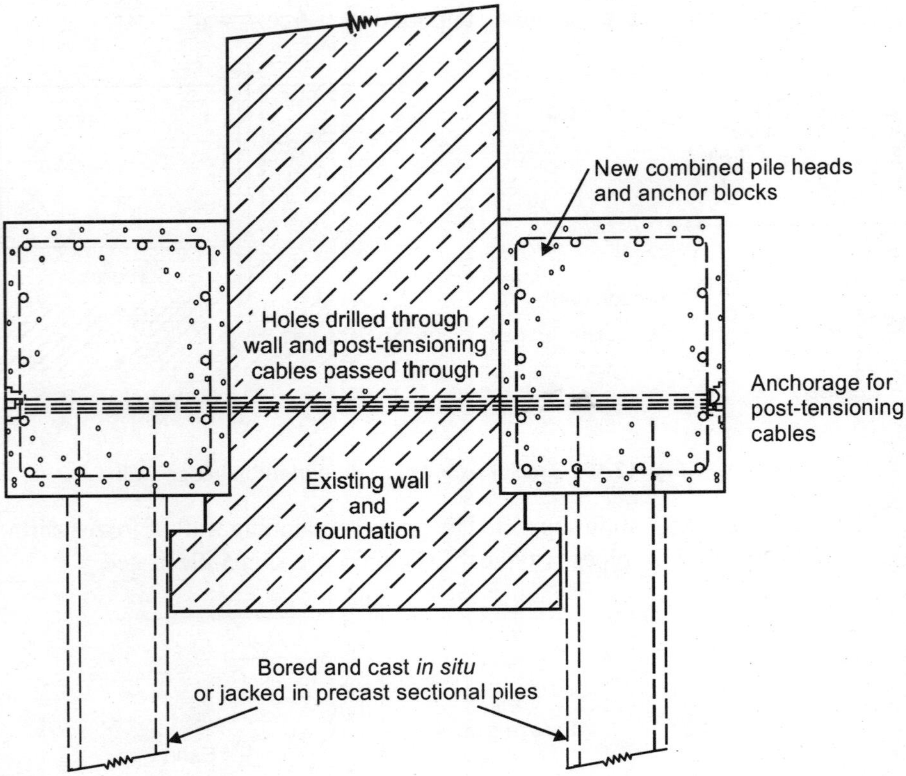

Fig. 12.12: Strengthening of old foundations: Paired piles with pile heads linked by post-tensioned cables through old wall

References

1. Coduto, P, Donald. (2002). "Geotechnical Engineering", Prentice Hall of India Pvt. Ltd., New Delhi.
2. Das, Braja, M. (2002). "Principles of Geotechnical Engineering", Library of Congress Cataloging-in-Publication Data.
3. Gulhati, S.K. & Datta, M. (2005). "Geotechnical Engineering", Tata McGraw Hill Publishing Co. Ltd.
4. Jones, C.J.F.P. (1981). "Earth Reinforcement and Soil Structures", Thomas Telford, ASCE Press.
5. Launder, V.C. (1975). "Foundations", English Language Book Society and Macmillan.

Appendix

Some Applications of Ground Improvement Techniques at Various Sites

A-1 (i) New road construction–Tire pick up.

A-1 (ii) New road construction with nanotac penetrative–No tire pick up.

A-2 Airfield using RBI-81.

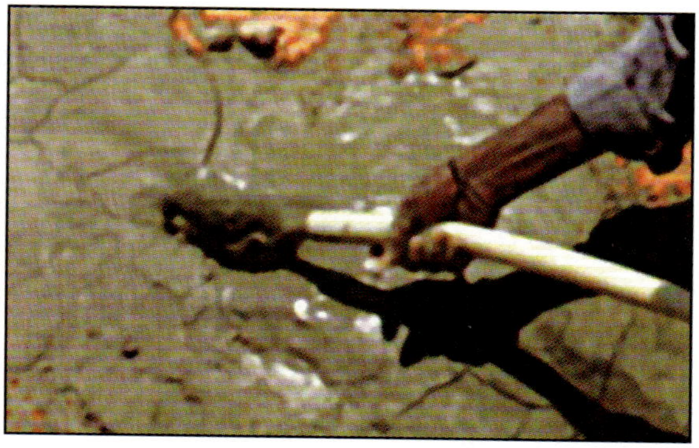

A-3 Grouting of the Cracks.

A-4 Reinforced earth (R.E.) wall construction with parawebs reinforcement.

A-5 Compaction of backfill after laying reinforcement.

A-6 R.E. wall for railway line in Holland.

A-7 R.E. wall with unique aesthetics (Terre Arme'e reinforced earth).

A-8 R.E. wall with unique aesthetics (Terre Arme'e reinforced earth).

A-9 R.E. wall with unique aesthetics (Terre Arme'e reinforced earth).

A-10 Roadway kind of structures. Retaining walls. Product: Reinforced earth, Facings, Reinforcements: Steel strips, Max. height 6 m, Overall length 500 m, Overall area 2000 m^2, Ghantoot Polo Club, UAE.

A-11 Roadway kind of structures. Retaining walls and abutments. Product: Reinforced earth, Facings: Terra Trel reinforcement HAR 45 × 5 steel strips, Max. height 7 m, Overall length 150 m, Overall area 1,000 m^2, North Annecy Bypass, France.

A-12 Kind of structures: Abutments and retaining wall. Product: Reinforced earth, Facings: Terraplus, Reinforcements: HA steel strips, Max. height 6.1 m, Overall length 2,649 m, Overall area 10,511 m^2, El Paso, Texas Loops 375, USA.

A-13 Kind of structures: Avalanche barrier. Product: Reinforced earth, Facings: TerraTrel (catching dam and splitters), Reinforcements: Galvanised HA steel strips, Max. height 14 m, Overall length 400 m, Overall area 7,930 m^2, Avalanche Barrier Neskaupstadur, Iceland.

A-14 Building with tiered retaining walls. Product: Reinforced earth, Facings: Square/Terraclass, Reinforcements: Galvanised steel strips 40 × 50 mm, Max. height 6 m, Overall length 3,200 m, Overall area 9,600 m^2, Lower California Pueblo Bonito Hotel, Mexico.

A-15 Application: Roadway retaining wall. Product: Reinforced earth, Facings: Terraplus with artistic graphics formed into the panels, Reinforcements: HA steel strips, Max. height 6.4 m, Overall length 914 m, Overall area 4,650 m^2, US 287 (Texas 021) Childress, Texas, USA.

A-16 Application: Roadway abutments and retaining walls. Product: Reinforced earth, Facings: Fish pattern on wall designed by port authority, Reinforcements: HA steel strips, Max. height 7.6 m, Overall height 152 m, Overall area 697 m^2, Owner: Port Canaveral Port Authority Engineer Aikenhead & Odom, King George Boulevard Port Canaveral, Florida, USA.

A-17 Marine structures with retaining walls. Product: Reinforced earth, Facings: TerraClass under water, TerraTrel above water, Reinforcements: H.A. 50 × 4 mm^2, Max. height 5 m, Overall length 450 m, Overall area 2256 m^2, Architect: Kuin & Kuin Architecten b.v., Bastionhouses, The Netherlands.

A-18 Roadway retaining wall and abutments, Product: Reinforced earth, Facings: Architectual Terraclass, Reinforcements : 45 × 5 HAR steel strips, Max. height 8.60 m, Overall length 613 m, overall area 2,430 m^2, Owner: Ministere de l' Equipment, des Transport et du logement, Boulogne, Sur-mer Harbour, France.

A-19 Roadway retaining wall, Product: Reinforced earth facings: Terraplus panels with stone form liner and artistic treatment to Texas map panel, Reinforcements: HA steel strips, Max. height 7.6 m, Overall length 865 m, Overall area 3,412 m^2, Owner : Texas Department of Transportation Texas, USA.

A-20 Application: Roadway, retaining walls parapets, Product: Reinforced earth, Facings: Mineral TerraTrel, Reinforcements: 45 × 5 HAR steel strips, Max. height 8.30 m, Overall length 786 m, Overall area 3,620 m^2, Owner Conseil General de l' Herault, France.

A-21 Retaining wall. Product: Reinforced earth and TechSpan arches, Facings: Special panels, Max. height 10 m, Overall length 208.80 m (2 walls), Overall area 1,067 m^2 TF-5 Santa Cruz De Tenerife–San Cristobal Tenerife, Spain.

A-22 Slope protection walls. Product: Reinforced earth, Facings: Plain, ribbed, architectural TerraClass, Reinforcements: Galvanised steel strips, Max. height 22 m, Overall length 8 walls, Overall area 15,000 m^2, Owner Putrajaya Holdings Engineers Perunding L & W Sdn Bhd, HS Liao Sdn Bhd IJM Bhd Putrajaya Convention Centre Putrajaya, Malaysia.

A-23 Application: Roadway, Kind of structures: Retaining walls, Product: Reinforced earth, Facings: TerraClass, Reinforcements: 45 × 5 HAR steel strips, Max. height 11.80 m, Overall length 1,955 m, Overall area 10,000 m^2, Saint-Astier A 89 – Toarcc 2.3 Mussidan-Perigueux, France.

A-24 Dynamic compaction (*Courtesy:* Menard)

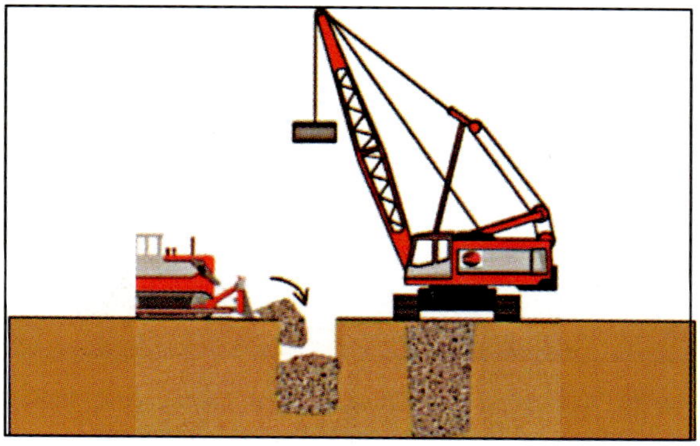

A-25 Dynamic replacement (*Courtesy:* Menard)

A-26 Geotubes for coastal protection (Maccaferri).

A-27 Groynes for river training (Maccaferri).

A-28 Rockfall and Avalanche barriers (Maccaferri).

A-29 Rockfall barriers (Maccaferri).

A-30 Application: Roadway, Ecobridge over A2 motorway (ecological connection safe passage of animals), Kind of structures: Mixed abutments. Product: Reinforced earth, Facings: TerraTrel, Reinforcements: 50 × 4 HA steel strips, Max. height 6 m, Overall length 300 m, Overall area 1,486 m^2, Rijssen Ecoduct De Borkeld over A2 motorway, The Netherlands.

A-31 Geogrid as basal reinforcement (Maccaferri).

A-32 Use of geomembrane for landfills (Garware wall ropes).

A-33 Geomembrane installation on slopes at HZL, Vizag (Garware wall ropes).

A-34 Bank protection of river Fulahar, Malda, W.B., India by use of jute geotextiles (National Jute Board).

A-35 Bio-engineering solution with use of jute geotextile for slope stabilisation in India (National Jute Board).

A-36 Sprouting of plant on jute geotextile in India (National Jute Board).

A-37 Control of railway track subsidence with jute geotextile in Eastern Railway, India (National Jute Board).

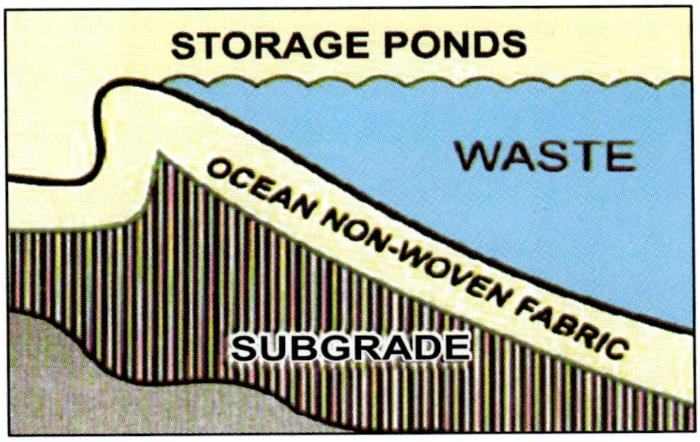

A-38 Non-woven fabric used as underliner for tailing ponds (Filter Fabs, Delhi).

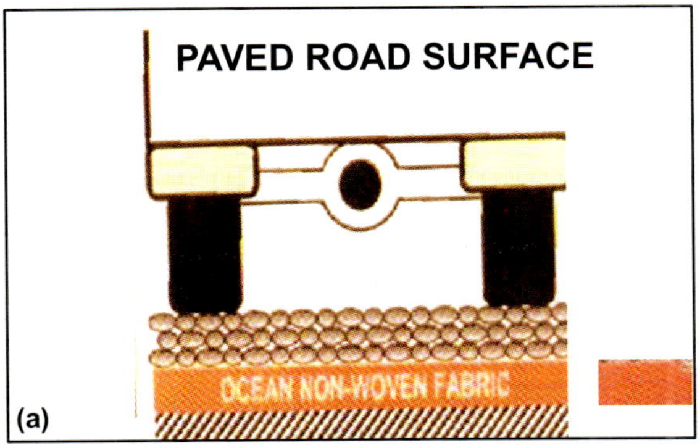

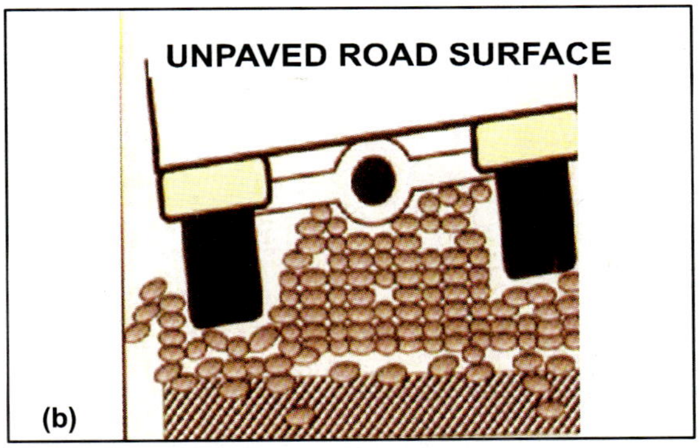

A-39 (a & b) On unpaved access roads for forestry/mining/industrial sites, heavy vehicles can't ply on wet weather. 'Ocean' made geotextiles have been used in converting unpaved roads to stable roads by placing it directly over soil. It permits water to drain, resulting the surface clean, dry and intact (Filter Fabs, Delhi).

A-40 Soil Nailing, designed by author for landslide control in Nainital
(*Courtesy:* Sukriti Infrastructure, Noida)

A-41 Soil Nailing, designed by author, for underground parking in Lucknow
(*Courtesy:* Sukriti Infrastructure, Noida)

A-42 Gabion box wall, designed by author, for DLF complex in Chandigarh
(*Courtesy:* Maccaferri, Panchkula)

A-43 Soil Nailing, designed by author, at Golden Temple, Amritsar
(*Courtesy:* AECS, New Delhi)

A-44 Slope stabilisation through geocells
(*Courtesy:* Strata Geosystems)

A-45 Cross-section of the soldier pile wall with anchors
(*Courtesy:* V.R. Raju, Keller)

Index

N

O

P

R

S